# Advances in Computational Intelligence and Robotics (ACIR) Book Series

Ivan Giannoccaro
University of Salento, Italy

ISSN:2327-0411
EISSN:2327-042X

## Mission

While intelligence is traditionally a term applied to humans and human cognition, technology has progressed in such a way to allow for the development of intelligent systems able to simulate many human traits. With this new era of simulated and artificial intelligence, much research is needed in order to continue to advance the field and also to evaluate the ethical and societal concerns of the existence of artificial life and machine learning.

The **Advances in Computational Intelligence and Robotics (ACIR) Book Series** encourages scholarly discourse on all topics pertaining to evolutionary computing, artificial life, computational intelligence, machine learning, and robotics. ACIR presents the latest research being conducted on diverse topics in intelligence technologies with the goal of advancing knowledge and applications in this rapidly evolving field.

## Coverage

- Evolutionary Computing
- Robotics
- Artificial Life
- Adaptive and Complex Systems
- Synthetic Emotions
- Artificial Intelligence
- Fuzzy Systems
- Automated Reasoning
- Brain Simulation
- Natural Language Processing

IGI Global is currently accepting manuscripts for publication within this series. To submit a proposal for a volume in this series, please contact our Acquisition Editors at Acquisitions@igi-global.com or visit: http://www.igi-global.com/publish/.

I0894909

The Advances in Computational Intelligence and Robotics (ACIR) Book Series (ISSN 2327-0411) is published by IGI Global, 701 E. Chocolate Avenue, Hershey, PA 17033-1240, USA, www.igi-global.com. This series is composed of titles available for purchase individually; each title is edited to be contextually exclusive from any other title within the series. For pricing and ordering information please visit http://www.igi-global.com/book-series/advances-computational-intelligence-robotics/73674. Postmaster: Send all address changes to above address.

## Titles in this Series

*For a list of additional titles in this series, please visit:*
*http://www.igi-global.com/book-series/advances-computational-intelligence-robotics/73674*

***Artificial Intelligence of Things (AIoT) for Productivity and Organizational Transition***
Sajad Rezaei (University of Worcester, UK) and Amin Ansary (University of the Witwatersrand, South Africa)
Business Science Reference • © 2024 • 368pp • H/C (ISBN: 9798369309933) • US $275.00

***Internet of Things and AI for Natural Disaster Management and Prediction***
D. Satishkumar (Nehru Institute of Technology, India) and M. Sivaraja (Nehru Institute of Technology, India)
Engineering Science Reference • © 2024 • 334pp • H/C (ISBN: 9798369342848) • US $345.00

***AI Applications for Business, Medical, and Agricultural Sustainability***
Arshi Naim (King Khalid University, Saudi Arabia)
Engineering Science Reference • © 2024 • 322pp • H/C (ISBN: 9798369352663) • US $315.00

***Innovative Machine Learning Applications for Cryptography***
J. Anitha Ruth (SRM Institute of Science and Technology, India) G.V. Mahesh Vijayalakshmi (BMS Institute of Technology and Management, India) P. Visalakshi (SRM Institute of Science and Technology, India) R. Uma (Sri Sai Ram Engineering College, India) and A. Meenakshi (SRM Institute of Science and Technology, India)
Engineering Science Reference • © 2024 • 294pp • H/C (ISBN: 9798369316429) • US $300.00

***The Ethical Frontier of AI and Data Analysis***
Rajeev Kumar (Moradabad Institute of Technology, India) Ankush Joshi (COER University, Roorkee, India) Hari Om Sharan (Rama University, Kanpur, India) Sheng-Lung Peng (College of Innovative Design and Management, National Taipei University of Business, Taiwan) and Chetan R. Dudhagara (Anand Agricultural University, India)
Engineering Science Reference • © 2024 • 456pp • H/C (ISBN: 9798369329641) • US $365.00

***Empowering Low-Resource Languages With NLP Solutions***
Partha Pakray (National Institute of Technology, Silchar, India) Pankaj Dadure (University of Petroleum and Energy Studies, India) and Sivaji Bandyopadhyay (Jadavpur University, India)
Engineering Science Reference • © 2024 • 314pp • H/C (ISBN: 9798369307281) • US $300.00

***Computational Intelligence for Green Cloud Computing and Digital Waste Management***
K. Dinesh Kumar (Amrita Vishwa Vidyapeetham, India) Vijayakumar Varadarajan (The University of New South Wales, Australia) Nidal Nasser (College of Engineering, Alfaisal University, Saudi Arabia) and Ravi Kumar Poluru (Institute of Aeronautical Engineering, India)
Engineering Science Reference • © 2024 • 405pp • H/C (ISBN: 9798369315521) • US $300.00

IGI Global
PUBLISHER of TIMELY KNOWLEDGE

701 East Chocolate Avenue, Hershey, PA 17033, USA
Tel: 717-533-8845 x100 • Fax: 717-533-8661
E-Mail: cust@igi-global.com • www.igi-global.com

# Exploring the Ethical Implications of Generative AI

Aftab Ara
*University of Hail, Saudi Arabia*

Affreen Ara
*Department of Computer Science, Christ College, Bangalore, India*

IGI Global
PUBLISHER of TIMELY KNOWLEDGE

A volume in the Advances in Computational Intelligence and Robotics (ACIR) Book Series

Published in the United States of America by
IGI Global
Engineering Science Reference (an imprint of IGI Global)
701 E. Chocolate Avenue
Hershey PA, USA 17033
Tel: 717-533-8845
Fax: 717-533-8661
E-mail: cust@igi-global.com
Web site: http://www.igi-global.com



Library of Congress Cataloging-in-Publication Data

Names: Ara, Aftab, editor. | Ara, Affreen, 1983- editor.
Title: Exploring the ethical implications of generative AI / edited by Aftab Ara, Affreen Ara.
Description: Hershey, PA : Engineering Science Reference, [2024] | Includes bibliographical references and index. | Summary: "The objective of the book Ethical Implications of the ethical challenges posed by generative AI, and to propose guidelines for its responsible development and use"-- Provided by publisher.
Identifiers: LCCN 2023055048 (print) | LCCN 2023055049 (ebook) | ISBN 9798369315651 (hardcover) | ISBN 9798369315668 (ebook)
Subjects: LCSH: Artificial intelligence--Moral and ethical aspects. | Artificial intelligence--Philosophy.
Classification: LCC Q334.7 .E97 2024 (print) | LCC Q334.7 (ebook) | DDC 174/.90063--dc23/eng/20240104
LC record available at https://lccn.loc.gov/2023055048
LC ebook record available at https://lccn.loc.gov/2023055049

This book is published in the IGI Global book series Advances in Computational Intelligence and Robotics (ACIR) (ISSN: 2327-0411; eISSN: 2327-042X)

British Cataloguing in Publication Data
A Cataloguing in Publication record for this book is available from the British Library.

All work contributed to this book is new, previously-unpublished material. The views expressed in this book are those of the authors, but not necessarily of the publisher.

For electronic access to this publication, please contact: eresources@igi-global.com.

# Table of Contents

# Detailed Table of Contents



This thought-provoking chapter is systematic research based on theories from literature review and empirical data exploring AI's potential impact on educational equity in higher learning. The atuhors examine how AI-powered systems might inadvertently perpetuate biases, affecting marginalized students disproportionately. This chapter discusses institutions' responsibility to implement AI in ways that support inclusivity and diversity. In addition, it highlights initiatives that prioritize fairness and transparency in AI algorithms. AI systems can perpetuate existing biases if not designed with the right in mind. Additionally, if an AI system is not transparent about its decision-making processes, identifying or rectifying potential biases will be nearly impossible. Readers will gain a deeper understanding of AI's challenges and opportunities in reshaping education, focusing on ensuring that no student is left behind.



Generative artificial intelligence (AI) refers to a form of AI technology that possesses the capability to produce a diverse array of content, encompassing text, images, artificially generated data, sound and chat responses. Generative AI is a complex branch of AI that allows systems to create new content by using current information such as text, audio, video. Due to the distribution of unlawful material, copyright violations, privacy of data violations, and the amplifying of preexisting biases, this powerful technology also raises serious ethical issues. Privacy, security issues and raising ethical concerns addressing the future of generative AI in different business domains and functions namely, Marketing, Human Resources, Finance, operations and information technology are discussed in this chapter.

This chapter's objective is to provide an overview of how artificial intelligence (AI) has become an essential part of human life. It explains the sources of bias and its types in AI technology. With the help of previous studies, the chapter elucidates the strategies that can be used to avoid decision-making as a source of bias in AI technology. It also talks about the importance of understanding how human bias can also cause AI systems to exhibit bias towards certain groups. Unfairness in AI is also one of the most common sources of biassed data, and it's explained with strategies for detecting and addressing unfairness. The chapter also covers the need for transparency in AI technology along with ethical considerations, as transparency in AI is essential to ensuring that AI systems operate in adherence to ethical standards.



This chapter provides an overview of the ethical considerations that should be taken into account while using generative AI technologies, specifically in the field of education, as well as concrete suggestions for programmers and end-users. Therefore, students, researchers, and academics in a variety of fields who are interested in the ethical aspects of generative AI will find this chapter useful, as it will also provide an overview of the existing ethical frameworks in the field of education. In that sense, this chapter can be viewed as a concise introduction to the current state of the ethical issues being studied and a proposal for balancing risks and opportunities.



In an era driven by digital innovation, artificial intelligence (AI), or Generative AI, emerges as a transformative force reshaping the landscape of higher education. Its potential to personalize learning experiences, bolster research capacities, and streamline administrative operations is revolutionary. However, the integration of Open AI into academia raises complex ethical issues for faculty and learners. The need for comprehensive ethical guidelines is imperative to ensure that the integration and utilization of AI in higher education are aligned with the core values of academic integrity and social responsibility. This chapter examines the ethical frameworks essential for governing the use of generative AI technologies in academia and provides practical recommendations for stakeholders involved. Additionally, emerging AI technologies such as experimental NotebookLM and Gemini will be discussed as future directions for AI use in teaching, learning, and research.

In the digital landscape, where innovation knows no bounds, generative artificial intelligence (Gen AI) emerges as a true virtuoso, orchestrating a seamless symphony of creative wonders. Artificial intelligence that can generate fresh and plausible images, texts, and motion graphics rapidly is called Generative AI. It possesses the remarkable ability to effortlessly transform prompts into a rich tapestry of visual art, evocative prose, dynamic videos, and an array of captivating media. This chapter embarks on an exploratory journey, delving deep into the profound influence that Generative AI wields across a diverse spectrum of industries. Through an immersive exploration of these remarkable used cases, this chapter offers a glimpse into the boundless potential and the promises that Generative AI holds, positioning itself as a versatile catalyst for progress within the expansive industrial landscape.



This chapter explores how the social implications of AI are being posited, often sensationalized as a threat to humanity, rather than being framed in something humanly designed that ought to remain within the control of its maker, transparent in terms of capacity to undertake complex decision making and which most importantly is accountable for every individual action made in terms of design and programming. The aims of the chapter are threefold, namely, to consider global ethics and the impact that AI could potentially have in terms of increasing societal inequalities in terms of existing infrastructure; to provide an insight into the developmental and progressive use of AI across organizational infrastructures such as global medicine and health and the military; finally, to embed the concept of ethical AI and the potential for its praxis across all areas of its integration.



Law enforcement is joining the fast-growing artificial intelligence (AI) research field. The chapter tries to fix that. This chapter utilized a "systematic literature review." The authors gathered research papers on using algorithms and AI in police work. This was done with Scopus, a fancy academic database. They searched for papers on "law enforcement," "policing," "crime prevention," "crime reduction," and "surveillance." Combine these terms with "algorithm" or "artificial intelligence." They found that AI has great potential to aid law enforcement. It can recognize faces, forecast crimes, and track people. These AI tools usually analyze photos, behavior, language, or a combination. However, there are significant "but" ethical issues that exist. AI can cause unjust treatment, confusion about responsibility, oversurveillance, and privacy invasion. AI's benefits and cool abilities are often highlighted over its drawbacks. Another observation is that writings on the same topics agree on what AI can achieve, its potential, and what we should explore next.

The rapid integration of generative AI technology across various domains has brought forth a complex interplay between technological advancements, legal frameworks, and ethical considerations. In a world where generative AI has transcended its initial novelty and is now woven into the fabric of everyday life, the boundaries of human creativity and machine-generated output are becoming increasingly blurred. The paper scrutinizes existing privacy laws and regulations through the lens of generative AI, seeking to uncover gaps, challenges, and possible avenues for reform. It explores the evolution of jurisprudence in the face of technological disruption and debates the adequacy of current legal frameworks to address the dynamic complexities of AI-influenced privacy infringements. By scrutinizing cases where personal data has been exploited by nefarious actors employing generative AI for malevolent purposes, a stark reality emerges: the emergence of a new avenue for privacy breaches that tests the limits of existing legal frameworks.



Generative AI systems have given incredible ability to independently produce a wide variety of content types, including textual, visual, and more. Complex issues with copyright protection and intellectual property rights have arisen as a result of this change. With a focus on fostering responsible global governance, this research delves into the complex legal and ethical considerations underlying Generative AI. The goal of this chapter is to take a look at the complicated legal issues that come up because of Generative AI's ability to generate material on its own. This chapter analyzes the current legal documents, legislation, and international treaties, focusing on ethical concerns. Ultimately, the authors want to have a positive impact on efforts to build responsible and efficient international frameworks for regulating Generative AI. This study provides an exhaustive case for the implementation of legal frameworks that can efficiently tackle the intricate legal and ethical quandaries posed by Generative AI, while simultaneously encouraging the progress of innovation and creativity.



This chapter dissects the proposal for an AI Liability Directive by the European Parliament and the Council. The Directive aims to adapt civil liability rules for artificial intelligence (AI) systems. Two main types of non-contractual liability are scrutinized: fault-based and strict liability. The core of the

# Foreword

The book was given while I am preparing for a lecture concerning "an ethic in artificial intelligence". It was like a gift for a child who eager for candy. As a colleagues in MIS department, I witness the effort of conducting research and organize topics by the author who has passion in Artificial Intelligence (AI). Dr Aftab are qualified to be the co-editor for the book based on her experience in the field. She conducted various research about a technology. Coming from an engineering background has given a tremendeous advantage for her in understanding the sophisticated algorithm laid in AI.

I was so impressed by the themes that cover a wide range of ethical issues of AI. Looking from the perspective of the editors' approach in navigating the issues of AI from the ethical perspectives has given me more exposure that indeed need to be discussed.

AI mimic human judgment by automates data in problem solving, learning, and reasoning. Algorithm developed in machine improve their ability to learn from patterns to generate a quick insight for that judgment. It maximizes the business operation, education etc. However, it is crucial for us to analyze the ethical concerns towards fast-paced technology to avoid any issues in the future.

Readers will come across a various viewpoint from professionals, academics, and practitioners who have devoted their careers to researching and influencing the ethical landscape of AI throughout these pages. Their observations will make you reevaluate your presumptions, widen your perspective, and motivate you to consider the implications of AI for society.

The book began on an exploration to investigate the relationship between ethics and artificial intelligence in education, which is a basis of ethical behavior. It explores the challenges of developing an inclusivity and diversity of artificial intelligence systems. The goal is to present a transparency and fairness issues that ought to guide the creation and application of AI technology.

The book talk about issues like issues in business domain that involves ethical dilemma such as copyright and privacy violations, data protection, the effect of AI on the workforce and employment and the convergence with product and service development.

Furthermore, the role AI plays in resolving global issues like the judiciary, rural populations, and healthcare are included in the chapters. Issues such as digital judiciary, global governance, and mental health effect. The analysis gains understanding of the effect of AI in the domain.

It is hoping that this book acts as a spark for important discussions and actions. The authors aim to find a convergence between a diverse sector and create a holistic environment for AI. By compiling all studies together, it can facilitate a further understanding of AI potential.

AI promise a lot for the future, and it is our duty to govern it with ethically so it could bring more benefits than drawbacks to society.

*Azira Ab Aziz*
*University of Ha'il, Saudi Arabia*

# Preface

Generative AI, with its boundless potential, stands as a beacon of innovation across diverse sectors. Its capacity to reshape healthcare, forge immersive virtual experiences, automate content generation, and more, underscores its profound impact on society. Yet, this rapid evolution also unveils a landscape fraught with ethical complexities demanding our keen scrutiny and deliberation.

In *Exploring the Ethical Implications of Generative AI*, we embark on a timely journey to dissect and decipher the ethical conundrums inherent in this transformative technology. Wang (2023) uncovers the shadowy underbelly of generative AI, unveiling its susceptibility to exploitation by illicit markets for counterfeit endeavors. Qadir (2023) highlights the nuanced constraints within educational spheres, exposing biases and misinformation perpetuated by systems like ChatGPT. Zohny (2023) provocatively probes into the realm of authorship, shedding light on the ethical dilemmas posed by AI-generated content in scholarly publications. Houde (2020) warns of the potential malevolent applications of generative AI in business, urging for heightened vigilance and deeper research.

"Ethical Implications of Generative AI: A Comprehensive Analysis" endeavors to furnish a holistic perspective on the ethical imperatives governing the deployment of generative AI. This compendium is meticulously crafted to cater to a broad spectrum of readers, including industry leaders, policymakers, scholars, and fervent enthusiasts alike.

Our objective is clear: to navigate the ethical labyrinth of generative AI by identifying key issues, reviewing extant research, proposing guidelines, and contextualizing ethical implications across varied domains. We aim to not only augment existing discourse but also offer actionable insights for responsible development and utilization.

This edited volume extends an invitation to engage with the ethical fabric of generative AI, fostering a collective dialogue that transcends disciplinary boundaries. Whether you're a seasoned executive seeking strategic integration or an inquisitive mind delving into the societal ramifications of technological innovation, this book beckons you to join the conversation.

Welcome to a comprehensive exploration of the ethical frontiers in the realm of Generative AI.

## ORGANIZATION OF THE BOOK

### Chapter 1: AI and Equity in Higher Education: Ensuring Inclusivity in the Algorithmic Classroom

Authored by Amdy Diene, this chapter dives into the intricate interplay between AI technology and educational equity within higher learning institutions. Drawing from both theoretical frameworks and empirical data, the chapter meticulously examines how AI systems, if not designed and implemented thoughtfully, can exacerbate existing biases, particularly affecting marginalized student populations. It stresses the imperative for institutions to adopt AI in a manner that fosters inclusivity and diversity, emphasizing initiatives geared towards transparency and fairness in algorithmic decision-making. Readers will gain insights into the challenges and opportunities presented by AI in reshaping education, with a focus on ensuring equitable access and outcomes for all students.

### Chapter 2: Artificial Intelligence in Different Business Domains- Ethical Concerns

Penned by Sam Paul B. and Anuradha A., this chapter delves into the intricate landscape of generative AI within various business sectors. It elucidates the multifaceted capabilities of generative AI and the ethical dilemmas it poses, ranging from copyright infringements to privacy violations and the amplification of biases. Through a comprehensive exploration, the chapter navigates the ethical implications of generative AI in domains such as marketing, human resources, finance, operations, and information technology. By addressing privacy and security concerns, this chapter offers valuable insights for businesses seeking to harness AI technologies responsibly while remaining ethically mindful.

### Chapter 3: Bias and Fairness in AI Technology

Authored by Muhsina P.R. and Zidan Kachhi, this chapter provides a sweeping overview of the integration of artificial intelligence (AI) into various facets of human life, with a specific focus on bias and fairness considerations. Through a meticulous examination of prior research, the chapter dissects the sources and types of bias prevalent in AI technology, alongside strategies to mitigate bias in decision-making processes. It underscores the critical importance of understanding human biases that may seep into AI systems, emphasizing the need for transparency and ethical considerations to ensure fairness. By shedding light on the complexities of AI fairness and transparency, this chapter offers practical insights for navigating ethical challenges in AI development and deployment.

### Chapter 4: Ethical Considerations in the Educational Use of Generative AI Technologies

Penned by Burak Tomak and Ayşe Yılmaz Virlan, this chapter offers a comprehensive overview of the ethical considerations surrounding the utilization of generative AI technologies in education. It provides concrete suggestions for programmers and end-users, catering to students, researchers, and academics interested in the ethical dimensions of AI. By synthesizing existing ethical frameworks and proposing strategies for balancing risks and opportunities, this chapter serves as a valuable resource for understanding the ethical landscape of generative AI in educational settings.

### Chapter 5: Ethical Navigations - Adaptable Frameworks for Responsible AI Use in Higher Education

Penned by Allen Farina and Carolyn Stevenson, this chapter delves into the ethical frameworks essential for governing the use of generative AI technologies in academia. It offers practical recommendations for stakeholders involved in higher education, while also discussing emerging AI technologies and their potential impact on teaching, learning, and research. By addressing the ethical dimensions of AI integration, this chapter aims to uphold the core values of academic integrity and social responsibility within higher education institutions.

## Chapter 6: Exploring Transformative Potential Generative AI's Multifaceted Impact on Diverse Sectors

Authored by Sadhana Mishra, this chapter embarks on an exploratory journey into the transformative potential of Generative Artificial Intelligence (Gen AI) across various industries. Through immersive case studies and analyses, the chapter illuminates Gen AI's ability to generate diverse content types and its profound influence on sectors such as media, arts, and entertainment. By showcasing its versatility and creative prowess, this chapter underscores the pivotal role of Gen AI as a catalyst for progress within the industrial landscape, offering valuable insights for stakeholders navigating the dynamic intersection of AI and diverse sectors.

## Chapter 7: For Better or For Worse?: Ethical Implications of Generative AI

Authored by Catherine Hayes, this chapter critically examines the social implications and ethical considerations surrounding the integration of generative AI across organizational infrastructures. It underscores the importance of ethical AI design and accountability in mitigating societal inequalities and ensuring transparency in decision-making processes. By delving into global ethics and progressive uses of AI in various domains, the chapter advocates for an ethical framework that upholds human values and fosters responsible AI deployment across sectors.

## Chapter 8: Harnessing the Power of Artificial Intelligence in Law Enforcement: A Comprehensive Review of Opportunities and Ethical Challenges

Penned by Akash Bag, Souvik Roy, and Ashutosh Pandey, this chapter provides a comprehensive review of the opportunities and ethical challenges associated with integrating artificial intelligence (AI) in law enforcement. Through a systematic literature review, the chapter highlights AI's potential to aid law enforcement efforts while addressing ethical concerns related to unjust treatment, surveillance, and privacy invasion. By offering a balanced perspective on AI's benefits and drawbacks, this chapter contributes to the ongoing discourse on responsible AI deployment in law enforcement.

## Chapter 9: Lensing Legal Dynamics for Examining Responsibility and Deliberation of Generative AI Tethered Technological Privacy Concerns: Infringements and Use of Personal Data by Nefarious Actors

Authored by (Dr.) Bhupinder Singh, this chapter scrutinizes the legal dynamics surrounding the responsibility and deliberation of generative AI technology, particularly concerning technological privacy concerns and the misuse of personal data by nefarious actors. Through a critical examination of existing regulations and proposed directives, the chapter highlights the complexities of AI-induced privacy infringements and calls for clarity in legal frameworks to address emerging challenges.

## Chapter 10: Navigating the Legal and Ethical Framework for Generative AI: Fostering Responsible Global Governance

Penned by Anuttama Ghose, S. M. Aamir Ali, and Sachin Deshmukh, this chapter delves into the intricate legal and ethical considerations underlying generative AI, with a focus on fostering responsible global governance. It analyzes current legal documents, legislation, and international treaties, offering insights into ethical concerns surrounding AI-generated content. By advocating for the implementation of efficient legal frameworks, the chapter aims to promote responsible innovation while safeguarding ethical principles in AI development and deployment.

## Chapter 11: Navigating the Legal Landscape of AI-Induced Property Damage: A Critical Examination of Existing Regulations and the Quest for Clarity

Authored by Akash Bag, Astha Chaturvedi, Sneha, and Ruchi Tiwari, this chapter examines the legal landscape surrounding AI-induced property damage, particularly focusing on existing regulations and proposed directives. Through a critical analysis of the AI Liability Directive, the chapter highlights the challenges and perspectives surrounding fault-based and strict liability in AI-related property damage cases. By dissecting key provisions and discussing potential impacts on future policies, the chapter contributes to the ongoing discourse on AI responsibility and legal frameworks.

## Chapter 12: Navigating the Quandaries of Artificial Intelligence-Driven Mental Health Decision Support in Healthcare

Authored by Sagarika Mukhopadhaya, Akash Bag, Pooja Panwar, and Varsha Malagi, this chapter explores the integration of artificial intelligence (AI) into mental health services, focusing on its potential benefits and ethical challenges. Through empirical data and a literature review, the chapter identifies five recurring ethical issues, including handling inaccurate suggestions and preserving patient autonomy. By thoroughly analyzing these challenges, the chapter sheds light on the complex ethical terrain at the intersection of AI and mental health, offering insights crucial for healthcare professionals and AI researchers.

## Chapter 13: Sustainable Islamic Financial Inclusion: The Ethical Challenges of Generative AI in Product and Service Development

Authored by Early Kismawadi, this chapter investigates the ethical consequences of integrating generative AI in the sustainable development of Islamic financial services. Through an interdisciplinary framework that integrates Islamic finance and technological ethics, the chapter explores the potential impact of AI on Islamic finance and offers insights for policymakers, technology developers, and financial institutions. By examining the convergence of AI technology and Islamic finance, the chapter seeks to foster a sustainable and inclusive ecosystem guided by ethical principles.

These chapters collectively offer a comprehensive exploration of the ethical implications and challenges associated with generative AI across diverse domains. From education and business to law enforcement and healthcare, each chapter provides valuable insights and recommendations for navigating the ethical complexities of AI integration while fostering responsible innovation and governance.

## IN CONCLUSION

As editors of "Ethical Implications of Generative AI: A Comprehensive Analysis," we are deeply gratified to present this meticulously curated collection of chapters, each offering profound insights into the ethical complexities surrounding generative AI. From the intricacies of algorithmic decision-making in educational settings to the transformative potential of AI across diverse sectors, the contributions within this volume represent a culmination of scholarly inquiry and interdisciplinary collaboration.

Through systematic research, empirical analyses, and critical examinations of existing frameworks, our esteemed authors have navigated the ethical labyrinth of generative AI with rigor and nuance. They have illuminated the challenges posed by biases, privacy infringements, and the amplification of societal inequalities, while also offering pragmatic solutions and ethical guidelines for responsible AI development and deployment.

As we reflect on the myriad dimensions explored within these chapters, it becomes evident that the ethical implications of generative AI extend far beyond technological innovation—they intersect with fundamental principles of fairness, transparency, and accountability. Whether in the realms of education, business, law enforcement, or healthcare, the ethical imperatives guiding AI deployment are paramount in shaping a future that prioritizes human dignity and societal well-being.

We hope that this volume catalyzes continued discourse and action, inspiring stakeholders from diverse fields to engage thoughtfully with the ethical quandaries posed by generative AI. By fostering collaboration, transparency, and ethical reflection, we can collectively navigate the complexities of AI-driven innovation while safeguarding fundamental human values.

We extend our deepest gratitude to the authors for their scholarly contributions and our readers for engaging with this vital discourse. May the insights gleaned from this volume inform ethical decision-making and pave the way for a future where generative AI is harnessed responsibly for the betterment of society.

*Aftab Ara*
*University of Hail, Saudi Arabia*

*Affreen Ara*
*Department of Computer Science, Christ College, Bangalore, India*

# Acknowledgment

Writing this book, *Exploring the Ethical Implications of Gen AI*, has been a journey of exploration and learning. This journey would not have been possible without the invaluable contributions of many others.

We want to acknowledge the countless researchers, policymakers, and thought leaders who have contributed to the field of AI ethics. Their work provides the foundation upon which this book builds.

Our immense gratitude goes to the entire team at IGI-Global, thank you for believing in this book and helping bring it to the world.

Lastly, we give our special thanks to our family for their unwavering encouragement and support throughout this process. Your unwavering encouragement was the fuel that kept us going. Thank you for believing in us throughout this intellectual odyssey.

Aftab Ara and Affreen Ara

# Chapter 1
# AI and Equity in Higher Education:
## Ensuring Inclusivity in the Algorithmic Classroom

**Amdy Diene**
https://orcid.org/0000-0002-1818-6195
*Liberty University, USA*

## ABSTRACT

*This thought-provoking chapter is systematic research based on theories from literature review and empirical data exploring AI's potential impact on educational equity in higher learning. The atuhors examine how AI-powered systems might inadvertently perpetuate biases, affecting marginalized students disproportionately. This chapter discusses institutions' responsibility to implement AI in ways that support inclusivity and diversity. In addition, it highlights initiatives that prioritize fairness and transparency in AI algorithms. AI systems can perpetuate existing biases if not designed with the right in mind. Additionally, if an AI system is not transparent about its decision-making processes, identifying or rectifying potential biases will be nearly impossible. Readers will gain a deeper understanding of AI's challenges and opportunities in reshaping education, focusing on ensuring that no student is left behind.*

## INTRODUCTION

The most significant disruption from technological development in this decade is Artificial Intelligence (AI), according to How et al. (2020), with the acknowledgment that these machines can solve many tasks requiring human intelligence, including facial recognition, natural language processing, and even driving a car. In this chapter, we will investigate the impact, negative or positive, AI has on higher education. Higher education has long stood as a beacon for progress and enlightenment, where knowledge, critical thinking, and diverse perspectives converge. However, as AI-powered tools weave their way into the fabric of academic institutions, we are met with new challenges that echo age-old issues of equity, fair-

DOI: 10.4018/979-8-3693-1565-1.ch001

ness, and inclusivity. The harnessing of algorithms in the classroom presents both a promise and a peril, with the potential to either uplift or further marginalize. Research in AI and equity in higher education has shown promising results. Cheddadi (2021) and Austin (2023) highlight AI's potential to promote equity by predicting student outcomes and detecting human bias in decision-making processes. However, (Holstein & Doroudi, 2021) warn that AIEd systems may inadvertently exacerbate existing inequities, requiring a more critical examination of their design and implementation. This thinking aligns with (Baldwin & James, 2010) emphasis on the need to reconceptualize equity in the context of massified, globalized higher education.

This chapter sheds light on these complexities, challenging educational stakeholders to ensure that the might of machines overshadows no student. As tech moguls have before it, generative AI enters the game; however, it faces a fierce battle in the business world, especially in areas like healthcare, HR, and academia. What are tomorrow's students ' expectations with artificial intelligence invited to dance on academia's floor? With (AI) making its way into the academic arena, students of tomorrow are bursting with expectations. However, not everyone is on board with this new technology, similar to when the telephone and television were invented. The invention of the phone was met with fear, and critics wondered if the telephone would disrupt face-to-face communication (Blinkoff & Hirsh-pasek, 2023). When television entered the arena, people fretted about potential harm. We are curious to know how ChatGPT or CLAUDE will impact students' ability to acquire knowledge. Some professors suggest an analogy from the calculator and math to ChatGPT and writing. In the same way that calculators became an essential tool for students, it is believed that ChatGPT has the same potential to become a necessary tool for writers and students alike, posit Blinkoff & Hirsh-pasek.

The Southeastern Conference (SEC) and its member universities focus on integrating artificial intelligence into the classroom and supporting workforce development training, The Chronicle (2020). As well as being known for its athletic prowess, the SEC is leading the way in transforming higher education by creating the SEC Artificial Intelligence Consortium. The consortium focuses on applying artificial intelligence and data science to workforce development because it acknowledges the rapid advancement and growing applications of artificial intelligence and data science in all sectors of society. It is important to them that their students are prepared for a future in which artificial intelligence is becoming an increasingly important part of the workforce. The universities embrace AI as a learning tool while acknowledging ethical concerns and hesitations with discussions around issues such as how AI works, effective software use, and academic integrity. The future landscape of academia will include automatic data collection, grading, creativity, feedback, automatic peer review, prompt engineering, contextualized citations, personal tutor, instant lit review, personal Teaching Assistant (TA), and Research Assistant (RA).

Adam Stevens notes that ChatGPT is only a threat to academia if our education system continues to "pursue rubric points and not knowledge" (Lambert & Stevens, 2023). Educating students the old-fashioned way will not prepare them for success or the jobs of tomorrow. It is time to let the old model die peacefully. Used rightly, ChatGPT can be a fantastic tool for students, just like the calculator is, and not something to be feared, according to Blinkoff and Hirsh-pasek. The authors think that in the coming years, this cutting-edge technology will continue to evolve and reshape various industries, including education because its integration into academia holds immense potential for revolutionizing learning methodologies and preparing future students. AI-enabled tools such as virtual tutors, peer reviewers, adaptive learning systems, and AI-enabled grading and feedback are expected to improve the learning experience for tomorrow's students.

*Figure 1. Courtesy of academic insight lab*

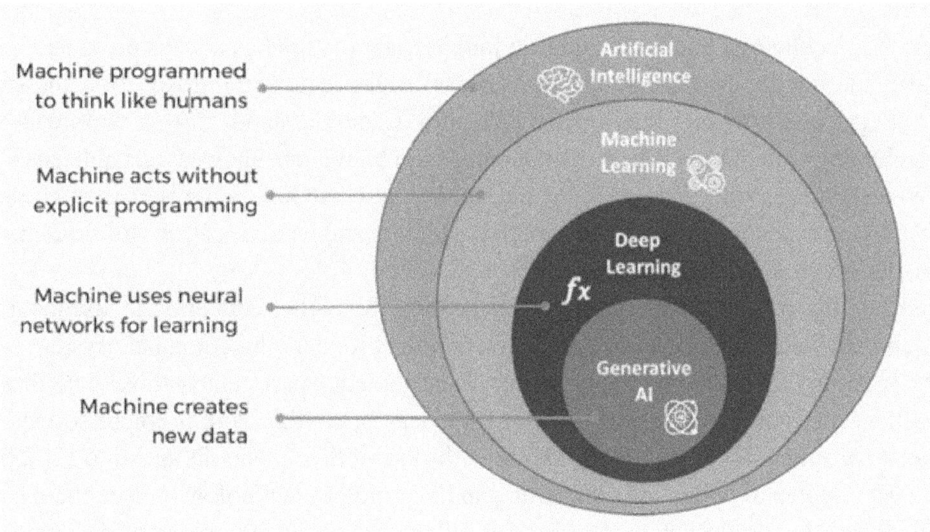


In addition, AI can provide personalized learning experiences to students, helping them stay engaged and improve their overall academic performance. For instance, AI-enabled learning systems can monitor student's progress and offer real-time, personalized support to help address any learning gaps or misconceptions. Although AI can provide customized learning experiences, there is also the potential for students to become too reliant on technology; students need to be allowed to learn how to solve problems independently and grow the critical thinking skills they need to succeed.

The abundance of data on education has created a strong link between data and learning analytics, according to (Cheddadi & Bouache, 2021). In their research within higher education (HE), they propose using deep neural networks to predict students' final results based on their prior knowledge that could be used in favor of underprivileged groups to assist and enhance their chances of success. Deep neural networks are one of the layers of artificial intelligence -when machines use neural networks for learning, called deep learning. The other layers comprise machine learning -when machines act without explicit programming, and generative AI -when machines create new data (as shown in the figure below. The results of their study show that deep learning could promote equity in higher education and is a powerful tool for exploring big data within learning analytics.

In the same vein, another study conducted by Austin et al. (2023) using deep learning algorithms found that AI can serve as an assessment tool to detect potential human bias in decision-making for students in HE. They uncovered insights with deep learning algorithms by comparing recipients awarded scholarships and identifying opportunities for HE institutions to implement a framework, a repeatable set of algorithms to help identify potential bias before granting future scholarship recipients. Like humans, AI algorithms may display biases based on who trained them. Just as humans are trained to become experts, so is artificial intelligence. It is forecast that 100 years from now, AI will possess more expertise than any individual human will ever possess and less bias than those who first trained them. As we move into the future of AI-driven education, we will examine how students will navigate this environment, how to empower them, and students' expectations vs. reality within the classrooms. We will also discuss inclusivity and equity within an AI classroom and students' employability and skills.

## BACKGROUND

The evolution of technology and its integration into various sectors has always presented a dichotomy. While it paves the way for advancements, it poses challenges requiring introspection and strategic maneuvering. Integrating Artificial Intelligence (AI) into academia stands at this crossroads today. The history of technology adoption, such as the telephone and television, showcases a pattern of skepticism followed by acceptance, with generative AI and tools like ChatGPT being the latest entrants. As the penetration of AI in academia deepens, it is crucial to understand its implications for students, educators, and the broader educational ecosystem.

Artificial Intelligence (AI) is increasingly permeating the higher education landscape, offering opportunities and challenges for students, educators, and the broader educational ecosystem (D'agostino, 2023; Knox & Pardos, 2022). AI tools promise to assist students, particularly those with disadvantages, develop skills needed for success, bolster creative, research, writing, and problem-solving skills, and provide executive function support for students with disabilities, posits D'agostino (2023); however, these tools also threaten to widen the education gap like no other technology before them because AI's potential to reproduce systemic bias is a significant concern. Algorithms, if not designed and implemented ethically and equitably, risk replicating the biases of the past, said (Knox & Pardos, 2022), and this is particularly true for under-resourced institutions and students lacking college knowledge who may find navigating the complexities of higher education even more challenging. The AI divide, where some students possess sophisticated skills in prompt engineering while others have scant experience conversing with machines, is another critical issue, and this divide can lead to a situation where those who know how to use AI tools effectively benefit while those who do not, are left behind, according to (Complete College America, 2023).

Despite these challenges, AI has the potential to equalize and scale access to a college degree, increasing individual, economic, and societal benefits, Ascione (2024). However, this requires a commitment to ethical and equitable AI in higher education (Knox & Pardos, 2022); moreover, researchers and policymakers must proceed with caution and healthy skepticism, ensuring that these technologies are designed and implemented ethically and equitably because there is no easy task and will require sustained, rigorous research to complement the rapid technology advances in the field. AI's role in higher education is a double-edged sword. While it offers great promise in assisting students and enhancing learning, it also poses significant risks, particularly in terms of equity; therefore, as we move forward, it is crucial to ensure that AI is leveraged in a way that promotes inclusivity and equity in the algorithmic classroom with ongoing research, ethical considerations, and a commitment to ensuring that all students have the opportunity to benefit from these powerful tools.

## METHODOLOGY

A comprehensive review of historical and contemporary literature is conducted to draw parallels between past technological integrations and the current AI scenario. Cited works from (Blinkoff & Hirsh-pasek, 2023; Lambert & Stevens, 2023; Jirout et al., 2023), and others provide insights into the potential, challenges, and expectations surrounding AI in education. Additionally, real-world examples, such as the adoption or resistance by various school districts, were examined to grasp the on-ground sentiments and responses. The analyses explored the data to identify patterns, themes, and relationships. The analysis results were then used to understand the topic comprehensively. In order to organize the data collection, take notes, and keep track of citations, Citavi, a scholarly reference management software, was used, and to analyze and draw conclusions from the data, NVivo, another qualitative and mixed-methods research software, was used.

## LITERATURE REVIEW

Artificial intelligence (AI) is rapidly transforming academia and reshaping the student experience. As Pedro et al. (2019) and UNESCO (2019) highlight, governments and institutions worldwide are introducing AI across all aspects of education to prepare learners for an increasingly automated future aligned with sustainable development goals. However, integrating AI into education raises pressing questions about the technology's societal implications, including bias, transparency, and equitable access. This literature review synthesizes current scholarship at the intersection of AI and academia to analyze how emerging technologies could impact students, administrators, and educators.

Overall, research suggests AI holds immense promise in augmenting and enhancing education (Alsobeh & Woodward, 2023; Ayinla, 2022; Schwalbe & Wahl, 2020) with data-driven, personalized platforms that can provide real-time support, accurately assess student understanding, and offer tailored guidance. However, researchers also caution that if poorly implemented, AI risks exacerbating existing inequalities across race, class, gender, and other demographics (Jirout et al., 2023). Biased algorithms could cement unfair advantages, widen achievement gaps, undermine inclusion efforts, and disproportionately harm marginalized communities.

To tap AI's potential while safeguarding equity, scholars overwhelmingly emphasize the need for thoughtful governance, stressing principles of transparency, auditing, and ongoing stakeholder education (Emre et al., 2021; Fortney & Steward, 2017; Kazim et al., 2021; Mökander, 2023; Noble, 2018; Raji et al., 2020; Ugwudike, 2022). Algorithmic transparency enables the evaluation of results and identification of errors or manipulation. Regular bias audits by diverse experts can detect and rectify unfair preferences, while stakeholder education empowers administrators, educators, and students to make informed decisions regarding AI adoption. Such oversight prevents "black box" systems that lack explainability. Crucially, subject matter experts underscore the continued necessity of human judgment in AI-augmented education (Jirout et al., 2023), arguing that technology should supplement but not supplant experienced educators' nuanced understanding of diverse student needs. AI insights can inform but should not determine high-stakes decisions around assessment, advancement, admissions, or other areas with significant consequences for young lives.

This literature review examines such considerations, spotlighting both possibilities and challenges associated with onboarding AI in academia. The analysis aims to elucidate critical governance strategies institutions might implement to guide AI's integration responsibly amid rapidly evolving technological landscapes. It also seeks to clarify outstanding questions deserving of further research, like how to balance innovation and inclusion or how biased data may permeate learning algorithms. Ultimately, the review highlights how proactive, equitable policies surrounding nascent technologies in education could profoundly shape many students' lifelong trajectories.

## A FUTURE OF AI-DRIVEN ACADEMIA: HOW WILL STUDENTS NAVIGATE?

In this section, we explore the changing expectations of tomorrow's students and the transformative role of artificial intelligence in academia. In their paper (Pedro et al., 2019; UNESCO, 2019), Artificial Intelligence in Education anticipates how AI affects higher education to allow for informed and appropriate policy responses. They gathered data on introducing AI in education from mainly developing countries to accomplish sustainable development goals, which strive for equitable, quality education for all. They explored how governments and educational institutions are reworking educational programs to prepare learners for increased AI in all aspects of social activities. Their paper reflects on future directions for AI in education and new discussions around the possibilities and risks of AI in education for sustainability.

In the next era, employers are shifting away from traditional indicators of potential employee success, such as college degrees and years of experience, and prioritizing tangible skills, according to a Fortune Magazine headline, stating that skills will take precedence over pedigree in the next era of work, Jirout et al. (2023). Tomorrow's students are increasingly seeking out personalized learning experiences and relying on digital technologies such as artificial intelligence to provide them. AI-powered technologies can supplement traditional instruction, helping improve student outcomes and reduce institution costs. For example, AI-powered learning platforms can provide personalized learning experiences tailored to individual student needs, such as providing differentiated instruction for students with different levels of understanding. AI-powered technologies can potentially reduce academic bias by providing a more equitable learning environment and a learning platform to identify and address differences between student populations, such as those from different socioeconomic backgrounds or those with other learning disabilities.

Furthermore, AI-powered technologies can help students access the same resources and materials, ensuring everyone has the same learning opportunities. Personalized feedback and instruction ensure students receive the necessary support to thrive academically. However, some researchers have argued that AI technologies could exacerbate educational inequalities. They claim that if AI-powered technologies are only available to wealthy districts and schools, there could be a widening achievement gap between rich and poor students. Additionally, if these learning platforms are not properly calibrated, they could reinforce rather than reduce educational inequalities. We explore how AI-powered educational tools reshape the learning landscape, from personalized learning pathways to intelligent tutoring systems. This section also delves into the ethical considerations surrounding AI's integration into education and the need to balance technology and the human touch. Educators will gain insights into how educational institutions can effectively harness AI to enhance student experiences while maintaining the essence of traditional education. According to Forbes, "AI's intelligence led to several schools either indefinitely or temporarily banning the software, including New York Public Schools, Seattle Public Schools, and the Los Angeles Unified School District." The thought is that schools might be concerned about the privacy of students' data, bias, lack of transparency, reliability, ethical concerns, staff concerns about job security, or the schools' lack of technological know-how. However, while some schools have chosen to ban or limit the use of AI, others might decide to adopt it with guidelines in place.

### Empowering Students in the Age of AI

There is no question that artificial intelligence is here to stay, just as Microsoft Word and Excel are today. In March 2023, Microsoft announced that its upcoming AI feature, Copilot, would be available in business apps like Word, PowerPoint, and Excel. This feature will allow users to leverage AI to improve their documents and presentations with automated recommendations, natural language processing, and predictive analytics. This feature will be similar to what Grammarly or ChatGPT is doing today. Additionally, Copilot will enable users to access AI-driven insights from within their documents, making it easier to make decisions quickly and accurately. Copilot will shed light on the symbiotic relationship between artificial intelligence and the academic journey, emphasizing how AI can help students thrive in an increasingly automated environment, just like Microsoft Office did decades ago.

Artificial Intelligence (AI) has become a prominent and transformative force in various aspects of our lives, including education. With advancements in machine learning algorithms and computational power, AI has the potential to revolutionize the academic journey for students, ensuring their success in an increasingly automated environment (Schwalbe & Wahl, 2020). From predictive analytics for academic success to AI-driven career counseling, the authors highlight real-world examples of how AI prepares students for diverse future careers. They emphasize the importance of fostering adaptability and creativity alongside AI-driven tools, spotlighting that a well-rounded education remains crucial in an era of rapid technological advancements. According to (Alsobeh & Woodward, 2023), academia has much to gain from the rapid development of artificial intelligence, machine learning, and natural language processing (NLP). However, they note the essentiality of addressing technology's challenges and opportunities for diversity, equity, and inclusion (DEI) in teaching and learning. The potentials of AI in the educational system include personalized real-time support for students, accurately assessing students' understanding levels, and tailored advice.

## AI's Inadvertent Perpetuation of Biases

Though a marvel of technological advancement, artificial intelligence is no stranger to biases. The root cause is the data it is fed. When the data is tainted with societal prejudices based on race, gender, socioeconomic status, or other factors, AI systems can unintentionally adopt these biases; for instance, an AI-powered admission system might favor applicants from specific backgrounds if its training data is predominantly sourced from those groups. Similarly, a personalized learning tool might make recommendations that do not cater to the diverse learning needs of all students if it was primarily trained on data from a narrow demographic.

Moreover, marginalized students might find themselves disproportionately affected. If an AI system is unaware of cultural norms or linguistic nuances, it will likely misinterpret students' responses, leading to inaccurate assessments of their knowledge or capabilities. AI systems rely on algorithms to understand the data they collect, and these algorithms rely on data collected in a certain way. If an AI system is not trained to consider the nuances of different cultural contexts, it will have difficulty interpreting the data accurately. To address this issue, AI systems must be trained to consider the cultural norms and linguistic nuances of the environment they are operating in. This will ensure the AI system can accurately interpret students' responses and assess their knowledge or capabilities.

## The Onus on Educational Institutions

In the wake of potential pitfalls, educational institutions have a paramount responsibility because they are the stewards of the academic journey; principles of equity and inclusivity should guide their adoption of AI. Firstly, institutions must be proactive in auditing their AI tools. This auditing could involve regularly scrutinizing the sources of data, the design of algorithms, and the outcomes they produce. Diverse teams of experts, including ethicists, sociologists, and educators, should be consulted to ensure that a holistic view is taken. Secondly, there should be an emphasis on 'human-in-the-loop' systems. Even the most advanced AI cannot understand the nuances and intricacies of human behavior like a seasoned educator can. Therefore, while AI can provide valuable insights and augment teaching, final decisions should often rest in human hands.

Fairness and transparency must be emphasized for AI to champion educational equity truly. AI algorithms must be developed with input from stakeholders from different backgrounds, including students, teachers, and parents. The algorithms must also be tested for accuracy and bias to ensure they are not creating unfair outcomes for specific groups. Additionally, the algorithms must be monitored and evaluated to ensure they function correctly. AI should only supplement existing educational practices and not replace human judgment. AI systems must be designed with fairness in mind and should include safeguards to ensure that any bias is not perpetuated. Additionally, AI systems should be transparent in their operation so that any potential issues can be identified and addressed quickly. Some key considerations include the following: Algorithmic transparency, diverse data sources, bias detection and rectification, and stakeholder education.

## Algorithmic Transparency

Institutions must understand the inner workings of their AI tools; a 'black box' approach that hides the decision-making process is contrary to academic openness. State-of-the-art AI tools are increasingly used in academic research, making research accessible and enjoyable for scholars, Ayinla (2022); therefore, institutions need to understand how they work to evaluate their output and ensure accuracy. Hiding the decision-making process makes it impossible to verify the accuracy and reliability of the results (Fortney & Steward, 2017). These institutions must comprehensively understand the algorithms and models used in their AI tools to ensure transparency and prioritize systems that clearly explain their decisions. Without a clear explanation of how the AI tool arrived at a particular decision, it becomes difficult to identify and address any potential biases, errors, or manipulation. Furthermore, without a comprehensive understanding of the algorithms and models, it would be difficult to assess the results' accuracy and determine their reliability.

## Diverse Data Sources

Considering a wide range of student demographics ensures that a well-rounded AI system can accommodate a broader range of learning needs, which will lead to AI tools that can better address overall learning needs and avoid narrow or biased viewpoints. AI tools that are tailored to specific demographics, such as gender or race, may miss out on important nuances and considerations that are essential for providing equitable learning. By considering the full range of student demographics, AI systems can better evaluate the nuances of students' different learning styles and needs, allowing for a more comprehensive and equitable learning experience. As a result, AI tools tailored to specific demographics should be avoided to guarantee an honest learning experience for all students. However, there are potential drawbacks to this approach. For example, AI systems that have been tailored to specific demographics may not take into account the needs of students outside of that demographic. Additionally, it may be difficult for AI systems to keep up with the ever-evolving needs of students, which could result in an inequitable learning experience over time.

## Bias Detection and Rectification

Regularly scheduled audits can help detect biases in AI outputs. Tools and methodologies that actively hunt for preferences and rectify them should be integrated into the system's lifecycle. Several authors including, (Kazim et al., 2021; Mökander, 2023; Raji et al., 2020; Ugwudike, 2022), suggest a framework for algorithmic auditing that supports artificial intelligence system development drawing on an organization's values or principles to assess the fit of decisions made throughout the process. According to Raji et al., these systems have the potential to replicate or amplify harmful social biases (Buolamwini & Gebru, 2018; Kiritchenko & Mohammad, 2018; Raji & Buolamwini, 2019). According to Noble's (2018) Algorithms of Oppression, recognizing that technology is built with and integrates bias is a model for many academic, critical, and social movements. According to Mökander, policymakers and technology providers must promote auditing as an AI governance mechanism. Academic researchers can play an essential role in how AI systems are designed and used and how they impact societies. In the same vein, Emre et al. (2021), in their study, systematizing audits in algorithmic recruitment, provide convincing facts about the development of artificial intelligence systems (AI) designed to measure individual differences. They are concerned about this high-impact technology use due to the potential for unfair hiring resulting from the algorithms used by these tools and how audits and assessments on AI-driven systems can be used to assure fair deployment and responsibility in the most appropriate manners. Ugwudike (2022) notes that AI systems generate outcomes through their inputs, which inform the design, operations, and outputs. The author used predictive policing algorithms as a case study to illustrate how assumptions can pose adverse social consequences and should be systematically evaluated during audits to detect unknown risks, avoid AI harms, and build ethical systems.

## Stakeholder Education

In order to create a diverse learning community, educators, students, and administrators must be made aware of both the potential and the pitfalls of AI, providing them with the tools to make informed decisions, voice concerns, and contribute to developing an inclusive environment. AI has the potential to level the playing field for students of all backgrounds, but it also has the potential to perpetuate existing biases and inequalities. Therefore, ensuring all stakeholders have the information they need to make the best decisions is critical. To ensure that AI is used effectively and responsibly, it is essential to provide all stakeholders with comprehensive and accessible information that enables them to make informed decisions. However, some argue that AI could also actually lead to more inequality.

# FINDINGS

The current generation is increasingly leaning towards personalized learning experiences facilitated by digital technologies. AI's capacity to offer customized learning pathways and reduce institutional costs positions it as a valuable educational asset. AI holds the potential to democratize education by providing equal learning opportunities to students from diverse backgrounds, bridging the knowledge gap that often exists due to socioeconomic disparities. Like all data-driven systems, AI is susceptible to inherent biases based on its training data. Unchecked, these biases can perpetuate societal prejudices, leading to unfair outcomes in educational settings. Institutions bear a significant responsibility to ensure the ethical deployment of AI, such as including regular audits, consultations with diverse experts, and a balanced human-AI approach. To foster trust and inclusivity, AI systems used in education must maintain algorithmic transparency to enable stakeholders to understand and question the decisions made by these systems.

## LIMITATIONS

While the findings aim to capture a holistic view, individual experiences with AI in education can vary. The outcomes might differ based on geographical location, institutional policies, and specific student demographics. The pace at which AI technology evolves might lead to certain aspects of this analysis becoming outdated shortly. Continuous updates are essential to maintain relevance. While the chapter acknowledges potential challenges, there might be unforeseen negative implications of AI in education that are not captured here.

## FUTURE RESEARCH

Integrating AI in academia is an ongoing journey warranting continuous exploration. Future research areas could include AI's impact on educator roles. As AI assumes a more significant role, how does it impact educators? Are their roles diminished, or have they become more of a guide or mentor? How do students perceive AI-driven learning tools? How does the teacher-student relationship change in an AI-dominated environment? A longitudinal study examined AI-driven education's long-term effects on students' career success, adaptability, and overall well-being. They are crafting comprehensive ethical guidelines for AI deployment in educational settings, ensuring fairness, inclusivity, and transparency. The ever-evolving landscape of AI in academia promises both advancements and challenges. The goal of fostering enriched and equitable learning experiences can be realized with balanced integration.

## CONCLUSION

The marriage of AI and higher education is undoubtedly transformative. However, as we stand at this juncture, we must reflect, reassess, and realign our strategies. The quest for knowledge should be as inclusive as the knowledge itself. By consciously crafting AI systems with fairness and transparency, higher education can ensure that no student is left behind in the algorithmic age. Integrating Artificial Intelligence (AI) into the academic realm is akin to opening a Pandora's box of potential and pitfalls. On the one hand, AI promises personalized education, efficiency, and inclusivity, while on the other, it raises concerns about biases, equity, and the essence of the human touch in education.

The historical skepticism surrounding technological innovations, like the telephone or television, mirrors the current hesitancy with AI tools such as ChatGPT. However, as with past innovations, adaptation and integration seem inevitable. As the exploration of AI's place in academia continues, the emphasis must remain on using AI to augment, not replace, human educators. Institutions bear the mantle of responsibility, ensuring that AI's adoption prioritizes every student's holistic development, fosters critical thinking, and maintains the core values of education. Equipped with lessons from the past, a cautious eye on the present, and a visionary gaze into the future, the academic community stands poised to navigate the complex dance of AI and education. By championing transparency, equity, and continuous learning, the true potential of an AI-integrated academic world can be harnessed to benefit all stakeholders involved.

## REFERENCES

Alsobeh, A., & Woodward, B. (2023). AI as a Partner in Learning: A Novel Student-in-the-Loop Framework for Enhanced Student Engagement and Outcomes in Higher Education. In *Proceedings of the 24th Annual Conference on Information Technology Education*, (pp. 171–172). Research Gate. 10.1145/3585059.3611405

Ascione, L. (2024). Here is how AI can increase equity in student success. e*Campus News*.

Austin, T., Rawal, B. S., Diehl, A., & Cosme, J. (2023). *AI for Equity: Unpacking Potential Human Bias in Decision Making in Higher Education*. AI, Computer Science and Robotics Technology.

Baldwin, G., & James, R. (2010). *Access and equity in higher education*. Elsevier. . doi:10.1016/B978-0-08-044894-7.00825-3

Blinkoff, E., & Hirsh-pasek, K. (2023). *ChatGPT: Educational friend or foe*? Brookings. https://www.brookings.edu/articles/chatgpt-educational-friend-or-foe/

Buolamwini, J., & Gebru, T. (2018, January). Gender shades: Intersectional accuracy disparities in commercial gender classification. In *Conference on fairness, accountability, and transparency*, (pp. 77-91).

Cheddadi, S., & Bouache, M. (2021, August). Improving equity and access to higher education using artificial intelligence. In *2021 16th International Conference on Computer Science & Education (ICCSE)* (pp. 241-246). IEEE. 10.1109/ICCSE51940.2021.9569548

Complete College America. (2023). *The AI Divide: Equitable Application of AI in Higher Education to advance the completion Agenda*. Complete College. https://completecollege.org/wp-content/uploads/2023/11/CCA_The_AI_Divide.pdf

D'agostino, S. (2023). *How AI tools both help and hinder equity in higher ed*. Inside higher Ed. https://www.insidehighered.com/news/tech-innovation/artificial-intelligence/2023/06/05/how-ai-tools-both-help-and-hinder-equity

Fortney, C., & Steward, D. (2017). A Qualitative study of nurse observation of symptoms in infants at end-of-life in the neonatal intensive care unit. *Intensive & Critical Care Nursing*, *40*, 57–68. doi:10.1016/j.iccn.2016.10.004 PMID:28189383

How, M. L., Cheah, S. M., Chan, Y. J., Khor, A. C., & Say, E. M. P. (2020). Artificial intelligence-enhanced decision support for informing global sustainable development: A human-centric AI-thinking approach. *Information (Basel)*, *11*(1), 39. doi:10.3390/info11010039

Jirout, J., Hirsh-pasek, K., & Evans, N. (2023). What ChatGPT can't do: Educating for curiosity and creativity. *Fortune Magazine*. https://www.brookings.edu/articles/what-chatgpt-cant-do-educating-for-curiosity-and-creativity

Kazim, E., Koshiyama, A. S., Hilliard, A., & Polle, R. (2021). Systematizing audit in algorithmic recruitment. *Journal of Intelligence*, *9*(3), 46. doi:10.3390/jintelligence9030046 PMID:34564294

Kiritchenko, S., & Mohammad, S. M. (2018). *Examining gender and race bias in two hundred sentiment analysis systems*. arXiv preprint arXiv:1805.04508. doi:10.18653/v1/S18-2005

Knox, D., & Pardos, Z. (2022). *Toward Ethical and Equitable AI in Higher Education*. Inside Higher Ed. https://www.insidehighered.com/blogs/beyond-transfer/toward-ethical-and-equitable-ai-higher-education

Lambert, J., & Stevens, M. (2023). *AI ChatGPT: Academia Disruption and Transformation*. In T. Bastiaens (Ed.), *Proceedings of EdMedia + Innovate Learning* (pp. 1498-1504). Association for the Advancement of Computing in Education (AACE). https://www.learntechlib.org/primary/p/222671/

Mökander, J. (2023). Auditing of AI: Legal, Ethical and Technical Approaches. *Digital Society : Ethics, Socio-Legal and Governance of Digital Technology*, *2*(3), 49. doi:10.1007/s44206-023-00074-y

Noble, S. U. (2018). Algorithms of oppression. In *Algorithms of oppression*. New York university press. doi:10.2307/j.ctt1pwt9w5.11

Pedro, F., Subosa, M., Rivas, A., & Valverde, P. (2019). *Artificial intelligence in education: Challenges and opportunities for sustainable development*. Research Gate.

Raji, I. D., & Buolamwini, J. (2019). *Actionable auditing: Investigating the impact of publicly naming biased performance results of commercial ai products*. In Proceedings of the 2019 AAAI/ACM Conference on AI, Ethics, and Society, 429-435. 10.1145/3306618.3314244

Raji, I. D., Smart, A., White, R. N., Mitchell, M., Gebru, T., Hutchinson, B., & Barnes, P. (2020). Closing the AI accountability gap: Defining an end-to-end framework for internal algorithmic auditing. In *Proceedings of the 2020 conference on fairness, accountability, and transparency*, (pp. 33-44). ACM. 10.1145/3351095.3372873

Schwalbe, N., & Wahl, B. (2020). Artificial intelligence and the future of global health. *Lancet*, *395*(10236), 1579–1586. doi:10.1016/S0140-6736(20)30226-9 PMID:32416782

Ugwudike, P. (2022). AI audits for assessing design logics and building ethical systems: The case of predictive policing algorithms. *AI and Ethics*, *2*(1), 199–208. doi:10.1007/s43681-021-00117-5 PMID:35909984

UNESCO (2019). *Artificial intelligence in education: challenges and opportunities for sustainable development*. UNESCO.

# Chapter 2
# Artificial Intelligence in Different Business Domains:
## Ethical Concerns

**B. Sam Paul**
*Vellore Institute of Technology Business School, Vellore Institute of Technology, Chennai, India*

**A. Anuradha**
*Vellore Institute of Technology Business School, Vellore Institute of Technology, Chennai, India*

## ABSTRACT

*Generative artificial intelligence (AI) refers to a form of AI technology that possesses the capability to produce a diverse array of content, encompassing text, images, artificially generated data, sound and chat responses. Generative AI is a complex branch of AI that allows systems to create new content by using current information such as text, audio, video. Due to the distribution of unlawful material, copyright violations, privacy of data violations, and the amplifying of preexisting biases, this powerful technology also raises serious ethical issues. Privacy, security issues and raising ethical concerns addressing the future of generative AI in different business domains and functions namely, Marketing, Human Resources, Finance, operations and information technology are discussed in this chapter.*

## INTRODUCTION

Traditional AI is focused on recognizing patterns, making decisions, enhancing analytics, categorizing data, and detecting fraud. Generative AI goes beyond the limitations of AI by attempting to generate wholly new data that mimics human-created material. Early adopters in a variety of sectors, including banking, retail, automotive, health care and agriculture are fusing fast computing with Generative artificial intelligence to alter corporate operations, service offerings, and productivity. These sectors include drug development, financial services, retail, and telecommunications. Although there are many advantages to generative AI, possible risks and privacy concerns still pose difficulties.

DOI: 10.4018/979-8-3693-1565-1.ch002

## OBJECTIVES OF THIS CHAPTER

- ❖ Providing a comprehensive overview of Artificial Intelligence application in different domains of business management namely marketing, human resources, finance, operations and information technology.
- ❖ Review the current potential of Artificial Intelligence in management of business with ethical implications.

## GENERATIVE ARTIFICIAL INTELLIGENCE IN DIFFERENT BUSINESS DOMAINS/FUNCTIONS

### AI in Marketing Domain/Function

The impact of artificial intelligence on several industries is undeniable, positioning it as a potentially ground breaking technology of the 21st century. One such industry that stands to be significantly transformed is marketing, with the anticipated integration of generative AI poised to reshape its landscape. The act of generating content constitutes merely a single component within the overall process. The efficacy of AI-generated content is contingent upon its ability to effectively engage the intended audience at the appropriate juncture. Incorporating the produced content into an automated marketing pipeline that not only identifies the client profile but also delivers a customized experience at the suitable stage of interaction is crucial for prompting the intended response from the consumer.

The usage of artificial intelligence has the potential to not only automate and optimize processes, but also deliver tailored and appealing information to customers. Generative AI is increasingly crucial in a wide range of industries, including retail, healthcare, finance and education.

With the increased usage of Generative AI in marketing, there are persistent ethical concerns. It includes everything from privacy and security concerns to discrimination against those who are marginalized and even gender prejudices. Marketing is one key sector where AI has had an influence.

#### Building Generative AI Into Marketing

**❖ Audience Segmentation**

The system offers the capability to analyze client data and behavior with the purpose of identifying patterns and trends that may subsequently be leveraged to generate more effective and focused marketing campaigns.

**❖ Tailoring Marketing**

The adoption of generative AI has the potential to automate the development of marketing content. Various forms of content, including textual materials for posts on social media, blogs, and emails, as well as multimedia content such as images and videos, are integral components of this phenomenon.

*Figure 1. Top domain of generative AI exploration*

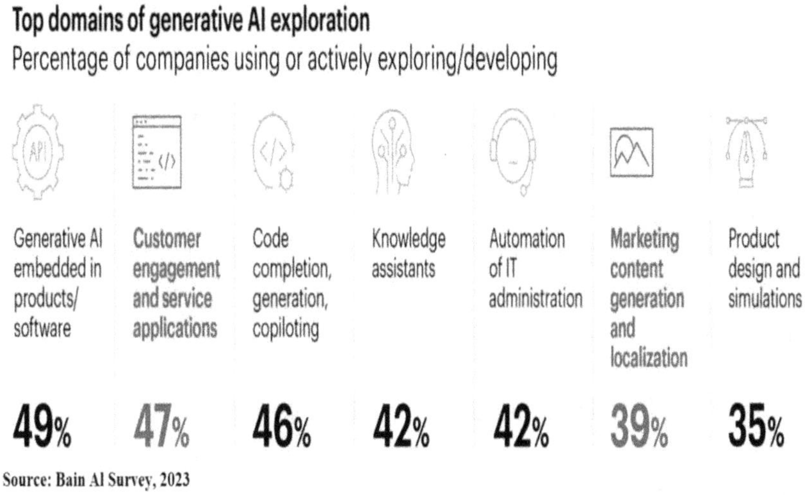


**❖ Marketing Automation**

The implementation of generative AI holds promising prospects for the automation of several aspects within the field of marketing, including email advertising, social networking marketing, and search engine advertising. This includes the implementation of automated processes for generating and disseminating advertisements, along with the evaluation of marketing campaign effectiveness.

## Positive Aspects of Artificial Intelligence in Marketing

**❖ Customization message**

Companies should try to use statistical analysis to establish consumer preferences and make recommendations based on that information like Amazon prime video recommends shows and flipkart recommends product for customers.

**❖ Cost reductions and increased marketing efficiency**

Deep learning, delivered through Artificial Intelligence, allows computers to more precisely evaluate consumer behaviours and predict which groups are most likely to become buyers.

**❖ Customization products**

Identifying prospective customers using data such as by demographics, geography, purchase history and so on. Tracking and consumer data are also included with the products themselves. Generative AI helps customers connect in personalized ways.

Generative AI use case gaining most traction with companies is shown in the Figure 1 Top Domains of Generative AI Exploration.

## Negative Aspects of Artificial Intelligence in Marketing

❖ **Without humans, computers won't work**

Computers cannot change their minds, make innovative judgments, or use their imaginations. While artificial intelligence might help with some solutions.

❖ **Algorithms can be incorrect**

AI content technological advances should be carefully integrated with inbound marketing strategy. (Luis Arango et al., 2023).

## Ethical Aspect -Artificial Intelligence Usage in Marketing

❖ **Identify potential ethical concerns**

Prior to the integration of artificial intelligence (AI) into content creation processes for marketing, it is essential to conduct a thorough examination of the potential ethical issues that may emerge.

❖ **Evaluate the impact on stakeholders**

The potential impact of integrating artificial intelligence (AI) into content creation on various stakeholders, including customers, employees and the wider community will have to be examined.

❖ **Prioritize ethical values**

Prioritise ethical values such as impartiality, transparency, and privacy over profit and efficiency.

❖ **Involve diverse perspectives**

It is imperative to involve stakeholders from diverse backgrounds, especially those who may experience disparate impacts resulting from area of content creation.

❖ **Regularly review**

Regularly evaluate the usage of artificial intelligence in content generation for marketing to determine how it adheres to ethical norms and effectuate any requisite modifications.

❖ **Concerns about privacy and security**

AI may analyze data on the buying habits, preferences, behavior, and demographic data to produce more effective and relevant marketing messages. To maintain a safe image, businesses must be honest in collecting and utilizing customer data and provide customers with the choice to opt out of data gathering if necessary.

❖ **Overcoming the prejudice**

Another significant issue that should be addressed in relation to AI applications in marketing is the possibility of AI bias in the algorithms that power them. The data utilized to train AI algorithms frequently contains flaws and biases, resulting in contentious outcomes.

❖ **The Role of Ethics in AI Development**

AI technology is growing increasingly widespread in marketing and other industries with each improvement. However, it is critical to remember that ethics is the most important factor in designing and deploying AI. This entails recognizing and addressing the previously mentioned ethical challenges, as well as ensuring that these AI systems are developed and utilized in a responsible manner that respects persons' privacy and dignity and developers to be mindful of the potential effects of AI systems and to design them in an open and accountable manner.

❖ **Cultural Nuances and Insensitivity**

Generative artificial intelligence (AI) models may exhibit limitations in comprehending cultural nuances and sensitive topics, potentially resulting in the generation of incorrect or even harmful information.

❖ **Not a Substitute for Human Creativity**

Certain aspects of marketing campaigns can be automated by generative AI, but it can't substitute the imaginative thinking or feelings that marketers use for building compelling campaigns and "human touch."

## AI in Human Resource Domain/Function

Human resources (HR) function and processes can be significantly improved and streamlined with the help of artificial intelligence (AI).

- ❖ Human resources professionals have the opportunity to adopt artificial intelligence (AI) as a means to improve decision-making processes and deliver enhanced efficiency and cost-effectiveness to organizations.
- ❖ When incorporating generative AI into their HR and people responsibilities, organizations should exercise caution in prioritizing data protection and privacy.

Artificial intelligence (AI) refers to the discipline concerned with enabling machines to replicate human intelligence in order to execute various jobs. Voice assistants such as Siri and Alexa, along with customer service chatbots, are built upon artificial intelligence (AI) technology.

Generative artificial intelligence (AI) constitutes a distinct category within the broader field of AI. Human resources (HR) departments have the ability to employ applications in order to automate and enhance procedures, decrease expenses, enhance decision-making capabilities, and contribute to the enhancement of employee engagement.

*Figure 2. Primary goal of generative AI for HR processes*

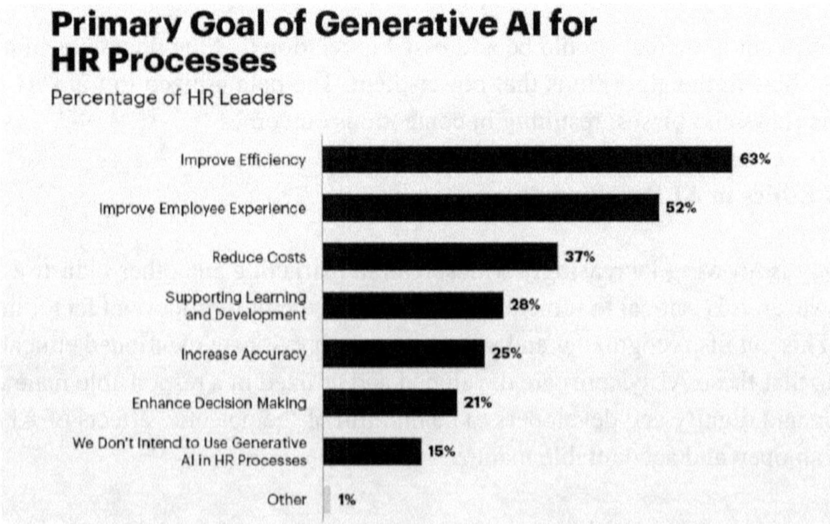


Source: Gartner,2021

Despite generative AI has shown beneficial impact on HR professionals and how individuals function, it is essential to understand the existence of some barriers. These obstacles include data security and privacy concerns, in addition to limitations and potential hazards. Algorithmic bias, the ethical and legal implications are among the various challenges that may arise. Shift has brought about several advantages for employees and has improved the utilization of resources, decision-making processes, and problem-solving capabilities inside these firms. Nevertheless, despite an increasing inclination towards scholarly pursuits, the body of research on artificial intelligence (AI)-based solutions for human resource management (HRM) remains restricted and lacks cohesion. (Pawan Budhwara et al 2022)

The primary goal of Generative AI for HR Process with percentage of HR leaders is shown in the above Figure 2 Primary Goal of Generative AI for HR Processes

## Building of Generative AI on HR

### ❖ Analyzing and interpreting information obtained through people analytic

Generative artificial intelligence (AI) systems possess the capacity to endure training by utilizing past human resources (HR) data, enabling them to generate projections regarding approaching outcomes. As an illustration, the models possess the capability to make predictions regarding attrition rates or discern possible high-performing applicants in the context of the hiring process.

### ❖ Resourcing and Talent Acquisition

Artificial intelligence (AI) algorithms have the capability to analyse curriculum vitae (CVs) and job applications in order to identify the individuals who possess the highest qualifications based on the requirements given in the description. The tool facilitates recruiters in improving their hiring decisions by furnishing them with relevant information.

❖ **Onboarding**

One of the key responsibilities of an organization is to impart information to newly hired personnel regarding its goal, values, and culture with support to newly hired personnel in comprehending and effectively manoeuvring through the policies and procedures of the organization.

❖ **Employee Engagement**

One way to enhance communication between employees and their bosses is by facilitating regular check-ins and feedback sessions. Providing aid for mental health and well-being, including strategies for managing stress, activities promoting mindfulness, and facilitating access to employee assistance programs.

❖ **Learning and Development**

Artificial intelligence has the capability to adaptively modify the difficulty level, tempo, and substance of learning materials based on the analysis of employee interactions, quiz outcomes, and assessment findings

❖ **Decision-Making**

The integration of artificial intelligence (AI) technologies empowers human resources (HR) professionals to augment their decision-making process through the utilization of accurate and unbiased data. As a result, this phenomenon contributes to enhanced outcomes for both individual employees and the overall organization.

## Positives Aspects of Artificial Intelligence in Human Resource

Artificial Intelligence Chatbots as a Complement to Human Training Efforts

One prevalent fallacy surrounding artificial intelligence (AI), computers, and automation at large pertains to the ultimate displacement of human labor by latest technologies. Although technology has the capacity to replace employees, it also holds significant potential to complement rather than entirely replace human labor.

## Negative Aspects of AI in Human Resource

❖ Excessive dependence in Generative AI might lead to dearth significant interpersonal relationships among colleagues, thus causing adverse effects on employee morale and satisfaction.
❖ ChatGPT cannot be considered a definitive and authoritative source, and it has a deficiency in empathetic capabilities. Generative artificial intelligence (AI) systems may have challenges in comprehending the subtleties of language or cultural variations, potentially resulting in instances of miscommunication or misrepresentation of conveyed messages.

- ❖ Generative artificial intelligence (AI) is dependent on technological infrastructure, which is susceptible to technical malfunctions, service interruptions, and various other challenges that might impede seamless communication and perhaps lead to user dissatisfaction.
- ❖ If Generative AI is not appropriately developed and trained, it has the potential to sustain prejudices and discrimination towards specific groups, such as minorities, resulting in unjust treatment or exclusion from communication platforms.
- ❖ Generative artificial intelligence (AI) techniques have the potential to produce information that is derived from obsolete or outdated data sources.

Ethical Aspect-Artificial Intelligence Usage in Human Resource

❖ ***Dramatically Increased Self-Service***

The implementation of HR self-service has elicited varied responses from employees in past instances. However, Generative AI provides enhanced conversational processes and personalised information, which may potentially increase its adoption among employees who prioritise simplicity of use when fulfilling their requirements.

❖ ***Productivity and Experience Enhancements***

The utilisation of advanced automation and data analysis in contemporary Generation AI (Gen AI) applications has yielded notable outcomes. These include a threefold increase in the speed of content generation and visualisation, the automation of over 50% of tasks involved in an on boarding process, and a doubling of engagement rates in recruitment when personalised messages are composed using Gen AI.

❖ **Truly Personalized**

Technology plays a crucial role in enabling Human Resources (HR) departments to gain deeper insights into employees with facilitating the monitoring of employees' work patterns, identifying employees learning and development needs, predicts employees vacation requirements, and determining the potential benefits of providing reminders for annual goals or strategic programmes.

## AI and Finance Domain/Function

Finance is in our daily life. We invest, borrow, lend, budget, and save money. Finance also provides guidelines for corporation and government spending and revenue collection. Traditional statistical solutions such as regression, PCA, and CFA have been widely used in financial forecasting and analysis. With the increasing interest in artificial intelligence in recent years, the Artificial Intelligence (AI) techniques in the finance domain systematically attempts to identify the current AI technologies used, major applications, challenges, and trends in Finance (Xuemei Li et al 2023).

Various Application of AI in Banking and Finance in future is shown the above Figure 3 Application of AI in Banking and Finance.

*Figure 3. Applications of AI in banking and finance*

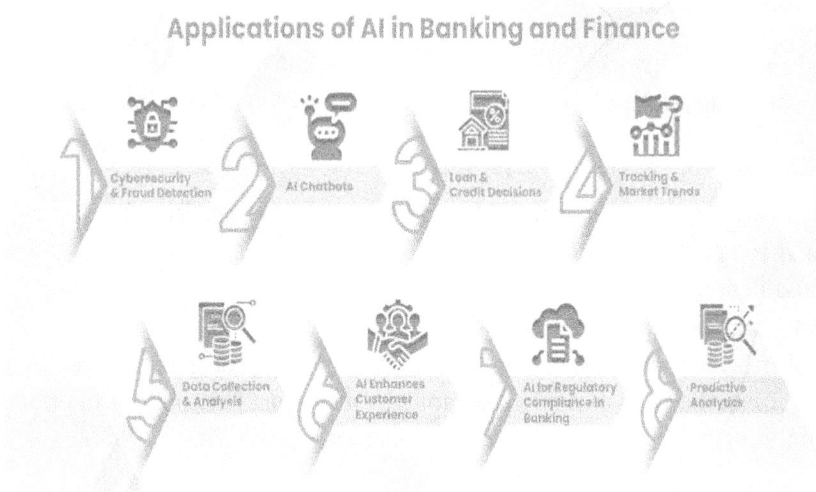


Source: esds.co.in, 2023

## Building Generative AI in Finance industry

Blockchain is an electronic, open ledger that records and stores transactions across the network of computers. Each block in the chain comprises a number of transaction records and is secured through cryptography by the previous block. This creates an immutable record of all transactions on the network. Although block chain technology is often associated with cryptocurrencies like Bitcoin, it has many other uses beyond digital currencies. It can create secure, decentralized systems for various applications, including supply chain management and voting systems. Generative AI models powered by block chain can predict financial market trends more accurately, leveraging historical financial data. This enables better investment decisions and risk management.

❖ ***Variational Autoencoders (VAEs)***

Variational Autoencoders (VAEs) are specifically designed to acquire knowledge about the fundamental input data and afterwards produce unique data that have a strong resemblance to the original data distribution.

**Optimization: VAEs** have the capability to acquire knowledge about the fundamental structure present in past market data, enabling them to construct novel investment portfolios.

**Anomaly detection: VAEs** possess the capability to detect anomalous patterns inside transactions.

**Modelling: VAEs** have the potential to be employed for the purpose of modelling and evaluating risks inside financial systems.

**Synthetic data generation: VAEs** have the capability to produce artificial financial data as a means of addressing constraints present in actual datasets.

**Options trading: VAEs** have gained significant popularity in the field of options trading due to their ability to produce synthetic volatility surfaces. This use of VAEs has proven to enhance the precision of options pricing with trading.

❖ **Applications of GANs**

Data generation

Generative Adversarial Networks (GANs) offer a viable solution to the issues posed by limited or biased availability of financial data by enabling the generation of synthetic financial data in an effective manner.

Financial Misuse

The utilization of Generative Adversarial Networks (GANs) has promise for enhancing fraud detection within the financial service sector, as they possess the capability to accurately distinguish between legitimate and fraudulent transactions.

Market simulation and scenario analysis

Generative Adversarial Networks (GANs) have the capability to generate synthetic market data, thereby enhancing comprehension of market dynamics, forecasting fluctuations in prices, and assessing the influence of multiple variables in financial organization.

❖ **Autoregressive models**

It is a particular type of time series models, are commonly utilized in the domain of finance for purposes of evaluation and prediction. Some of the applications of Autoregressive models are

## Time series forecasting

Models possess the capability to forecast forthcoming values of financial variables by leveraging their historical values. These models are employed in the prediction of stock prices, rate of interest, rate of exchange, and various other financial indicators.

**Optimization:** It is utilized to effectively capture the volatility and correlations exhibited by asset returns, hence facilitating the process of risk evaluation and portfolio optimization.

❖ **Virtual assistants**

Virtual Assistants in Financial Sector

Virtual assistants, Chatbots have become increasingly popular in Financial Sector. With their ability to improve client support and engagement, engage with clients through an interface, offering support and assistance around the clock.

Conversational Ability of Agents

Virtual assistants have the capability to produce solutions to consumer queries that are both contextually relevant and exhibit human-like qualities through the generative AI models. These models possess the capability to comprehend and grasp the underlying purpose or objective behind inquiries posed by customers, so enabling them to furnish precise and significant replies.

Trading Strategies

Trading methods play vital roles within the financial domain. Financial companies endeavour to optimize their profits with effectively mitigating dangers through the implementation of diverse trading and investment strategies.

Trade signals

Models play a pivotal role in the generation of trading signals and the identification of investment possibilities. Through the examination of extensive historical market data, generative AI algorithms have the capability to discern patterns.

AI-powered chatbots

Chatbots have the capability to provide personalized and customized responses, hence enhancing the level of engagement and customer-centricity in interactions. Through the comprehension of particular consumer preferences and histories, chatbots possess the capability to offer recommendations, suggestions, and solutions that are in accordance with the specific wants of the customer.

Compliance

Financial institutions are required to comply with an intricate network of regulations and norms enforced by regulatory bodies. Compliance refers to the act that transactions adhere to relevant regulations.

Ensuring compliance

Artificial intelligence has the capability to automate intricate regulatory evaluations, hence enhancing the efficiency and precision of compliance procedures. Through the utilization of sophisticated algorithms, generative artificial intelligence (AI) possesses the capability to evaluate extensive quantities of data, interpret regulatory frameworks, and detect probable instances of non-compliance.

Predicting risks

Through the examination of historical cyber occurrences and the utilization of threat intelligence, Generative AI algorithms possess the capability to discern prospective forthcoming hazards and vulnerabilities. These models have the capability to offer timely alerts and valuable perspectives on emerging hazards, enabling financial institutions to take proactive measures in mitigating risks prior to their actualization.

## ❖ Transformer models

The attention mechanism plays a crucial role in the architecture of a transformer model enables to dynamically allocate weights or levels of significance to specific segments of the input sequence while generating representations.

Applications of transformer models in finance:

Sentiment analysis

The application of contextual analysis and the examination of word relationships allows for the extraction of significant insights pertaining to market sentiment, hence facilitating the process of making well-informed investment decisions.

Document classification

Transformer models are employed for the purpose of categorizing preset classes of textual materials, such as financial reports and research papers. This process facilitates the organization and filtration of substantial volumes of financial data.

Financial text generation

Transformer models have the capability to produce synthetic financial statements, market research, and other significant textual content. The acquisition of written content production skills by individuals is facilitated through the identification and utilization of discernible patterns and structures within financial data.

### Positive Aspects of AI in Finance

- Budgeting and forecasting are greatly more efficient, providing business more time to create business partnerships while providing advise services.
- Human beings can be outperformed by artificial intelligence on fraud detection and prevention with Apps powered by artificial intelligence to eliminate bias from metrics.

Negative Aspects of AI in Finance

- Advanced technology currently prohibitively expensive for the majority of fintech organizations. Once computers develop their own capabilities, they are expected to be unstoppable. A lack of control over regulations could be a source of concern in years to come. Cybercrime risks can be fine-tuned by regulations.
- Data misuse in the financial technology sector often ends in large losses, which is why the AI should develop protocols and privacy laws.

Ethical Aspect -Artificial Intelligence Usage in Finance

- Artificial intelligence (AI) systems possess the ability to make choices at a faster pace compared to human beings. Consequently, they are being employed more frequently in the automation of trading processes, risk assessment procedures, and loan approval systems.
- An important ethical consideration revolves around the possibility of artificial intelligence (AI) systems exacerbating pre-existing biases. Artificial intelligence (AI) algorithms have the capacity to undergo training through the utilisation of data sets that may possess inherent biases.

## AI and Operations Domain/Function

The application of generative AI, which is capable of generating novel solutions and processes, empowers an operations manager to significantly transform their operational practices. One of the most efficacious strategies for an operations manager to optimize the potential of Artificial Intelligence(AI) is to use it for the purpose of scrutinizing prevailing procedures, detecting bottlenecks and other forms of inefficiencies, and devising novel methods and resolutions for extant predicaments.

### Building Generative AI in Industrial Operations

❖ **Supply Chain Management**

The utilization of AI-driven processes facilitates a more efficient workflow by delegating tasks such as data collecting, analysis, and decision-making to external sources. The utilization of this technology enables the ability to make inventory and order fulfilment decisions that would be unattainable with traditional approaches.

### ❖ Production Performance

AI technology identifies potential enhancement areas, allowing efficiency and operation with profit.AI can also be used to predict performance of the future, enabling managers with information and decisions. In future AI Technology will be able to rapidly identify bottlenecks for production process and make decisions.

### ❖ Quality Control

Artificial intelligence (AI) has the capability to analyze large volumes of data and offer valuable insights that support decision-making processes grounded in evidence derived from data.

Generative AI assists operations managers in comprehending trends over time, enabling to make more accurate future predictions and decisions (Abolghasemi, et al;2015)

### ❖ Staffing

By analysing employee performance data, AI tools can help detect trends and make decisions regarding resource allocation. The constant high performance of an employee enables the organization to make informed decisions regarding the potential promotion of such employee or the allocation of additional resources to support their work.

## Positive Aspects of AI in Operations

### ❖ Management Capabilities

Artificial intelligence (AI) has the potential to proactively mitigate these risks and offer insights into patterns that could otherwise remain undetected. Alternatively, AI can be focused on addressing certain areas of risk with capability to identify fake documents that may have eluded human detection.

### ❖ More Streamlined Compliance Processes

The potential for modification is inside generative AI. By integrating suppliers into a unified management system with other business platforms, the necessary data to demonstrate a company's adherence to compliance regulations becomes readily available to artificial intelligence.

### ❖ Focusing on Complex Tasks

Generative artificial intelligence demonstrates proficiency in comprehending and analyzing predetermined parameters, evaluating the supply chain information of an enterprise, and generating reports.

### ❖ Demand Forecasting and Planning

By considering multiple variables concurrently, Generative AI models can identify intricate patterns and correlations that conventional forecasting techniques may neglect. This enables businesses to anticipate demand fluctuations and align production and inventory levels accordingly, resulting in increased operational efficiency and cost savings.

### ❖ Inventory Optimization

Managing inventory levels is a delicate equilibrium between avoiding stockouts and minimising holding costs. Using historical data, demand patterns, and external factors, generative AI can help businesses optimise their inventory management by predicting optimal supply levels. Companies can reduce excess inventory, prevent overstocking, and improve supply chain responsiveness by utilising Generative AI.

### ❖ Manufacturing and Supply Chain

A Generative AI application can assist firms choose the most cost-effective and free of risk logistic strategies. The system can create best routes and timetables based on the company's fleet, as well as climatic and traffic patterns, enabling for timely and safe material delivery.

### ❖ Supplier Selection and Relationship Management

Selecting reliable suppliers and maintaining strong relationships is crucial for an uninterrupted supply-chain operation. By analysing a vast array of data, such as supplier performance metrics, quality records, pricing, and geographical considerations, generative AI can aid in supplier selection. By leveraging Generative AI, businesses can identify the most suitable suppliers based on predefined criteria and improve supply chain resilience.

### ❖ Predictive Maintenance

Generative AI models can predict when maintenance is required by identifying anomalies and patterns in the data, allowing organisations to schedule repairs and replacements in advance. This reduces downtime, extends the equipment's lifespan, enhances operational efficiency, and minimises maintenance costs.

### ❖ Route Optimization and Logistics

AI has the potential with efficiency of various operational tasks such as route planning, delivery scheduling, and resource allocation. This is achieved by taking into account factors such as traffic conditions, weather forecasts, vehicle capacity, and client needs.

### ❖ Supply Chain Optimization

AI has the potential to enhance multiple facets of the supply chain, including but not limited to optimising inventory levels, production schedules, and delivery routes.

### ❖ Fraud Detection

Generative artificial intelligence (AI) possesses the capability to analyse financial data and discern patterns that can be crucial in the detection of fraudulent activities. The models can also be trained to make predictions on the probability of fraudulent activities by utilising past data.

**❖ Product Design and Innovation**

The utilisation of generative artificial intelligence (AI) holds potential in facilitating product design and fostering creativity through its ability to generate novel concepts, optimise product configurations, and simulate diverse situations. It has the potential to facilitate the development of inventive and tailored goods that align with precise customer demands, all the while taking into account limitations within the supply chain and cost considerations.

**❖ Sustainability and Environmental Impact**

The utilisation of generative AI has the potential to make a significance contribution to the field of sustainability supply chain management through the optimisation routes with reducing fuel consumption emissions with optimization of packaging materials, reduction of waste across supply chain.

## Negative Aspects of AI Operations

**❖ Data Availability and Quality**

Generative AI models require vast quantities of high-quality data to learn and produce accurate outputs. Obtaining sufficient and reliable data in the supply chain can be difficult, especially when dealing with complex and dynamic data sources such as customer demand, production parameters, and logistics data. To ensure the efficacy of generative AI models, data collection, cleansing, and integration become essential steps.

**❖ Interpretability and Explainability**

In supply chain decision-making, it is crucial to maintain transparency as well as understand the rationale behind generated results. Assuring the interpretability and explainability of generative AI models is crucial for obtaining the confidence and acceptance of stakeholders.

**❖ Real-time Adaptation and Dynamic Environments**

Generative AI models may struggle to adapt to real-time data and respond to abrupt shifts in demand, disruptions, or supply chain changes. To successfully integrate generative AI into the supply chain, it is necessary to incorporate real-time data sources and develop models capable of rapid adaptation. The design challenge pertains to establishing a dependable randomized supply chain network while accounting for unpredictable disturbances in the placement of distribution centres (DCs) and transit modes. It is postulated that a disrupted data center does not inherently result in a complete failure of its entire capacity, but rather may experience a partial loss, with the remaining demand being accommodated by neighboring data centers. (Azad et al;2014)

❖ **Ethical and Legal Considerations**

The use of generative artificial intelligence in the supply chain raises ethical and legal concerns. For instance, the generation of synthetic data or content resembling real data may raise concerns regarding privacy or intellectual property. When deploying generative AI in the supply chain, ensuring compliance with data privacy regulations, intellectual property rights, and ethical guidelines is crucial.

❖ **Deployment and Scalability**

Compatibility issues, system outages, and the need for additional computational resources can all contribute to deployment complexities. To address these obstacles, a multidisciplinary strategy involving AI, supply chain management, data governance, and legal and ethical considerations is necessary.

### Ethical Aspect in Operations functions

Developers who employ generative AI bear the onus of upholding ethical principles through various measures, including but not limited to: carefully curating unbiased datasets, closely monitoring the generated outputs, establishing clear guidelines, refraining from incorporating harmful content, ensuring transparency in the AI system's operations, integrating human oversight into the development process, safeguarding user privacy, adhering to relevant regulations, conducting regular audits, and actively engaging with relevant stakeholders.

## AI in Information Technology

Nowadays, generative AI is mostly used to generate information in response to natural language queries; it doesn't need to know how to write code. However, there are many enterprise use cases for generative AI, including advances in material science and drug and chip design. The advancement generative AI platforms are becoming more and more popular among businesses because of their high success rate and ability to completely change industries.

### Building Generative AI in Information Technology

Computational methods that can produce seemingly original, meaningful content—like writing, images, or audio—from training data are referred to as generative AI. The broad adoption of this technology, exemplified by products like GPT-4, Copilot, and Dall-E 2, is presently transforming how we collaborate and conduct business. In addition to being utilized creatively to produce fresh text that mimics authors or new visuals that mimic artists, generative AI systems can and will help humans by acting as intelligent question-answering systems (Feuerriegel, S., *et al.* (2023). Applications in this context include IT help desks, where generative AI assists with routine chores like food preparation and medical advice, as well as transitional knowledge work duties.

## Positive Aspects of AI in Information Technology

- **Faster Adoption**

According to experts, businesses may adopt generative AI more quickly if they demonstrate intelligence, creativity, and context awareness.

- **Technology Budgets**.

The prospects for discretionary spending have been clouded by the economic downturn in major economies, although companies may decide to raise their budgets in an effort to boost output and provide greater value to clients.

- **Deflationary Risks**

Although generative AI systems can be a great help with software development-related tasks, there is a risk that they will have a deflationary effect on IT firms.

- **Ensuring quality and development time**

AI drives down the time taken to perform a task. It enables multi-tasking and eases the workload for existing resources.AI enables the execution of hitherto complex tasks without significant cost outlays and operates 24x7 without interruption or breaks and has no downtime with facilitating decision-making by making the process faster and smarter.

## Negative Aspects of AI in Information Technology

- **Job Displacement:** AI automation may cause job losses in specific industries, which would have an impact on the labor market and labor force.
- **Ethical Concerns:** Data privacy, algorithm bias, and potential misuse of AI technologies are just a few of the ethical concerns that AI brings up.
- **Lack of Creativity and Empathy:** AI is limited in its capacity to comprehend emotions and generate novel ideas because it lacks human traits like creativity and empathy.
- **Cost and Complexity:** AI system development and implementation can be costly and need specific skills and resources.
- **Reliability and Trust:** AI systems might not be entirely dependable all the time, which could cause people to doubt their ability to make decisions.
- **Dependency on Technology:** People who rely too much on AI risk becoming technologically reliant and losing their ability to think critically.

Ethical Aspects of AI in Information Technology

❖ **Bias and Discrimination**

Societal biases are present in the vast volumes of data that AI systems are trained on. Thus, these prejudices may become deeply embedded in AI systems, thereby sustaining and exacerbating unjust or discriminatory results in vital domains like recruitment, credit, criminal justice, and assignment of resources.

❖ **Transparency and Accountability**

Transparency is essential in critical domains such as autonomous vehicles and health care to determine decision-making processes and accountability. To enable appropriate corrective action to be taken in the event that AI systems make mistakes or cause harm, it is especially important to make accountability clear.

❖ **Social Manipulation and Misinformation**

Fake news, misinformation, and disinformation are commonplace in politics, competitive business, and many other fields. AI algorithms can be exploited to spread this misinformation, manipulate public opinion, and amplify social divisions. Vigilance and countermeasures are required to address this challenge effectively.

❖ **Privacy, Security, and Surveillance**

In AI, protecting people's privacy and human rights becomes critical, requiring strong defenses against data breaches, illegal access to private data, and shields against widespread surveillance.

## Future of Generative AI in Business Management: A General Discussion

❖ **Guidelines For Trusted Generative Artificial Intelligence**

- **Long-term viability**: While creating accurate models, right-sized models should be created to reduce our carbon footprint.
- **Integrity**: gathering data to train and assess models with permission to access data.
- **Precise:**Produce verifiable results that maintain precision, efficacy, and retention in systems by allowing clients to train systems on their own data.
- **Legitimize**: Artificial intelligence should only be used as a complement to humans or where human judgement is necessary.
- **Secured**: In order to mitigate potential harm, it is imperative for organizations to safeguard the confidentiality of personal information embedded within training data.

❖ **Generative AI-Specific Principles**

- **Limitations and suggestions for execution and use:**

To limit the implementation and application of Generative artificial intelligence technology where it is essential to prevent harm, law and regulation should be assessed and applied as stated, or updated in collaboration with all stakeholders

- **Holding ownership:** Intellectual property (IP) legislation and regulation do not yet sufficiently account for inherent elements of how Generative artificial intelligence systems are designed and function.
- **Personal data management:** Generative artificial intelligence system providers ought to maintain publicly accessible repositories where system flaws may be documented and, if required, fixed.
- **Authenticity:** Any kind of application or device that employs Generative artificial intelligence ought to make this clear to the right stakeholders.
    - ❖ **Enhanced security and privacy:** The recommendations are vital, but we also need to develop a community of politicians, researchers, and corporate leaders who will work together in the public interest to grasp Generative artificial intelligence limitations and hazards, as well as its advantages.

❖ **The Positive Effects of Generative Artificial Intelligence**

- Increases productivity through task automation or acceleration.
- Allows advanced data analysis and investigation.
- Eliminates skill and time constraints for content creation and creative uses.
- Generate artificially generated information to train and enhance other AI systems.

❖ **The Negative Aspects of Generative Artificial Intelligence**

- **Figment**: Certain artificial intelligence models have a proclivity to create mistakes that do not conform to reality, practical terms, or sensible thinking reasoning.
- **Data labelling reliance**: Despite the fact that many generative artificial intelligence models may be trained unsupervised using unidentified data, both data quality and authenticity remain a concern.
- **Challenges with managing content:** AI models have the ability of recognizing and filtering out undesirable information
- **Concerns about ethics:** Algorithms have been shown to magnify or repeat pre-existing prejudicial views and flaws in the data used for training.
- **Reservations about the law:** Many of the consequences of emerging artificial intelligence techniques are now unresolved by the legal system, which lacked a sufficient structure.
- **Concerns over privacy:** Concerns have been expressed concerning acquiring, preservation, utilization, and confidentiality of personal and business data.
- **Impact of politics:** Generative artificial intelligence raises the issue of erroneous or incorrect data, as well as the legitimacy of media such as realistic photos.

## Ethical Best Practices for Generative AI

When applied in a controlled and regulated manner, the technical advances brought about by the creation of Generative AI models have a larger potential to help society. This involves the adoption of ethical best practices in order to reduce the possible dangers that come with its benefits.

Align with global standards

The UNESCO AI ethics guidelines, which was adopted by all 193 Member States in November 2021.

The guidelines emphasize four core values:

- Human rights and dignity
- Peaceful and just societies

- Diversity and inclusiveness
- Environmental flourishing

They also outline ten principles, including proportionality, safety, privacy, multi-stakeholder governance, responsibility, transparency, human oversight, sustainability, awareness, and fairness. The guidelines also provide policy action areas for data governance, environment, gender, education, research, health, and social wellbeing.

## CONCLUSION

Although Generative Artificial Intelligence is becoming more popular, its usage by businesses poses significant ethical concerns. Companies can enhance the precision, security, and reliability of their deployed products by diligently adhering to established criteria and proactively implementing protections. However, sticking with a strong framework for ethics can help businesses navigate this period of rapid expansion. Many present legal and regulatory frameworks have not completely come up with generative artificial intelligence and its possible uses because it is a unique technology that is constantly evolving. Chatbots and personal assistants are examples of how artificial intelligence may successfully penetrate and modify our society. Businesses must emphasise the appropriate application of generative artificial intelligence by making sure that it is reliable, secure, authentic and productive. Organisations must comprehend the ethical consequences and institute appropriate measures to avoid hazards. In order to build a more reliable artificial intelligence, both society and industry will additionally establish enhanced systems for tracing the provenance of information. Generative AI holds immense potential to revolutionize various sectors, from healthcare to education, by creating new content and enhancing productivity. Although, the utilisation of GenAI technologies in the financial industry has significant potential it is necessary to exercise precaution when adopting such technologies. The implementation of GenAI has the potential to yield notable gains in operational efficiency by enhancing the quality of client interactions, and strengthen risk management through compliance measures.

## LIMITATIONS OF THE STUDY

This study only focuses on the review of existing literature in journals and websites and does not carry out an empirical analysis.

## REFERENCE

Abolghasemi, M., Khodakarami, V., & Tehranifard, H. (2015): A new approach for supply chain risk management: Mapping SCOR into Bayesian network. *Journal of Industrial Engineering and Management (JIEM), ISSN 2013-0953, Omnia Science, Barcelona, 8*(1), 280-302. doi:10.3926/jiem.1281

Arango, L., Singaraju, S. P., & Niininen, O. (2023). Consumer Responses to AI-Generated Charitable Giving Ads. *Journal of Advertising, 52*(4), 486–503. doi:10.1080/00913367.2023.2183285

Azad, N., Davoudpour, H., Saharidis, G. K. D., & Shiripour, M. (2014). A new model to mitigating random disruption risks of facility and transportation in supply chain network design. *International Journal of Advanced Manufacturing Technology*, *70*(9-12), 1757–1774. doi:10.1007/s00170-013-5404-0

Budhwar, P., Malik, A., De Silva, M. T. T., & Thevisuthan, P. (2022). Artificial intelligence – challenges and opportunities for international HRM: A review and research agenda. *International Journal of Human Resource Management*, *33*(6), 1065–1097. doi:10.1080/09585192.2022.2035161

Feuerriegel, S., Hartmann, J., & Janiesch, C. (2023). Generative AI. *Business & Information Systems Engineering*. doi:10.1007/s12599-023-00834-7

Li, X. (2023) Artificial intelligence applications in finance: a survey. *Journal of Management Analytics*. ACM. . doi:10.1080/23270012.2023.2244503

*Top Strategic Technology Trends for 2022: Generative AI.* (2021). Gartner. https://www.gartner.com/en/documents/4006921.

# Chapter 3
# Bias and Fairness in AI Technology

**Muhsina**
*PES University, India*

**Zidan Kachhi**
https://orcid.org/0000-0002-8317-6356
*PES University, India*

## ABSTRACT

*This chapter's objective is to provide an overview of how artificial intelligence (AI) has become an essential part of human life. It explains the sources of bias and its types in AI technology. With the help of previous studies, the chapter elucidates the strategies that can be used to avoid decision-making as a source of bias in AI technology. It also talks about the importance of understanding how human bias can also cause AI systems to exhibit bias towards certain groups. Unfairness in AI is also one of the most common sources of biassed data, and it's explained with strategies for detecting and addressing unfairness. The chapter also covers the need for transparency in AI technology along with ethical considerations, as transparency in AI is essential to ensuring that AI systems operate in adherence to ethical standards.*

## INTRODUCTION

Artificial intelligence (AI) has become an integral part of our lives and has transformed the way we interact with technology. With the advancement of AI, there has also been a growing concern about the potential for bias and unfairness in AI technology. In recent years, there have been several high-profile cases where AI systems have demonstrated bias and unfairness, which has led to a growing concern about the ethical implications of AI. This chapter will examine the issue of bias and fairness in AI technology, including its causes, effects, and potential solutions. The chapter will also explore the ethical implications of biased AI and provide recommendations for addressing this issue.

AI bias often comes from the data used in its training process. The factor that majorly affects AI performance is data quality. Like this, if an AI is trained with biased data, it will mirror that bias. Let's

DOI: 10.4018/979-8-3693-1565-1.ch003

say, if an AI is programmed mainly using data from a specific race or gender, it may struggle to correctly identify individuals of different races or genders. Lack of diversity in the groups responsible for AI development and training also adds to AI bias.

A team mainly having people from one type of background can lead to a biased AI system. When this happens, the bias can impact a lot of people and society as a whole. A big issue is when biased AI keeps old social inequalities going. For instance, an AI system might not fairly assess job candidates if it has a bias against women or certain racial or ethnic groups. This could lead to uneven job chances. Biased AI can also lead to unfair treatment of individuals. For example, an AI used for deciding if someone should be given credit might not be fair if it biases towards certain neighborhoods or income ranges. People could be refused credit or asked to pay more interest.

Bias in AI can be addressed in a number of ways. One strategy is to make sure that the AI systems' training data is representative of the whole population and diversified. To do this, gather data from multiple sources and make sure it is fair across demographics like gender and ethnicity. Diversifying the personnel in charge of creating and educating AI systems is another way to find a solution. A diverse team diminishes the likelihood of individual biases affecting policy creation. Last, uphold ethical benchmarks. Recently, a significant problem regarding AI fairness and equality has emerged, catching scholars, researchers, and decision-makers' attention. Studies have made progress in onset, outcome, and potential solutions.

Research from Buolamwini and Gebru (2018) revealed an interesting point. Major tech companies like IBM, Microsoft, and Face++ have developed facial recognition software. But they found a problem. The software shows more mistakes for women and people of color. Surprising, right? Their conclusion provides a clue. A lack of diversity in training data causes bias in the software. This means the software isn't as reliable as we'd expect!

The authors suggested companies making face recognition software need to have varied, representative training data to stop bias in their software. Mittelstadt et al. (2016) study, looked at how biased AI could impact ethics. They said biased AI might make social unfairness worse, even causing some people to face discrimination. The authors said that AI tech creators should include fair and unbiased aspects in their AI design and development stages.

Artificial Intelligence technology has advanced significantly, but it still displays prejudice and often falls short when it comes to fairness. This presents some significant challenges that demand urgent solutions. Firstly, it's crucial that we produce more diverse data sources for training AI systems. The lack of diversity in the current data often results in biased outcomes and increased inequality. Secondly, there's no standard methodology to detect and mitigate bias across all fields of AI. This hinders the creation of effective, versatile solutions. Lastly, in dealing with the complex social and technical problems of AI bias, interdisciplinary research that encompasses fields like ethics and sociology is needed more than ever. In fact, the everchanging character of AI systems makes it tough to consistently oversee and renew algorithms for lasting fairness. Likewise, little investigation exists on the lifelong effects of partial AI systems on society, such as its consequences on democracy, social unity, and human rights. It is crucial to bridge these gaps in research to promote ethical AI tech growth that fetches benefits for every person in society.

Another need to explore bias and fairness in AI is transparency and accountability. AI can often be puzzling, making it hard to see and fix biases. By digging into these topics, we can make sure AI is not only transparent and answerable but also impartial. Focusing on bias and fairness when creating AI helps make them more ethical and responsible. As AI gets a bigger role in our regular activities, we need to make sure it's suitable for everyone. This means designing AI that respects privacy, safeguards personal data, and upholds human rights. Studying bias and fairness in AI can also help to promote diversity and inclusion in these systems. When the people who develop AI systems are not diverse enough, they may not be able to identify the biases in the data they are using to train the machines.

Examining the fairness of AI is crucial for several reasons. One significant reason is that it can help to curb any harmful implications this technology might have on individuals and our society. Let's say, for instance, AI in hiring can unknowingly favour certain genders or races. This happens when candidate screening relies heavily on patterns in resumes. To fix this, we need to understand and address these biases. This way, everyone, no matter their race or gender, gets a fair shake at jobs.

After careful analysis, it can be concluded that the chapter provides a comprehensive and detailed background information and literature review regarding its topic of bias and fairness in AI technology. The chapter discusses the causes, effects, and potential solutions to address the issue of AI bias, citing studies from Buolamwini and Gebru (2018) and Mittelstadt et al. (2016) to support its claims.

Furthermore, the chapter delves into the ethical implications of biased AI and provides recommendations for addressing this issue. The author(s) suggest diversifying the personnel in charge of creating and educating AI systems and upholding ethical benchmarks to ensure fairness and equality in AI technology.

However, to further strengthen the chapter's impact, it is recommended that the author(s) expand their discussion of the literature by including more recent studies and research on the topic. Additionally, a more in-depth analysis and discussion of the potential solutions to address AI bias and their effectiveness should be included.

Overall, the chapter provides a well-researched and detailed foundation for understanding the issue of bias and fairness in AI technology, and with some modifications to the literature review, it could provide a stronger and more descriptive analysis of the topic.

## Sources of Bias in AI Technology

AI technology's significant source of bias is the absence of diversity in the data sets utilized to train AI models. Typically, AI models are trained on extensive data sets intended to represent real-world scenarios. However, if the training data is biased towards a particular group of people, the resulting model may show discrimination towards other groups. For example, if a dataset of facial recognition system contains mainly images of white males, it may not be as precise in recognizing women and people of colour (Buolamwini and Gebru 2018). The absence of diversity in the training data may result in prejudiced outcomes in AI models.

Additionally, the selection of data by humans is a crucial factor that can introduce bias into AI systems. Human biases, whether conscious or unconscious, can impact the curation and selection of data, leading to biased AI outcomes. Biassed human input can result in unfair and discriminatory outcomes from AI systems. For instance, if the selection of data is based on prior decisions or outcomes, the AI system's results will also reflect similar biases, perpetuating discrimination. The AI industry lacks diversity, with individuals from specific ethnic and cultural backgrounds dominating the industry, leading to potential biases in AI systems towards those groups. Furthermore, the homogeneity of the industry results in a limited representation of perspectives and experiences, which leads to inadequate development of AI systems. A study conducted by researchers at the AI Now Institute found that women only accounted for 15% of AI research staff at Facebook and 10% at Google, while black researchers only made up 1.5% of Facebook's research staff and 2.5% of Google's research staff. To address the lack of diversity in the

AI industry, it is crucial to increase the representation of women and minorities through scholarships, mentorship programmes, and partnerships with organizations that promote diversity in the tech industry.

The tech industry's lack of diversity is a leading cause of AI technology's homogeneity. According to the National Centre for Women & Information Technology report, women constitute merely 26% of the computing workforce. In addition, ethnic and racial minorities are also underrepresented in the tech industry. The scarcity of diversity in AI technology has extensive and potentially dangerous consequences. One of the most significant implications is the perpetuation of biases and discrimination. For instance, Amazon discontinued an AI recruitment tool in 2018 that favored men and discriminated against women. The tool was programmed using resumes submitted to Amazon over a decade, primarily from men. As a result, the tool favoured male candidates and penalised resumes with words like "female" and "women" (Dastin, 2018). The lack of diversity in the data used to train AI algorithms is another reason for the lack of diversity in AI technology. AI algorithms are as good as the data used to train them, and if the data is not diverse, the algorithms will not reflect the diversity of the real world, leading to biased algorithms that perpetuate stereotypes and discrimination.

After careful analysis, it can be concluded that the chapter provides a comprehensive and detailed background information and literature review regarding its topic of bias and fairness in AI technology. The chapter discusses the causes, effects, and potential solutions to address the issue of AI bias, citing studies from Buolamwini and Gebru (2018) and Mittelstadt et al. (2016) to support its claims.

Furthermore, the chapter delves into the ethical implications of biased AI and provides recommendations for addressing this issue. The author(s) suggest diversifying the personnel in charge of creating and educating AI systems and upholding ethical benchmarks to ensure fairness and equality in AI technology.

However, to further strengthen the chapter's impact, it is recommended that the author(s) expand their discussion of the literature by including more recent studies and research on the topic. Additionally, a more in-depth analysis and discussion of the potential solutions to address AI bias and their effectiveness should be included.

## Decision-Making in AI Technology

Defining the issue and goals holds substantial importance in preventing bias in AI technology. By clearly outlining the problem and objectives, the AI system can be tailored to fulfill precise aims without introducing extraneous biases. This approach also guarantees that the AI system is crafted with inclusivity, transparency, and accountability in mind. A study conducted by Barocas and Selbst (2016) emphasizes that defining the problem and objectives plays a pivotal role in ensuring that the AI system aligns with specific goals and avoids unnecessary biases. In the quest for unbiased AI, identifying and mitigating potential sources of bias is pivotal. Crawford and Calo (2016) emphasize the significance of defining the problem and objectives, fostering inclusivity. Considering diverse user needs and preferences ensures fairness. Moreover, transparency and accountability are achieved by articulating the problem and objectives, fostering trustworthy decision-making processes. This can foster a deeper level of trust and confidence in the system's decision-making abilities. It's absolutely crucial for those developing the system to articulate the problem and objectives clearly. By doing so, they can make sure the AI system is tailored to fulfil specific goals, without adding any needless biases. The system also needs to be inclusive, transparent, and accountable. In this way, developers can bring to life AI systems that are precise, fair, and ensure equal treatment for everybody using it.

The scarcity of diversity in AI technology has extensive and potentially dangerous consequences. One of the most significant implications is the perpetuation of biases and discrimination. For instance, Amazon discontinued an AI recruitment tool in 2018 that favoured men and discriminated against women. The tool was programmed using resumes submitted to Amazon over a decade, primarily from men. As a result, the tool favoured male candidates and penalized resumes with words like "female" and "women" (Dastin, 2018). The shortage of varied information for guiding AI algorithms contributes to a similar lack in AI technology itself. If the data used for teaching AI lacks variety, the resulting algorithms won't mirror our diverse real world. This can cause algorithms that continue to promote stereotypes and foster discrimination. AI effectiveness, like life, thrives on variety. Without diverse data, we risk creating biased, stereotypical systems. One key reason for bias in AI technology lies in the training data. The precision of AI systems can greatly depend on data quality and volume. When the data has bias, the AI system may learn and replicate that bias, resulting in unfair and inaccurate choices.

For instance, research carried out by Buolamwini and Gebru (2018) found that facial recognition tech had trouble accurately identifying those with darker skin and females. This was because the data used to teach these systems had bias in it. This highlights why it's crucial to use a wide variety of representative data to instruct AI systems. This helps in diminishing bias. One reason AI can show partiality in its choices is the formula it uses to dissect information. These formulas might have bias if they are crafted on certain beliefs that mirror common societal biases, or are built on incomplete or inaccurate data. Obermeyer et al. (2019) conducted a study that found an algorithm used by hospitals to identify patients who need extra care was biased against black patients. The algorithm failed to identify sicker black patients than white patients, leading to unequal access to healthcare. To minimize bias in AI algorithms, it is crucial to ensure that they are designed to be transparent, auditable, and free from assumptions that reflect societal stereotypes.

AI technology's decision-making process is prone to biases due to the involvement of humans. In the development, training, and deployment of AI systems, humans have a significant role and can introduce their biases into the decision-making process. For instance, as per the study conducted by Dastin (2018), Amazon's AI recruiting tool had a bias against women because of the biased data used to train the system. The system was trained on resumes submitted mostly by men, and hence, the AI system favoured male candidates over female candidates. To prevent this type of bias, it is crucial to ensure that diverse teams are involved in the development, training, and deployment of AI systems

Some strategies that can be used to avoid decision-making as a source of bias in AI technology.

AI technology requires defining the problem and objectives to avoid bias. This helps ensure that the system achieves specific objectives and operates without introducing avoidable biases. Additionally, it is a means of guaranteeing that the AI system is transparent, accountable, and inclusive. According to a study by Barocas and Selbst (2016), defining the problem and objectives can prevent potential sources of bias and ensure the accuracy, efficiency, and effectiveness of the AI system

When creating an AI system that is fair and inclusive for everyone, it is essential to define the problem and objectives, as indicated by Crawford and Calo (2016) study. To ensure that the system is designed reasonably and equitably, it is crucial to consider the distinctive needs and preferences of various user groups. This step is also instrumental in making the AI system more transparent and accountable. A clear definition of purpose is key for AI design. This clear roadmap helps make the AI process easy to follow, which increases trust. The design team handles this task carefully. The aim? To make decision-making crystal clear. Trust in system decisions grows this way. Defining purpose and being concise benefits developers when creating AI. This step aids in the alignment of the system with specific goals, sans any unnecessary bias. Accessible, upfront, and responsible these are the pillars of a great AI system. By adhering to these, developers can assure AI that it remains accurate, just, and works for everyone.

It is crucial to use diverse, representative, and unbiased data when training AI systems. Biased data can lead to inaccurate and unfair decision-making, as AI systems tend to replicate the bias present in the training data, which can cause significant disparities in accuracy for underrepresented groups, according to Buolamwini and Gebru (2018). As such, it is imperative to thoroughly evaluate the training data used in AI systems to identify and address any potential sources of bias. For instance, when training an AI system to recognize faces, it is necessary to include faces from different races, ages, and genders to ensure inclusivity and accuracy for all users. It is also essential to ensure that the data used in AI systems is representative of the real-world context to guarantee that the AI system can work effectively for all users, regardless of their background or demographic. Biased data is a significant source of bias in AI systems, which is why it is crucial to evaluate the training data meticulously to ensure that it is diverse, representative, and unbiased. By doing so, developers can create AI systems that are equitable, fair, and accurate for all users.

AI technology can be made less biased by using transparent algorithms. Transparent algorithms ensure that AI systems base their decisions on accurate and impartial information, which can help build trust in the decision-making process. According to Mittelstadt et al. (2016), transparent algorithms can increase accountability and trust in AI systems' decision-making process. Therefore, designing algorithms that are transparent and easy to understand is crucial. If an algorithm used in AI systems is designed to favour specific groups or individuals, it can lead to bias against low-income families or minorities, for example, an algorithm designed to predict loan eligibility. It is crucial to evaluate AI algorithms carefully to identify and mitigate any potential sources of bias as per Caliskan et al. (2017). Algorithmic bias can arise from various factors such as biased feedback, biased design, and biased data. So, it's important to think about all of this to make sure AI systems don't have any bias. According to Machlev et al. (2022), developers can use things like explainable AI to make algorithms clear. That way, you can see how an AI system decides. Explainable AI lets you see and understand why AI systems make the decisions they make. We can then make sure they're right and fair. And making algorithms that are easy to understand is key, this means explaining things in a simple, clear way. Then, people can trust AI systems and get them more. It's also real important to have clear algorithms in AI. This can stop bias from happening. Having clear algorithms can make sure AI systems decisions are based on what's right and fair. The use of techniques such as XAI and the creation of easily understandable algorithms can establish trust in the decision-making process of AI systems among both developers and users.

AI systems can mirror human bias. This might happen during their production, use, or maintenance. People who create these systems can unintentionally let their opinions influence the AI. For example, if one cultural group is heavily represented in the AI's creators, that bias might appear in the system's actions. A study from Carnegie Mellon University shows this occurrence. It tells us that gender bias is common in the creation of language processors, which are part of AI systems. This study pointed out that words like "programmer" are linked to men, while words like "nurse" are mostly associated with women (Zhao et al., 2017).

Increasing AI industry diversity by hiring various individuals is vital to fight human bias. This tactic makes sure AI system development includes many perspectives and experiences, lowering the chance of bias. Also, setting up and applying ethical rules for the development, use, and upkeep of AI systems is vital. These rules should focus on fairness, transparency, and responsibility when making AI systems.

To find any possible bias sources, AI systems need constant checks. This can happen through routine audits and system behaviour evaluations. Quick finding and reducing of any bias can happen this way.

Nevertheless, the problem of partiality in AI technology has emerged as a major worry. Partiality in AI technology takes place when AI systems or algorithms display discriminatory actions towards specific groups, individuals, or communities. In this section, we will examine the origins of partiality in AI technology and investigate a few of the methods to alleviate these partialities.

## Unfairness in AI

Artificial intelligence, or AI, has become a huge part of our life. It's in the devices we use every day, like phones and streaming apps. But there are concerns about AI being biased or causing unfairness. AI bias happens when the programming behind the tech copies the unfairness already in our society. For example, some facial recognition technology makes more mistakes with people of colour. This can cause concerns about racial bias. Unfairness in AI happens when the tech is used, on purpose or not, to keep unfairness going. Like, if someone can't get a loan because of their race or where they live. These problems can make existing unfairness even worse and make it harder for people who are already struggling

AI's efficiency leans heavily on its training data quality. If this data shows bias, the AI will mirror that bias too. Bias might sneak into data due to past prejudice or sketchy sampling. Take an AI recruitment system trained on data once slanted towards guys. It might continue to favor males. This is despite the present-day workforce being different. Algorithms also trigger unfairness. They can lean towards certain groups or outcomes. An algorithm predicting criminal returns might use race. This leads to unjust treatment of some racial groups. Apart from algorithms, bias in data is a major suspect behind unfairness in AI tech. An unfair benefit to men in a job selection process could be perpetuated by AI systems, for instance. If the training data leans toward men, the AI might continue to favor males even if later data no longer mirrors the workforce. It's critical to spot biased data to uphold fairness in AI systems. A good way to spot skewed data is to examine the information used to teach an AI system. This examination can spotlight any groups that are not enough or too much represented in the data. Plus, it can highlight any elements or factors that show a strong link with the end result being guessed.

If any of these features or variables are highly correlated with the outcome, they could potentially serve as proxies for other factors, such as race or gender. This approach is backed by research on detecting and mitigating bias in AI systems and can help create a fairer, more unbiased AI technology.

Buolamwini and Gebru (2018) conducted a study on commercial gender classification systems and discovered that these systems were less accurate for darker-skinned individuals and women. The research emphasizes the significance of scrutinizing the data used to train AI systems to ensure that they are just and impartial. In a similar vein, Caliskan et al (2017) employed an algorithm to analyze language corpora, revealing that the algorithm associated certain professions such as "programmer" and "engineer" with men, while connecting other jobs such as "homemaker" and "nurse" with women. This research highlights the likelihood of bias in data employed to train AI systems, indicating the need for thorough analysis to detect and address this bias.

Kleinberg et al. (2016) conducted a study on the use of risk assessment algorithms in the criminal justice system. The authors found the algorithms were biased. More African American defendants were flagged as high-risk than white defendants with similar histories. Hence, we must seriously examine the data used to train AI systems. The goal is to make sure they're fair and don't reinforce social inequities. One way to find prejudiced data is to check the fairness of the AI system using different criteria. For example, when an AI system is set to predict loan defaults, it should be checked by looking at the number of loans approved or turned down for various race or ethnic groups. Noteworthy differences mean it's key to investigate more. We need to find out if the differences come from biased data or unfair algorithms.

Detecting and fixing unfairness in AI tech is key to making these systems fair and trustworthy. Biased data, or data that isn't balanced, causes the most unfairness in AI tech. This can cause unfair algorithms, causing unequal treatment and bias. So, it's crucial to spot this biased data to make AI systems fair. Checking data and judging AI systems for fairness using different measures are effective ways to find biased data. One important way to find biased data is by testing AI systems for fairness using different measures.

A study by Buolamwini and Gebru (2018) showed why it's important to test AI for fairness with different standards. They found that commercial gender identification systems had lower accuracy for those with darker skin and females. They pointed out we need to inspect and handle biases in AI to get fair and trustworthy results. The research showed how biased computer programs could impact

Underrepresented groups, meaning we should always be careful and improve how we test AI for fairness. They discovered the AI favored certain groups. This was due to the fact the data they learned from wasn't unbiased, which led to unfair and biased results.

Think of an AI system developed to predict repeated offenses. We could evaluate this by seeing how many folks from different races and ethnicities are tagged as high-risk. If there's a clear disparity, we must check if biased data or algorithms generated these differences. If we test AI fairness using multiple ways, we might identify any neglected or overly highlighted groups. This can aid in constructing strong methods to reduce bias. Knowing whether artificial intelligence (AI) is biased is crucial for ensuring its fairness and dependability. Any bias in data or programming may result in unfairness and bias; recognizing these is vital.

When data is scrutinized, it unveils insights into fairness. It's essential to rigorously evaluate AI systems for fairness and comprehend their forecasting methods. Detecting and rectifying unfairness in AI ensures equitable development of AI systems, promoting societal justice. Biased algorithms pose a significant challenge in AI technology. They tend to favor specific groups or outcomes, potentially resulting in unjust discrimination, such as using race as a determining factor in loan default rate predictions, leading to unfair treatment of certain racial or ethnic groups.

Spotting biased algorithms in AI technology is essential to guarantee fairness. This ensures that the system operates in a fair and trusted manner. Biased algorithms pose a major challenge. They can be manipulated to show favouritism towards certain groups, leading to unfair outcomes. Therefore, it's crucial to detect and address biased algorithms to uphold fairness in AI systems. Testing AI systems for fairness is a significant approach to achieve this objective. For instance, an AI predicting chances of reoffending could be evaluated in a unique way. We could compare how many individuals from various races or ethnicities get labelled as high-risk. If differences are discovered, more investigation is needed. We need to figure out are the differences due to biased information or biased algorithms?

Using interpretability techniques is a key approach in uncovering biased algorithms. For example, when an AI system is built to forecast recidivism, interpretability methods can reveal the specific features driving its predictions. If these features serve as proxies for factors like race or gender, the AI system might manifest bias. For example, A study conducted by Kleinberg et al. (2016) uncovered troubling biases within algorithms intended to forecast recidivism, particularly in relation to African American defendants. The research revealed that these algorithms, prioritizing accuracy over fairness, utilized factors like zip codes and education levels, inadvertently becoming proxies for race. This underscores the critical need to identify and rectify biases within algorithms with effective solutions. Testing the fairness of AI systems is crucial. It involves assessing various metrics to unearth any unfairness or discrimination within the decision-making process. Detecting notable disparities prompts thorough inquiry to pinpoint the underlying causes of such unfairness. This exploration paves the way for the creation of viable remedies to mitigate bias.

Detecting and rectifying injustices within AI technology plays a pivotal role in guaranteeing the fairness and reliability of AI systems. Biased algorithms stand out as a prominent culprit in fostering unfairness within AI technology. These algorithms can be intentionally programmed to show favouritism towards specific groups or outcomes, resulting in unjust treatment and prejudice. It's indispensable to detect biased algorithms as an essential measure in ensuring fairness in AI systems. The deployment of interpretability techniques serves as a viable approach to comprehending the decision-making process of AI systems and pinpointing any underlying biases. For example, in the context of an AI system aimed at predicting recidivism, interpretability techniques can be leveraged to identify the specific features or variables influencing its predictions. A different method for identifying biased computer patterns is by using understanding techniques. This helps us see how the AI system decides. For instance, if an AI system aims to forecast repeat offenses, we can use understanding techniques. These techniques reveal the elements or factors that the AI system uses for its forecasts. If any of these elements or factors are stand-ins for other things like race or gender, the AI system can be biased. Suppose there's a feature or variable that's really just representing something else, like race or gender. This could mean the AI system has a bias. A method called LIME (Local Interpretable Model-Agnostic Explanations) could help us understand this better. LIME isn't tied to any specific model and it gives explanations for single predictions an AI system makes. So, how does LIME work? It creates a local model to mimic the actions of the AI system around a certain prediction.

The local model can be utilized to recognize the crucial features or variables influencing the prediction process. Furthermore, another interpretability method known as SHAP (Shapley Additive Explanations) comes into play. SHAP serves as a comprehensive technique applicable to elucidate the outcome of any machine learning model. It functions by calculating the contribution of each feature to the prediction, thus providing a unified framework for interpreting the model's output. Effectively using SHAP allows the identification and assessment of the pivotal prediction-influencing features or variables, also considering whether they serve as proxies for other factors, such as race or gender. For instance, a research conducted by Buolamwini and Gebru (2018) utilized various interpretability strategies to pinpoint the origins of bias in commercial gender classification systems. Their findings revealed that these systems were less precise when it came to darker-skinned individuals and women. They determined that the root cause of this unequal AI performance was the biased data used for training the systems, leading to a perpetuation of unfairness and discrimination.

Moreover, we can probe AI systems for fairness using varying measurements to reveal if these systems are fostering any inequality or prejudice in their decision-making process. If we find notable divergencies, thorough research can help uncover the root of this inequality and pave the way for creating effective resolutions to reduce bias. Conclusively speaking, it's crucial to spot inequality within AI technology to ensure that the systems we put in place are fair and dependable. Biased algorithms can result in prejudice and discrimination; hence their detection is a priority. We can sniff out these biased algorithms through strategies like fairness tests on AI systems using varied measurements, and employing interpretability methods to comprehend how AI systems are forming their predictions. Through the identification and management of skewed algorithms, we can assure the development of AI systems that are equitably weighted, benefiting everyone and advocating social fairness.

Figuring out if there's any unfairness in artificial intelligence can be quite tough, given that these systems can get pretty complex and hard to understand. Yet, there are several methods that we can use to uncover any bias. One way is to closely look at the data that was used to train the AI. We do this by checking the spread of the data and looking for any groups that are either too prominent or too scarcely represented. Another way is to conduct fairness checks on the AI system itself by using varied measures. For example, let's say we're testing a recruitment AI system for fairness. We could assess it by contrasting the number of chosen candidates from varying demographic backgrounds. As soon as we spot any sign of injustice, multiple tactics can be applied to alleviate it. One such method is to enhance the caliber of data deployed for AI system training. This can be accomplished by guaranteeing that the data is a true reflection of the population and devoid of any prejudiced elements. Another method is to tweak the algorithm to eradicate any biases. If, for instance, an algorithm is found to exhibit prejudice against particular groups, we can modify it to eliminate such bias. In addition, incorporating elements of transparency and accountability into AI systems ensures that any injustice is identified and rectified swiftly.

## NEED FOR TRANSPARENCY IN AI TECHNOLOGY

A report from the European Union Agency for Fundamental Rights (2018) emphasizes the importance of having clear and transparent decision-making processes in AI systems. This transparency is crucial in preventing these systems from infringing on people's fundamental rights or causing discrimination against certain groups. Moreover, keeping the workings of AI systems open and clear to stakeholders upholds accountability. It helps people understand the mechanics of these systems and how decisions are made. Openness in AI processes can also cultivate greater public trust and acceptance by addressing concerns related to privacy, security, and fairness. Transparency, thus, contributes significantly to the advancement and acceptance of AI systems. In a different research study, a team from the University of Oxford and the Alan Turing Institute discovered just how vital transparency is in the decision-making processes of AI.

In addressing concerns about bias in AI systems, the researchers found that adopting transparent and open practices plays a crucial role in identifying and rectifying these biases. This transparency ensures the ethical and fair utilization of AI. It is instrumental in addressing issues related to accountability, bias, and public perception, all of which are essential for the successful implementation and application of AI systems. Transparent decision-making processes are key in ensuring the ethical use of AI and are pivotal in managing public perception and accountability, thereby guaranteeing fairness. In the rapidly growing landscape of artificial intelligence, there is a prevailing unease regarding its reliability. Trust holds a crucial role in shaping the dynamics between humans and AI, exerting its influence on the utilization and acceptance of AI systems. The absence of trust could potentially lead to apprehension towards engaging with AI and even result in total avoidance.

The intricacy of AI systems often eludes human comprehension since they are typically intricate. This lack of clarity can lead to a lack of responsibility and openness, potentially eroding trust. Further, AI systems can be susceptible to bias, resulting in possibly unjust or prejudiced decisions, further undermining trust. Therefore, establishing a trusting rapport between humans and AI systems is paramount. This can be achieved through two pivotal approaches: transparency and accountability. Transparency involves elucidating the computational decision-making process and the data employed. This empowers individuals to comprehend and exert influence over the system. The subsequent strategy entails rendering AI systems accountable. This encompasses providing avenues for individuals to challenge the system's decisions or assign blame in cases of malfunction. Incorporating a sense of responsibility is pivotal for establishing trust as the system's adherence to performance benchmarks is imperative (Nissenbaum, 1996). The ethical integrity of AI systems is closely tied to honesty. Inaccurate training data can lead to unjust or detrimental conclusions, eroding trust (Liao & Hirschberg, 2018). Prevention of such issues involves meticulous data selection for system training and the development of equitable algorithms. Additionally, prioritizing user experience in AI system design, encompassing user-friendly interfaces and transparent system operations, is crucial for fostering user confidence. Ultimately, the ethical framework governing AI greatly influences public confidence in artificial intelligence. Making rules about right and wrong can help. This can include creating moral rules for building and using AI 'thinkers'. If we show that we're committed to using AI in a right and responsible way, we can make people trust AI more (Mittelstadt and others, 2016).

AI (Artificial Intelligence) decision making has gotten quite a bit of attention lately. People are worried. They believe AI should be transparent, able to be understood. Can it be fair? Can it treat all without bias? This is important. We need to examine AI closely. We must dig deep into the issue of AI legitimacy. Transparency and explainability are key. What about potential problems with AI decisions? We ought to investigate that too. Also, how do we ensure AI fairness in designing? We need to keep a clear eye on transparency. A major issue with AI making choices is the absence of clearness and understanding. AI designs can be tricky, leading to confusion in figuring out how they make certain picks. This could be tough in spotting mistakes or biases when making decisions, causing the public to doubt these systems. Hence, the demand for AI decisions to be clear and understood is increasingly important (Hajian, 2016). Understanding AI means knowing how it makes a choice. Think of it like this, "Transparency" is like seeing through an AI's thoughts. "Explainability" is like someone telling you how AI thinks. If we don't have both, people might not trust AI (Meuwissen & Bollen, 2021). Several things can put algorithmic decision-making's integrity at risk. A top concern is the ability for these systems to echo current biases and inequalities. Unfair data used to educate the AI system could result in biased outcomes, leading to unequal treatment and discrimination. To illustrate, an employment algorithm that uses data weighted against women might overlook capable female applicants (Lepri et al., 2017).

AI-based decision-making sometimes faces credibility challenges due to the absence of human supervision. Even though AI can flawlessly process enormous data volumes quickly, they can't grasp data's real meaning. Thus, they may reach conclusions that don't consider wider societal, cultural, and ethical aspects which humans would inherently reflect upon. This absence of human supervision can prompt decisions that may be unfair or potentially damaging (Lepri et al., 2017). To uphold the integrity of algorithmic decision-making, several measures can be implemented. An essential step involves developing AI systems with fairness and transparency. This entails using unbiased datasets, conducting regular testing, and monitoring the system. It's crucial for AI systems to be designed in a manner that is explainable and transparent, allowing humans to comprehend the decision-making process (Hajian, 2016). Another essential measure is to maintain human supervision over the decisions made by AI. AI systems are fantastic tools, but they shouldn't substitute human decision-making altogether. They should act as aids, bolstering and enriching human decision-making by offering valuable insights and suggestions. These can assist people in making more informed and effective decisions.

## Ethical Considerations in AI Systems

AI can change a lot of areas. Think healthcare, money matters, moving around, and learning. But, people make and train AI. So, AI can pick up the bad stuff from people, like bias. This has come up in a few big cases. It makes us question if we should use AI. One big story about AI bias happened at Amazon. They made an AI tool to help pick people for jobs. The AI learned from resumes sent to Amazon over 10 years. Most of those resumes were from guys. The system didn't favour women. It sort of "punished" resumes with words like "women's," "female," and "diversity." After realizing it was unfair, this tool was discarded.

AI bias exists in facial recognition too. Systems tend to misjudge folks with darker skin. In 2018, the American Civil Liberties Union (ACLU) studied Amazon's Recognition. It made more mistakes with darker than lighter skin. It's worrisome, right? Law enforcement uses this tech more and more, and biased systems might cause unwarranted arrests or convictions. Even the criminal justice system has seen prejudiced AI. Back in 2016, ProPublica revealed an AI system used by judges. This system would try to guess the chances of defendants repeating their offenses. Shockingly, it was unjust towards black defendants. More often than not, it wrongfully marked them as high-risk. It was almost double the rate for white defendants. This bias could unjustly extend jail time for black defendants, thus worsening racial inequality within the justice system.

AI bias can impact the healthcare industry too. In 2019, some researchers shared a startling finding in the Science journal. They found an AI system, used to rank patients for extra attention, favoring others over black patients. The system seemed to undervalue the healthcare necessities of black patients, leading to fewer of them getting more care. The lowered care could affect black patients in severe ways, worsening their health situations. Health disparities already exist for these folks. These situations stir questions about AI ethics. If algorithms show bias, discrimination increases and society's fairness wanes. Some people, like women, colored folks, and other overlooked communities, might suffer more. This could mean fewer chances, unjust actions, or lack of access to key services and resources.

If people sense that AI systems hold biases against them, their trust may erode. This perceived unfairness could possibly put a strain on their relationships within their community. To rectify things, we need to acknowledge that AI systems might have biases. The critical solution lies in creating unbiased AI systems. This can be achieved by utilizing diverse training data, consistently reviewing for biases, and incorporating varied perspectives in the development process of AI systems.

AI is swiftly progressing, showing up more in areas like healthcare, the justice system, finance, and schools. Yes, AI brings lots of good things. But it also brings up ethics questions we need to figure out. This makes sure AI is used in a good, responsible way. We'll look at ethics questions about AI. These include being clear (or transparent), being fair, being responsible (or accountable), and how AI affects people and society. Transparency is a crucial ethical consideration in AI. It refers to the ability of stakeholders to understand how an AI system arrives at a decision. Transparency is vital to promote scrutiny of algorithms by stakeholders. This scrutiny enables the identification and addressing of biases in the algorithms, promoting fairness and justice in AI applications. In healthcare, AI systems are used for patient diagnosis and treatment. The decisions made by AI systems can have significant impacts on patients' health outcomes. Therefore, it is essential to ensure that AI decisions are transparent to ensure that they align with ethical standards (Huang et al., 2020). In criminal justice, AI systems are used for decision-making in various areas such as sentencing and parole. The decisions made by AI systems can have significant impacts on people's lives. Therefore, it is essential to ensure that AI decisions are transparent to ensure that they align with ethical standards (Angwin et al., 2016).

It's important to consider fairness in AI systems. We want systems that are fair to everyone, without any discrimination. Sometimes AI systems can be biased because of the data they are trained on or how they work. For example, facial recognition systems can be less accurate at recognizing people of colour, which is really not cool (Buolamwini & Gebru, 2018). To make sure AI systems are fair, we need to use diverse data that includes everyone. And, we need to design algorithms that don't unfairly discriminate against anyone. Responsibility is key in AI technology. AI systems need to answer for their choices and deeds. Responsibility makes sure AI systems own their actions and can be blamed for any harmful results. In areas like health, responsibility is crucial to make sure AI systems give correct illness diagnosis and guidance for treatment. Responsibility is also important in criminal justice. Here, AI systems are used for decisions these decisions could heavily affect people's lives, like in sentencing and parole. An example might be if an AI system suggests a sentence deemed later as unfair; the system needs to be held responsible for its decision (Angwin et al., 2016). AI's effect on society is an important ethical aspect to consider. AI can bring about many changes. These changes can be as diverse as making services more efficient or taking away jobs. We need to use AI in ways that helps society and avoids harm. For example, we can use AI in healthcare to help patients and cut costs. But we also need to think about the problems AI can create in healthcare. This can include patients losing trust in their doctors or worries over privacy. (Huang et al., 2020).

Ethical considerations in AI implementation hold significant importance, aiming to guarantee ethical and responsible use of AI. This encompasses transparency, fairness, accountability, and assessing AI's societal impact. Transparency ensures AI decisions align with ethical standards, while fairness prevents discrimination. Accountability entails AI systems taking responsibility for their actions, and considering societal impact ensures AI is used for societal welfare. These ethical considerations are fundamental in ensuring AI benefits society without causing harm.

## CONCLUSION

In simpler terms, biases hidden in AI decision-making can greatly change its correctness, fairness, and equal treatment. Spotting and battling these root causes of bias is truly crucial. These sneaky biases can hide in the data, the formulas we employ, and in the human role played in making decisions. By putting money into varied and representative data for AI learning, crafting clear and verifiable formulas and fostering diverse teams in creating, training, and sharing AI systems, we can minimize bias in AI. For an AI system to be precise, fair, and impartial, avoiding biases is utterly necessary. To reduce bias in AI systems, let's highlight the main steps. First, clearly outline the issues and goals. Next, bring together different teams for their ideas. Assess the data thoroughly. Then, create transparent algorithms. Keep checking the system regularly. These steps are crucial to keep AI technology fair and unbiased. The seriousness of these steps can't be overstated. We must ensure that everyone gets equal treatment from AI technology. In conclusion, it's important to carefully examine the data used to train AI systems. This is a key strategy for finding and removing biased data. Moreover, closely examining this data helps us see if some groups are underrepresented or overrepresented. It also helps us identify if certain factors, like a person's race or gender, are being given undue weight. This way, we can ensure fairness and impartiality in AI systems. Research supports this method for finding and minimizing bias in artificial intelligence systems. It's essential in crafting AI technologies that are fair for all, contributing to a just society. Ultimately, it's critical to spot any unfairness in AI technology to guarantee that these systems are trustworthy and equitable. After all, if algorithms are biased, they may lead to discrimination. Identifying these algorithms is vital.

Strategies to spot biased algorithms involve examining AI systems for fairness using various metrics and employing interpretability methods to comprehend their decision-making process. Detecting bias is essential to ensure the equitable development of AI systems, fostering societal justice for all. Importantly, trust between humans and AI hinges on transparency, accountability, fairness, user-centered design, and ethical considerations. Embracing these principles solidifies the reliability and societal benefits of AI systems.

The legitimacy of AI-driven decision-making is a pivotal concern within the technology domain, prompting earnest deliberation among professionals and scholars. A noteworthy threat to their credibility is the opaqueness and lack of elucidation within AI systems, potentially eroding public faith. Moreover, there is apprehension regarding the reinforcement of existing biases and disparities by these systems, exacerbated by insufficient human oversight. To uphold the authenticity of algorithmic decision-making, endeavours should be directed towards crafting AI frameworks that embody fairness and lucidity while ensuring human supervision. Notably, AI systems are increasingly woven into critical domains like healthcare and criminal justice, warranting meticulous ethical deliberation. In the realm of morality and artificial intelligence (AI) there are pivotal concerns encompassing aspects like openness, justice, responsibilities, and the influence of ai system on the community. The transparency of ai driven decision making is fundamental as it guarantees compliance with ethical norms, allowing stakeholders to examine algorithms, acknowledge biases and advocate impartiality and equity in AI implications. It's crucial to design AI system with equity and responsibility, ensuring their societal influence uphold ethical benchmarks.

## REFERENCES


Angwin, J., Larson, J., Mattu, S., & Kirchner, L. (2016). *Machine bias: There's software used across the country to predict future criminals. And it's biased against blacks*. ProPublica. https://www.propublica.org/article/machine-bias-risk-assessments-in-criminal-sentencing

BarocasS.SelbstA. D. (2016). Big data's disparate impact. *Social Science Research Network*. SSRN. doi:10.2139/ssrn.2477899

Buolamwini, J. (2018). *Gender Shades: Intersectional accuracy Disparities in commercial gender classification*. PMLR. https://proceedings.mlr.press/v81/buolamwini18a.html.

Caliskan, A., Bryson, J. J., & Narayanan, A. (2017). *Semantics derived automatically from language corpora contain human-like biases*. ResearchGate. https://www.researchgate.net/publication/316973825

Crawford, K., & Calo, R. (2016). There is a blind spot in AI research. *Nature*, *538*(7625), 311–313. doi:10.1038/538311a PMID:27762391

Dastin, J. (2018). *Amazon scraps secret AI recruiting tool that showed bias against women.* Reuters. https://www.reuters.com/article/us-amazon-com-jobs-automation-insight/amazon-scraps-secret-ai-recruiting-tool-that-showed-bias-against-women-idUSKCN1MK08G

EU Agency for Fundamental Rights. (2018). *Fundamental Rights and Artificial Intelligence: Key Fundamental Rights considerations for the future regulatory framework.* FRA. https://fra.europa.eu/sites/default/files/fra_uploads/fra-2018-focus-ai-rights_en.pdf

Hajian, S., Bonchi, F., & Castillo, C. (2016). Algorithmic Bias. *Algorithmic Bias: From Discrimination Discovery to Fairness-aware Data Mining*. ACM. . doi:10.1145/2939672.2945386

Huang, J., Li, J., Huang, X., & Cai, W. (2020). Ethical considerations in artificial intelligence and healthcare. *Journal of Healthcare Engineering, 2020*, 1–10. oi:10.1155/2020/8845224

Kleinberg, J., Mullainathan, S., & Raghavan, M. (2016). Inherent Trade-Offs in the Fair Determination of Risk Scores. https://doi.org//arXiv.1609.05807. doi:10.48550

Lepri, B., Oliver, N., Letouzé, E., Pentland, A., & Vinck, P. (2017). Fair, transparent, and accountable algorithmic decision-making processes. *Philosophy & Technology, 31*(4), 611–627. doi:10.1007/s13347-017-0279-x

Liao, L., & Hirschberg, J. (2018). Towards Understanding and Mitigating Bias in Neural Machine Translation. https://doi.org//arXiv.2106.13219. doi:10.48550

Machlev, R., Heistrene, L., Perl, M., Levy, K. Y., Belikov, J., Mannor, S., & Levron, Y. (2022). Explainable Artificial Intelligence (XAI) techniques for energy and power systems: Review, challenges and opportunities. *Energy and AI, 9*, 100169. doi:10.1016/j.egyai.2022.100169

MeuwissenM.BollenL. (2021). Transparancy versus Explainability in AI. *ResearchGate*. https://doi.org/doi:10.13140/RG.2.2.27466.90561

Mittelstadt, B., Allo, P., Taddeo, M., Wachter, S., & Floridi, L. (2016). The ethics of algorithms: Mapping the debate. *Big Data & Society, 3*(2), 205395171667967. doi:10.1177/2053951716679679

Nissenbaum, H. (1996). Accountability in a computerized society. *Science and Engineering Ethics, 2*(1), 25–42. doi:10.1007/BF02639315

Obermeyer, Z., Powers, B., Vogeli, C., & Mullainathan, S. (2019). Dissecting racial bias in an algorithm used to manage the health of populations. *Science, 366*(6464), 447–453. doi:10.1126/science.aax2342 PMID:31649194

Veale, M., & Binns, R. (2017). Fairer machine learning in the real world: Mitigating discrimination without collecting sensitive data. *Big Data & Society, 4*(2), 205395171774353. doi:10.1177/2053951717743530

Zhao, J., Wang, T., Yatskar, M., Ordóñez, V., & Chang, K. (2017). Men Also Like Shopping: Reducing Gender Bias Amplification using Corpus-level Constraints. *Men Also Like Shopping: Reducing Gender Bias Amplification Using Corpus-level Constraints*. ACL. . doi:10.18653/v1/D17-1323

# Chapter 4
# Ethical Considerations in the Educational Use of Generative AI Technologies

**Burak Tomak**
https://orcid.org/0000-0001-6678-431X
*Marmara University, Turkey*

**Ayşe Yılmaz Virlan**
https://orcid.org/0000-0002-6839-5745
*Marmara University, Turkey*

## ABSTRACT

*This chapter provides an overview of the ethical considerations that should be taken into account while using generative AI technologies, specifically in the field of education, as well as concrete suggestions for programmers and end-users. Therefore, students, researchers, and academics in a variety of fields who are interested in the ethical aspects of generative AI will find this chapter useful, as it will also provide an overview of the existing ethical frameworks in the field of education. In that sense, this chapter can be viewed as a concise introduction to the current state of the ethical issues being studied and a proposal for balancing risks and opportunities.*

## INTRODUCTION

A few years ago, the concept of a machine generating art, telling engaging stories, or digging into the complexities of medical and legal papers seemed like a science fiction fantasy—an almost unattainable ideal. However, the advent of the Generative AI age has profoundly changed our knowledge, bringing these seemingly insane ideas within reach. It is widely believed that the forthcoming significant digital revolution in several aspects of human existence, including lifestyle, communication, employment, commerce, and education, would be primarily driven by the advancements in AI (Zemel et. al., 2013). Yet, because of the increased employment of AI-powered ed-tech tools and the emphasis on distance

DOI: 10.4018/979-8-3693-1565-1.ch004

learning during the pandemic, the risks and ethical issues in AI has emerged as a significant concern for professionals and researchers (Chaudhry & Kazim, 2022).

According to Pecorari (2001), plagiarism is the act of taking content from a source without giving credit, either with or without the aim to deceive. Traniello and Bakker (2016) refer to it as "intellectual theft". With the help of the Internet, the availability of lots of information and access to it have both made it possible for everyone to use it as their own. However, Li and Casanave (2012) warn that this "unconventional and interactive" source of information must be detected (p. 166). Therefore, as Davies and Howard (2016) state, there is a need for a reconceptualization of plagiarism and recommendations for coping strategies especially after the introduction of the artificial intelligence tools. This idea is also supported by carried on by Flowerdew and Li, (2007). However, in their study Gao et al., (2022) assert that certain artificial intelligence systems, such as ChatGPT and Quillbot, have already integrated a variety of strategies to avoid detection of plagiarism. Furthermore, Yan (2023) shows in his research that students are limiting the odds of being detected as "cheating or plagiarizing" using present means of detection, which poses a higher risk to academic integrity and educational fairness. This is highlighted in the context of the fact that students are minimizing the likelihood of getting caught (Haque et al., 2022) as the AI tools can help them to easily get away from the risk of being detected (Susnjak, 2022).

According to Bostrom (2014), the management of artificial intelligence systems has to be carried out by setting constraints on their utilization following our value systems. This would guarantee that we have control over these systems. Two distinct points of view are taken into consideration in Nyholm's (2023) report on the subject. According to the author, when we consider human control over artificial intelligence to be a form of self-control, this might mean having control over AI is initially advantageous, as it has the ability to serve both as a means and possibly even as an end goal. The argument made here is that if control over artificial intelligence may be viewed as a type of "control over another person or a representation of a human", then such control might be perceived as something that is negative or something that is in incorrect assessment (Nyholm, 2023, p. 1230). It is a dilemma for some people, as some of them support the use of it while others reject the idea of accepting it as a useful tool. However, there are a number of factors that influence this process, including technological, social, political, and economic aspects, each possessing the capacity to hinder the effective execution of policy and design-oriented computer ethics (Jacobs & Simon, 2023). As a result, this decision is not a simple one to make. In that sense, it crucially important to be knowledgeable about the principles and the frameworks that are set by the leading institutions and organizations and policy makers.

## Current Ethical Frameworks

Several ethical and legal difficulties may be encountered while using the AI tools, including questions of responsibility and the risk of biased decision-making. For this reason, the European Union, UNESCO, and the World Economic Forum are just a few of the institutions and researchers that have carried out in-depth research on ethical implications of AI in our lives. Recently, there has been an obvious increase in the number of ethical projects that aim mainly at addressing concerns associated with data in AI (Huang, et. al., 2023). These efforts mainly focus on obtaining informed consent, protecting data privacy, and addressing biases that are present in datasets. What is more, these efforts try to point out different yet important concerns about the use of AI with regards to transparency, statistical pattern recognition, and the reduction of prejudiced assumptions (Huang, et. al., 2023). Although there has been such efforts made, AI technology and its tools have been implemented in many different fields without much concern. On top of it, there has not been a strong official announcement or statement that has particularly targeted the wide variety of concerns that have been brought up by AI especially in the field of education except from some frameworks and guiding principles that have been announced by some leading organizations like UNESCO or the European Union.

Now, it has become even more important for researchers from all around the world to figure out the capacity of the implementation of AI tools in their research to revolutionize education, by enhancing its inclusivity and equity at all levels. In order to ensure equitable access to artificial intelligence, it is a must to establish policies that actively empower marginalized populations, including women, girls, and economically disadvantaged communities (Fischer et. al., 2023). In order to improve educational standards, the integration of AI into educational fields must be considered in terms of human needs and ethical norms. For this reasons, it is also recommended that some measures should be taken to improve the overall educational experience for all learners and educators. These measures, or guidelines, could include the improvement of learning management systems, granting educators increased autonomy, and leveraging artificial intelligence. Thus, as Fischer et. al. (2023) suggest, the paramount objective should be given to equipping the learners and instructors with necessary knowledge and skills to effectively and safely employ artificial intelligence in their studies within a defined framework despite the considerable challenges it might present.

This being the case, it is for sure that each individual needs to obtain basic AI abilities to be adequately equipped for the future, which requires incorporating AI skills into educational fields (Fischer et. al., 2023). A wide range of skills, such as protecting data and guaranteeing its security, as well as understanding the processes involved in acquiring and manipulating data for AI could be considered here. Moreover, these skills also need an understanding of the complex mechanisms that make up AI. For this reason, to develop an effective strategy that integrates AI and education, it was imperative to actively involve and communicate with parties from all sectors and businesses. This was because the influence of artificial intelligence goes beyond any specific business or field. As a result, multiple organizations and institutions have attempted to create frameworks regarding artificial intelligence, and some studies in terms of creating frameworks have been conducted such as the, The Beijing Consensus on Artificial Intelligence (2019), The Ethical Framework for AI in Education (2020) and Artificial Intelligence and Education: Guidance for Policy-makers (2021). Similarly to their presence in various other domains, these frameworks follow to a set of principles linked to artificial intelligence. To conform to AI standards in various fields, they have strived to create a consistent and standardized guide as well as produce guidelines also in education.

## The Beijing Consensus on Artificial Intelligence

UNESCO also released another document which advocates for taking measures to ensure the smooth and constructive integration of AI into education (UNESCO, 2019). "The Beijing Consensus on AI and Education" was the first document to provide guidelines on how to employ AI technology in teaching and learning practices. Written in a plurilingual format, the Consensus was embraced by more than 50 government ministers and delegates from over 105 member States. The consensus follows the Qingdao Declaration of 2015, which pledged UNESCO member states to make effective use of emerging technologies for Sustainable Development Goal 4, was adopted during the International Conference on Artificial Intelligence and Education in Beijing. The final document included twelve subtopics under which 44 items related to AI usage in education were discussed and adopted as guidelines on the employment of AI technology to fulfill the Education 2030 Agenda.

As was stated in the "Beijing Consensus", the progression of AI has had a substantial impact on human civilizations, economies, and the labor market, in addition to education and learning systems that are designed to last a lifetime. Not only does it have the ability to transform education, teaching, and learning, but it also has enormous consequences for development of skills and employment opportunities. In the field of AI, there are numerous understandings, broad definitions, and diverse applications in a variety of contexts. This topic is complex and quickly evolving. Concerns of an ethical nature are also raised.

According to the Consensus, the implementation of AI can improve the capabilities of instructors from different fields and make the process of teaching more efficient. Nevertheless, it is of the utmost importance to ensure that human connection and collaboration continue to play a fundamental role in the field of education. As the Consensus imply, there is no way that machines could ever take the place of teachers, and it is crucially important to protect the educators' rights and the conditions under which they work (Beijing Consensus, 2019). Another important item discussed in the Consensus is about the duties and talents of instructors which should be evaluated and established in a flexible manner. Additionally, according to the guidelines accepted, the capabilities of teacher training institutions should be improved, and programs should be developed to improve the instructors' abilities to teach effectively in educational settings that use artificial intelligence (Beijing Consensus, 2019).

As was stated in the Consensus, AI has the capacity to enhance both learning and assessment, and teachers should be aware of trends and modify their lesson plans to encourage the incorporation of AI. In order to assist well-defined learning objectives across a variety of courses and to create skills that are applicable across disciplines, AI techniques can be utilized (Beijing Consensus, 2019). In addition, schools should provide assistance for pilot studies in AI to foster innovation in teaching and learning as these experiments can help extract valuable insights from successful cases and facilitate the expansion of evidence-based approaches. The Consensus also declares that AI approaches can support applications such as adaptive learning processes, the assessment of students' competencies, and facilitating large-scale and remote evaluations and it should be noted that these examples represent only a fraction of the potential capabilities that artificial intelligence can offer (Beijing Consensus, 2019).

The Consensus also aims to enable UNESCO to promote the funding, collaborative efforts, and global cooperation in the domain of AI in education. Here, the idea recommends the creation of a dedicated website exclusively focused on AI for educational objectives, which also offers accessible consultation resources, encompassing AI courses, tools, policies, and exemplary approaches to AI. Last but not least, the problem of variations in the development of artificial intelligence across different countries and regions is addressed in the framework of the Consensus. In order to prevent the occurrence of division, the job of monitoring and studying the impact of these differences has been transferred to international organizations and partners. UNESCO's goal is to strengthen its important position in the field of AI in education. The achievement of this could be attained through collaborative works of partners, and organizations, and the implementation of basic principles.

## The Ethical Framework for AI in Education

Founded in 2018 with an objective that aimed to establish a comprehensive framework, The Institute for Ethical AI in Education focuses on maximizing the educational advantages of AI in classrooms, while also protecting learners against the possible risks related to this technology. The Institute released a preliminary research report in February 2020, focusing on the topic of ethical artificial intelligence in education. The primary goal of this inquiry was to examine the potential risks and advantages of implementing artificial intelligence technologies in educational environments. The report offered an ethical framework with six major problems to guide future research and address any discrepancies between the educational advantages and risks associated with artificial intelligence (The Institute, 2021). This framework was provided as a method to direct future study. They released a framework with nine main objectives and 33 criteria with accompanying checklist items for each objective and the related criteria to have the highest level of authority when it comes to allocating resources (The Institute, 2021). In educational institutions, the framework is designed for use by decision-makers who are responsible for obtaining and implementing artificial intelligence technologies. Since educational leaders and practitioners have a crucial role in both maximizing the advantages of AI and mitigating its potential risks, through implementing this approach, providers became obligated to give priority to ethical design that aligns with the welfare of individuals who are acquiring knowledge.

In that sense, the Framework for Artificial Intelligence in Education empowers educational leaders and practitioners to improve the educational experience of students by facilitating the development, acquisition, and implementation of AI software. In addition, their responsibility includes the eradication of bias and user manipulation, while also ensuring that the design of materials aligns with educational principles. Furthermore, they are responsible for ensuring that the design of resources is fair and just. The Institute also emphasizes the necessity of educational improvements to enable *equal* access to the benefits of AI which can effectively address obstacles like limited curriculums and barriers to socio-economic mobility. It can also enable widespread access to lifelong learning with an advanced quality. For this reason, implementation of reforms is necessary to achieve this objective.

All in all, The Institute's objective is to advocate for government intervention in ensuring universal access to essential infrastructure, technology, and connectivity for all students. In other words, the pandemic, for example, could be a critical moment for education, as it presents an opportunity for artificial intelligence to address educational inequalities in an ethically acceptable manner, thereby enabling children from different socioeconomic groups to reach their full potential. Adhering to the suggestions outlined by the Framework is crucial to fully capitalize on the benefits that artificial intelligence can offer to the field of education.

## Artificial Intelligence and Education: Guidance for Policy-makers

Another important document including guidelines for the utilization of AI in education is "Artificial Intelligence and Education: Guidance for Policy-makers," proposed by UNESCO in 2021. According to the framework there raises several ethical questions, some of which are as follows:

- Defining ethical boundaries for the collection and use of data
- Limiting representation in large datasets
- Examining AI decision-making
- Considering the short-term nature of students' interests and the learning process
- Taking into account the responsibilities of private organizations and public authorities.

One of the most important elements in education is to use approaches that correspond to ethical standards, which places an emphasis on the significance of educational data and algorithms in a way that is open to inspection, and in accordance with ethical values. This addresses the inequalities that are present in AI systems, as well as the difficulties that might appear when finding a compromise between protection of data and allowing unrestricted access to information (UNESCO, 2021). In that sense, according to the guide, the initiative aims to achieve the following objectives: promoting the advancement of innovative AI technologies, conducting in-depth research on the ethical implications of AI, and establishing comprehensive data protection frameworks that cover all relevant stakeholders. In addition to these, it encourages research on several themes, including data privacy and security; concerns regarding the influence of AI on human rights, gender equality, and ethical considerations in this field (UNESCO, 2021).

Additionally, some regulations have been set on the use of AI tools, and the European Commission (2021) has also set some rules on their use. Cossette-Lefebvre and Maclure (2023) claim that the use of AI should be strictly regulated because it leads to social inequalities and disregards individual autonomy. It is quite natural because when a student writes the prompt about a question that s/he needs an answer to, the tools provide him/her with a standard answer without considering the context in which s/he lives. Thus, this diminishes the creativity and personification that a student can make with his own writing, though some scholars claim that the output taken from ChatGPT is qualified enough to satisfy the needs of the students (Gao et al., 2022). What is more, Wenzlaff and Spaeth (2022) confirm that the answers of ChatGPT are equivalent to those of a human.

Keeping these in mind, it should be noted that the following concept serves as the overarching idea for AI and education policies: In addition to ensuring that artificial intelligence is regulated by humans and centered on helping people, it should also be implemented to improve the capabilities of both students and teachers. The frameworks focuses on creating ethical, fair and open AI applications and monitoring and analyzing how AI affects people and society (UNESCO, 2021). Also, encouraging the human values that are necessary for the development and implementation of AI is another important issue discussed in the framework. In the context of AI technologies that boost productivity, it is also important to investigate the potential conflict that may arise between the market and human values, skills, and social well-being (UNESCO, 2021).

One way to solve the major societal challenges that are brought up by artificial intelligence technology is to encourage widespread corporate and civic responsibility in order to address concerns such as justice, transparency, accountability, human rights, democratic principles, bias, and privacy. It is imperative that people continue to be at the center of education as an integral component of the design of technology. Additionally, it is essential to prevent the automation of work without first identifying and compensating for the values that are currently being practiced.

As Akgun and Greenhow (2022) say, artificial intelligence systems pose ethical problems and dangers. It seems that marketing campaigns that show algorithms to the public as objective and value-neutral tools are going against this. So, when it comes down to it, algorithms are a reflection of the values that their creators, who are in positions of power, embrace (Akgun & Greenho, 2022). At the same time that individuals are developing algorithms, they are also developing a collection of data that is representative of the historical and systemic prejudices that exist throughout society (Hrastinski, 2019). This data eventually becomes algorithmic bias. It is possible to observe a variety of gender and racial biases in various AI-based platforms, despite the fact that the bias is ingrained in the computational model without any intentional attempt to sustain it (Stahl & Wright, 2018).

Finally, as Stephen Hawking states at the launch of the Leverhulme Centre for the Future of Intelligence in 2016:

*"Success in creating AI could be the biggest event in the history of our civilisation, but it could also be the last – unless we learn how to avoid the risks. Alongside the benefits, AI will also bring dangers like powerful autonomous weapons or new ways for the few to oppress the many."*

So now, we would like to continue with the recommendations that could be taken into consideration when dealing with the ethical issues that may arise while we are employing AI tools in our works, as Professor Hawking points out.

## Recommendations on the Ethics of Artificial Intelligence

To deal with the ethical issues raised in the field of education, policymakers established more than just frameworks. Institutions or organizations like UNESCO have also made and published their recommendations at several occasions. At this point, it's crucial to emphasize that "Recommendation on the Ethics of Artificial Intelligence" was the first-ever global standard on AI ethics which was adopted by all 193.

The framework suggests that states should collaborate with international organizations, educational institutions, and private entities to provide AI literacy education to reduce digital divides and access inequalities (UNESCO, 2022). This includes promoting skills like basic literacy, numeracy, coding, media literacy, critical thinking, teamwork, communication, and AI ethics. Awareness programs should be promoted about AI developments, and research initiatives should be encouraged on responsible AI use. AI education should empower students and teachers, and best practices should be shared.

UNESCO (2023) also proposes that educational institutions provide explicit guidelines to students and instructors on how and when ChatGPT can (and cannot) be utilized. Such counsel should be agreed upon with students and teachers rather than forced on them (p. 13). Lodge (2023) assures this kind of awareness will lessen the possibilities of cheating from the sides of students.

UNESCO (2021) comes up with five different recommendations for the better use of AI in education (p.31):

### Interdisciplinary Planning and Inter-Sectoral Governance

Just like every other field, education will be more qualified and stronger with cooperation and collaboration with other majors. Thus, the use of AI in an efficient and ethical way requires cooperation of educators with some experts in other fields because this situation can entirely be handled with neither the teacher of a classroom nor the curriculum developer that has facilitated the flow and the content of the lesson or course. Therefore, instructors, learning scientists, and AI developers from diverse fields such as neuroscience, cognitive science, social psychology, and the humanities must collaborate to create user-centered and result-based AI solutions for for both the efficient and ethical use of AI. This can be accomplished through the use of consistent system-wide methods and evidence-based inclusive methodologies such as participatory design and co-creation frameworks (Pobiner & Murphy, 2018). What is more, authorities from the governments should also take part in the negotiations because their political power and eligibility for the implementation of the law will make the proper use of AI both easier and possible. Furthermore, these negotiations should be arranged both in local and international contexts, which will make the information share more likely among all the stakeholders whose opinions are so valuable that their contributions can be great asset to the solution of the problem. Additionally, different cases in different contexts are different pieces of the puzzle and they will be perfect examples for the different scenarios of the situation. They can all be discussed among the experts in international conferences and seminars in which people of interest will gather to share their valuable ideas about the topic and the collection of different opinions from different sources will surely broaden our minds about this vital issue.

## Policies on Equitable, Inclusive, and Ethical Use of AI

One of the most important things to consider in the use of AI like any other facilities is the equality, the absence of which will lead to unfair consequences. Thus, equality is something that must be considered for the efficient and ethical use of AI. This can be achieved with the availability of the necessary software, hardware, and the Internet connection for everyone in the society. Otherwise, there will be double standard and the ones who already have these things listed will be more advantageous than the others who do not have them. Governments must have a key role in the supply of these for the learners at any level of education from primary school to the university. What is more, governments should also keep an eye on the protection of data services because the data created by the learners can either be open-access or private. Thus, the balance between open-access and data privacy must be enabled by the authorities. García-Peñalvo (2023) warns that intellectual property, privacy, and recognition of authorship are the sensitive issues that should be paid attention in the use of AI tools. Furthermore, the bias included in the corpus of AI must be eliminated. This is something unwanted but unfortunately it can be easily detected. Especially some of the data available on the web is sexist showing the women inferior to the men. For instance, when you write a prompt about a successful life of a CEO on ChatGPT, it gives you a sample from a male person, which implicitly implies that it will surely be someone from a male dominant world. These biases must be invalidated. Pennington (2018) warns that AI tools must be tested and confirmed that they are free of biases.

## Develop a Master Plan for Using AI for Education Management, Teaching, Learning, and Assessment

New trends always bring some challenges along with them so people have the inclination to object to them without carefully thinking about the benefits. AI is one of these new changes in education that might be found something challenging and new by most of the people. However, AI technologies can be highly benefitted if they are used properly and on purpose. Therefore, educators must be enthusiastic for the encouragement of their use in education. Surely, this requires a careful and detailed plan. First of all, almost all educational institutions make use of learning management systems (LMS). These systems can be integrated with AI technologies for the benefit of both learners and teachers because AI will surely make the task of both teachers and learners easier. For instance, they can be utilized by the curriculum developers to create more personalized content for students with different proficiency levels. AI will organize activities for the students who need more language support as well as for the ones whose proficiency levels are much better than their peers by providing them with extra materials and sources. Kostka and Toncelli (2023) show in their study that students' enthusiasm for the course materials and their interactions with their peers have both increase with the integration of AI tools in the class implementations. What is more, the data that can be collected from these channels will give a great insight into how AI technologies should be utilized and integrated in the school curriculums. Thus, teachers and students must be open for cooperation, which means that they must give their consent for the collection of the data that will reveal the exploitation of AI technologies in the educational context. Abramson (2023) emphasizes that including learners in discussions about academic integrity will make it easier for the adaptation of teaching and learning to the age of AI. However, both students and teachers must be trained on their use because they must consider this as a “facilitative” source instead of something that does all the things on behalf of them. Therefore, students must be guided on how to make use of them. There should be a balance between what they do on their own and what ChatGPT does for them. In fact, they should not feel the pressure to use it to do the tasks just to complete but to create their own learning product. As students are encouraged to use the AI tools in facilitative role in their learning, they must be integrated in the assessment process, as well. AI-based formative assessment, when integrated into an AI-powered LMS, improves the evaluation of students’ learning with greater accuracy and efficiency. This integration removes human bias and yields superior results for learner improvement, as stated by Abramson (2023).

Not only the students but also the teachers must be encouraged to make use of AI properly. Teachers must use this technology for their own benefit. Honigsfeld and Dove (2023) suggest that teachers must collaborate with AL tools not only to develop their teaching skills but also to improve themselves in terms of curriculum development, material development, and assessment. For instance, they can make the AI tools prepare worksheets for their students, which will take their hours. Thus, they will benefit from it both for their and students’ own good. However, the important point here is that teachers should be trained on how to use it efficiently. Teacher training programs should be updated according to UNESCO ICT Competency Framework for Teachers (UNESCO, 2018). This surely does not mean that the role of the teachers will be subordinated to the AI tools. Teachers will play an active role in teaching but they will make of AI tools to increase the efficiency of their teaching. They will be the ultimate guide and counsellor in the language learning process of their learners because they will be the only reliable source that will determine the path that goes to higher language proficiency of the learners. AI should be integrated into every field in a human-centric manner, according to Pasquale (2020). This means that in order to increase the productivity of AI, control and the ultimate decision should remain in the hands of humans.

Scholars in the academia are also one of the groups that can benefit from the AI tools. While they are reviewing the literature, they can make use of AI tools to summarize the articles that they have to read and cite. Though AI tools cannot make citations, they can be used to give a compact summary of the articles that will be used in the literature review part of the articles written by the scholars. Rahman et al. (2023) propose that scholars can utilize this technology to condense extensive text and discern crucial discoveries “from the literature” (p. 10). They think that AI tools are appropriate to use as a co-writer of an academic text.

The use of AI technologies should not be restricted to the educational contexts like classrooms but they should be integrated in our lives forever because AI tools must be used for lifelong purposes, as well. This means that AI tools can be so beneficial that we can make use of them even when we do not actively attend a school or an educational institution. We can benefit from them for our personal development. Therefore, everyone in the society must know how to benefit from AI tools. In other words, everyone should be trained on the proper use of them because every citizen of the world has something to learn from AI tools, which should not be restricted to the educational institutions. In fact, training is necessary to develop AI literacy skills (Tlili et al., 2023).

These master plans for the use of AI tools will encompass several fields. Even though it may seem the use of AI tools will boost the possibility of the extinction of some jobs and vocations, it will facilitate any kind of job and make it grow faster. However, authorities should take into account the prediction models in order to identify trends in employment and skills, as well as design retraining programs for individuals in jobs that are at risk of AI automation. If they do it so, they will make arrangements accordingly. For instance, people in certain sectors will be trained considering both strengths and weaknesses of AI tools and their contribution to their field. If they become aware of these, they will improve themselves in this sense. What is more, older people who are called as seniors in certain professions should be protected as they might find the AI tools challenging to use but they can be adapted to the new conditions with some in-service training programs.

Considering all these, the authorities should make applicable plans for the future in terms of the integration of AI tools with our lives. Therefore, it is significant that skilled and qualified workers who are capable of using the AI tools efficiency are in need in any kind of sector in any field. Thus, they should use them in accordance with the needs of the sectors in which they are working. Thus, to prepare them for their future professions, they must be skilled enough to meet the demand or the requirements properly. As a matter of fact, these master plans should be carefully made considering all sectors in the market.

## Pilot Testing, Monitoring, and Evaluation, and Building an Evidence Base

To better serve the diverse learning environments, it is suggested that we promote the use of AI-enhanced personalized learning simulations, chatbots, augmented and virtual reality tools, language learning resources, AI-based artwork and music generators, exploratory learning systems, and automatic writing evaluation systems. It is crucial to place the system in its proper context, as the implementation of any new element necessitates adjustment to the new surroundings. Therefore, it is imperative to design and carry out local large-scale pilot evaluations of AI systems provided by external vendors. These evaluations will determine the relevance and effectiveness of the systems in relation to educational practices, objectives, diversity, culture, and demographics specific to the local context. These findings will be utilized to tailor the data, design, and integration of the artificial intelligence system to the specific requirements of the local community. In addition, it is vital to monitor and evaluate whether the implementation of the system will result in conflicts of interest or partnerships, as well as contradictions about the security of data or ownership, in order to ensure that the appropriate actions are done. These processes will build trust for the AI tools, of whose benefits people will be convinced. Thus, more research should be done especially in several parts of the world in order to determine their efficient and ethical use because the results of these studies, which will enable the scholars to better understand the consequences of AI tools in certain regions, will be exploited for a better use.

## Fostering Local AI Innovations for Education

Even though the previous four recommendations are equally significant for the contribution of the efficient and ethical use of AI tools, "fostering local AI innovations for education" should also be taken into consideration as the last but not the least important one because this is the last step that should be taken seriously. This can only be achieved with more investments and funding in the educational institutions in local context. Therefore, the process of determining market failures, complex educational systems around the world, and the problems about scaling programs should therefore involve participation from a variety of stakeholders, including funders, commercial developers, educators, and learning scientists. As a matter of fact, more research should be conducted to detect the strengths and weaknesses of the AI tools in certain contexts. Thus, more collaborations are needed.

## CONCLUSION

There are significant psychological consequences for both teachers and students if AI is used improperly in the classroom. What is more, it has been mentioned throughout the chapter that the use of AI tools affects every field. Thus, both the efficient and ethical use of AI tools is significant in this sense because it affects all parts of society not only at present but also in the future. Therefore, more effort must be put forward to guarantee reliable and secure AI products for all parties involved (Chaudhry & Kazim, 2022). As a matter of fact, people from the different stages of society should be trained on the proper use of them so that they will be aware of the efficient and ethical use of AI tools both for the benefit of their own profession and society itself. In other words, awareness raising activities such as trainings, seminars, or conferences on the ethical use of AI tools must be organized and consistently planned all around the world. Otherwise, the consequences can be detrimental. Hence, we need to weigh the risks against the benefits and smartly use this powerful technology. For this reason, actionable recommendations that can be utilized to develop policies and programs tailored to local contexts should be taken into consideration not only by the policymakers who will have the eligibility to make laws but also by the educators who will facilitate the implementations of these laws in the educational contexts all around the world. What is more, in an individual basis, the importance of raising awareness for the ethical pitfalls of AI technologies in terms of human emotions has been mentioned by Ghotbi (2023) who claims that human emotions also play an important role in the decisions we make.

As the above-mentioned institutes and organizations have emphasized, the active participation of governments around the world, the United Nations, and those who are actively involved in education at all levels is an absolute necessity. What is more, each and every country must do its best to consider these ethical issues in terms of the use of AI tools in order to better apply the rules and make them serve for the benefit of its citizens without creating any inequalities and unfairness. Younas and Zeng (2024) warn that "many countries worldwide have recognized the need to establish ethical principles to guide the development and deployment of AI systems with a formal law or policy" (p.3). Therefore, they claim that every country including the ones in Central Asia is aware of the potential influence of AI on their societies, economies, and their governance system; hence, they suggest that government bodies, academia, industry experts, civil society organizations, and the public must cooperate with each other to better benefit from AI tools considering the ethical issues. Chakrabarti and Sanyal (2020) state that India is doing its best to promote and regulate the use of AI properly but it is still behind the countries such as China and the United States of America. However, these scholars claim that the situation is even worse for the underdeveloped countries where algorithms have a bag impact on the socio-economically disadvantaged people. Thus, it is claimed in this article that these algorithms must be monitored by some people who are eligible and qualified enough to eliminate the unfairness and inequality created by the unethical use of AI tools.

## REFERENCES

Abramson, A. (2023). How to use ChatGPT as a learning tool. *Monitor on Psychology*, *54*(4). https://www.apa.org/monitor/2023/06/chatgpt-learning-tool

Akgun, S., & Greenhow, C. (2022). Artificial intelligence in education: Addressing ethical challenges in K-12 settings. *AI and Ethics*, *2*(3), 431–440. doi:10.1007/s43681-021-00096-7 PMID:34790956

*Beijing Consensus on artificial intelligence and education.* (2019). Paris: UNESCO. https://unesdoc.unesco.org/ark:/48223/pf0000368303

Bostrom, N. (2014). *Superintelligence: Paths, Dangers, Strategies.* Oxford University Press.

Chakrabarti, R., & Sanyal, K. (2020). Towards a 'Responsible AI': Can India take the lead? *South Asia Economic Journal, 21*(1), 158–177. doi:10.1177/1391561420908728

Chaudhry, M. A., & Kazim, E. (2022). Artificial Intelligence in Education (AIEd): A high-level academic and industry note 2021. *AI and Ethics, 2*(1), 157–165. doi:10.1007/s43681-021-00074-z PMID:34790953

Cossette-Lefebvre, H., & Maclure, J. (2023). AI's fairness problem: Understanding wrongful discrimination in the context of automated decision-making. *AI and Ethics, 3*(4), 1255–1269. doi:10.1007/s43681-022-00233-w

Davies, L. J. P., & Howard, R. M. (2016). Plagiarism and the Internet: Fears, Facts, and Pedagogies. In T. Bretag (Ed.), *Handbook of Academic Integrity* (pp. 591–606). Springer., doi:10.1007/978-981-287-098-8_16

European Commission. (2021). *Proposal for a regulation of the European Parliament and of the Council laying down harmonized rules on artificial intelligence (artificial intelligence act) and amending certain union legislative acts.* European Commission. https://eur-lex.europa.eu/resource.html?uri=cellar:e0649735-a372-11eb-9585-01aa75ed71a1.0001.

Fischer, I., Mirbahai, L., Beer, L., Buxton, D., Grierson, S., Griffin, L., & Gupta, N. (2023). *Transforming Higher Education: How we can harness AI in teaching and assessments and uphold academic rigour and integrity.* Warwick International Higher Education Academy (WIHEA), University of Warwick. https://warwick.ac.uk/fac/cross_fac/academy/activities/learningcircles/future-of-learning/ai__education_12-7-23.pdf

Flowerdew, J., & Li, Y. (2007). Plagiarism and second language writing in an electronic age. *Annual Review of Applied Linguistics, 27*, 161–183. doi:10.1017/S0267190508070086

GaoC. A.HowardF. M.MarkovN. S.DyerE. C.RameshS.LuoY.PearsonA. T. (2022). Comparing scientific abstracts generated by ChatGPT to original abstracts using an artificial intelligence output detector, plagiarism detector, and blinded human reviewers (p. 2022.12.23.521610). bioRxiv. doi:10.1101/2022.12.23.521610

García-Peñalvo, F. J. (2023). The perception of Artificial Intelligence in educational contexts after the launch of ChatGPT: Disruption or panic? *Education in the Knowledge Society, 24*, 1–9. doi:10.14201/eks.31279

Ghotbi, N. (2023). The ethics of emotional Artificial Intelligence: A mixed method analysis. *Asian Bioethics Review, 15*(4), 417–430. doi:10.1007/s41649-022-00237-y PMID:37808444

Haque, M. U., Dharmadasa, I., Sworna, Z. T., Rajapakse, R. N., & Ahmad, H. (2022). *'I think this is the most disruptive technology': Exploring Sentiments of ChatGPT Early Adopters using Twitter Data* (arXiv:2212.05856). arXiv. https://doi.org//arXiv.2212.05856 doi:10.48550

Hawking, S. (2016). *The best or worst thing to happen to humanity. The launch of the Leverhulme Centre for the Future of Intelligence.* CAM. https://www.cam.ac.uk/research/news/the-best-or-worst-thing-to-happen-to-humanity-stephen-hawking-launches-centre-for-the-future-of

Honigsfeld, A., & Dove, M. G. (2023). *5 collaborative teaching practices for teacher learning.* TESOL International Association. https://www.tesol.org/blog/posts/5-collaborative-teaching-practices-for-teacher-learning

Hrastinski, S., Olofsson, A. D., Arkenback, C., Ekström, S., Ericsson, E., Fransson, G., Jaldemark, J., Ryberg, T., Öberg, L., Fuentes, A., Gustafsson, U., Humble, N., Mozelius, P., Sundgren, M., & Utterberg, M. (2019). Critical imaginaries and reflections on artifcial intelligence and robots in post-digital K-12 education. *Post Digit. Science Education., 1*(2), 427–445. doi:10.1007/s42438-019-00046-x

Huang, R., Liu, D., Chen, Y., Adarkwah, M. A., Zhang, X. L., Xiao, G. D., Li, X., Zhang, J. J., & Da, T. (2023). *Learning for All with AI? 100 Influential Academic Articles of Educational Robots.* Beijing: Smart Learning Institute of Beijing Normal University. https://sli.bnu.edu.cn/uploads/soft/230413/1_1744328351.pdf

Jacobs, M., & Simon, J. (2023). Reexamining computer ethics in light of AI systems and AI regulation. *AI and Ethics, 3*(4), 1203–1213. doi:10.1007/s43681-022-00229-6

Kostka, I., & Toncelli, R. (2023). Exploring applications of ChatGPT to English language teaching: Opportunities, challenges, and recommendations. *The Electronic Journal for English as a Second Language, 27*(3), 1–19. doi:10.55593/ej.27107int

Li, Y., & Casanave, C. P. (2012). Two first-year students' strategies for writing from sources: Patchwriting or plagiarism? *Journal of Second Language Writing, 21*(2), 165–180. doi:10.1016/j.jslw.2012.03.002

Lodge, J. M. (2023). *Cheating with generative AI: Shifting focus from means and opportunity to motive [LinkedIn page].* LinkedIn. https://www.linkedin.com/pulse/cheatinggenerative-ai-shifting-focus-from-means-motive-lodge/

Nyholm, S. (2023). A new control problem? Humanoid robots, artificial intelligence, and the value of control. *AI and Ethics, 3*(4), 1229–1239. doi:10.1007/s43681-022-00231-y

Pasquale, F. (2020). *New laws of robotics.* Harvard University Press.

Pecorari, D. (2001). Plagiarism and international students: How the English-speaking university responds. In D. D. Belcher & A. R. Hirvela (Eds.), *Linking Literacies: Perspectives on L 2 Reading- Writing Connections* (pp. 229–245). University of Michigan Press.

Pennington, M. (2018). Five tools for detecting Algorithmic Bias in AI. *Technomancers – Legal Tech Blog.* https://www.technomancers.co.uk/2018/10/13/fivetools-for-detecting-algorithmic-bias-in-ai/

Pobiner, S., & Murphy, T. (2018). *Participatory design in the age of artificial intelligence.* Deloitte Insights. https://www2.deloitte.com/us/en/insights/focus/ cognitive-technologies/participatory-design-artificial-intelligence.html

Rahman, M., Terano, H. J. R., Rahman, N., Salamzadeh, A., & Rahaman, S. (2023). ChatGPT and academic research: A review and recommendations based on practical examples. *Journal of Education. Management and Development Studies*, *3*(1), 1–12. doi:10.52631/jemds.v3i1.175

Schiff, D. (2022). Education for AI, *not* AI for Education: The role of education and ethics in national AI policy strategies. *International Journal of Artificial Intelligence in Education*, *32*(3), 527–563. doi:10.1007/s40593-021-00270-2

Stahl, B. C., & Wright, D. (2018). Ethics and privacy in AI and big data: Implementing responsible research and innovation. *IEEE Security and Privacy*, *16*(3), 26–33. doi:10.1109/MSP.2018.2701164

Susnjak, T. (2022). *ChatGPT: The End of Online Exam Integrity?* (arXiv:2212.09292). arXiv. https://doi.org//arXiv.2212.09292 doi:10.48550

Tlili, A., Shehata, B., Adarkwah, M. A., Bozkurt, A., Hickey, D. T., Huang, R., & Agyemang, B. (2023). What is the devil is my guardian angel: ChatGPT as a case study of using chatbots in education. *Smart Learning Environments*, *10*(1), 1–24. doi:10.1186/s40561-023-00237-x

Traniello, J. F. A., & Bakker, T. C. M. (2016). Intellectual theft: Pitfalls and consequences of plagiarism. *Behavioral Ecology and Sociobiology*, *70*(11), 1789–1791. doi:10.1007/s00265-016-2207-y

UNESCO. (2018). *ICT Competency Framework for Teachers*. UNESCO. https://unesdoc.unesco.org/ark:/48223/ pf0000265721

UNESCO. (2021). *Artificial intelligence in education: Guidance for policy makers*. UNESCO. https://en.unesco.org/artificial-intelligence/education.

UNESCO. (2022). *Recommendation on the Ethics of Artificial Intelligence*. UNESCO. https://unesdoc.unesco.org/ark:/48223/pf0000381137.locale=en

UNESCO. (2023). *ChatGPT and artificial intelligence in higher education: Quick start guide*. UNESCO. https://unesdoc.unesco.org/ark:/48223/pf0000385146

Wenzlaff, K., & Spaeth, S. (2022). Smarter than Humans? Validating how OpenAI's ChatGPT Model explains crowdfunding, Alternative finance and community finance. SSRN *Scholarly Paper*. doi:10.2139/ssrn.4302443

Yan, D. (2023). Impact of ChatGPT on learners in a L2 writing practicum: An exploratory investigation. *Education and Information Technologies*, *28*(11), 13943–13967. doi:10.1007/s10639-023-11742-4

YounasA.ZengY. (2024) *Proposing Central Asian AI ethics principles: A multilevel approach for responsible AI*. SSRN. https://ssrn.com/abstract=4689770 or doi:10.2139/ssrn.4689770

# Chapter 5
# Ethical Navigations:
## Adaptable Frameworks for Responsible AI Use in Higher Education

**Allen Farina**
*Purdue University Global, USA*

**Carolyn N. Stevenson**
https://orcid.org/0000-0001-7521-8133
*Purdue University Global, USA*

## ABSTRACT

*In an era driven by digital innovation, artificial intelligence (AI), or Generative AI, emerges as a transformative force reshaping the landscape of higher education. Its potential to personalize learning experiences, bolster research capacities, and streamline administrative operations is revolutionary. However, the integration of Open AI into academia raises complex ethical issues for faculty and learners. The need for comprehensive ethical guidelines is imperative to ensure that the integration and utilization of AI in higher education are aligned with the core values of academic integrity and social responsibility. This chapter examines the ethical frameworks essential for governing the use of generative AI technologies in academia and provides practical recommendations for stakeholders involved. Additionally, emerging AI technologies such as experimental NotebookLM and Gemini will be discussed as future directions for AI use in teaching, learning, and research.*

## INTRODUCTION

The meteoric rise of Artificial Intelligence (AI) technologies has fundamentally reshaped various sectors, most notably in business, industry, and in particular, higher education. The release of ChatGPT to the masses by Open AI on November 30, 2022 is the genesis for this new age of AI (Marr, 2023). Throughout 2023, the influence of AI has been pervasive, with its tools increasingly being integrated into academic institutions and personal lives alike. However, this advancement has also presented unique challenges and opportunities for Higher Education Institutions (HEIs). The advent of ChatGPT, along

DOI: 10.4018/979-8-3693-1565-1.ch005

with the explosion of similar tools, has brought the potential of AI to be salient, prompting HEIs to explore its dual nature: its ability to enhance educational experiences and the ethical dilemmas surrounding its implementation. Central to these discussions are concerns about academic integrity, particularly cheating and plagiarism, along with broader ethical considerations regarding the appropriate use of AI. Additionally, safeguarding student privacy and fostering AI literacy remain top priorities. Such concerns are likely to continue in 2024 and beyond. In this rapidly evolving landscape, it becomes important for HEIs to establish comprehensive, ethical guidelines for AI use within their environments.

To effectively tackle the ethical challenges of Artificial Intelligence in HEIs, it is vital that these institutions adopt a proactive approach in policy making regarding AI usage. This process should involve all key stakeholders, including administrators, faculty, students, technology experts, and student advisors, to ensure comprehensive and inclusive strategies. This collaborative effort should focus on formulating a shared vision for the ethical application of AI in both physical and online educational settings. Policy development should be an iterative process, tailored to the specific needs of each HEI. Regularly revisiting and revising these policies will ensure their alignment with the evolving AI landscape and the unique needs of the college or university community.

As colleges and universities adjust policies due to the growing influence of Artificial Intelligence, it is also important to focus on the developing area of generative AI technology. New generative AI tools are constantly emerging. Google Lab's NotebookLM and Gemini are two newly launched products that while in the experimental stage hold the potential to streamline the research process and refine AI queries. NotebookLM is an experimental open AI tool introduced in July 2023 that is designed to use language models and an individual's content to synthesize information faster. In a general sense, NotebookLM is like a virtual research assistant that can summarize key points of curated content, identify themes and codes, generate guides, and serve as a virtual notebook. In December 2023, Google Lab's also introduced an experimental open AI tool known as Gemini. Gemini is Google's latest large language model (LLM). Gemini is a multimodal AI tool. In other words, it can deal with various forms of input and output, including text, code, audio, images, and videos.

As with any new technology, it is important for users to exercise critical thinking and ethics when using AI tools. Developing AI policy and practices grounded in an ethical framework for use is essential. Regardless of industry, or personal or professional use, AI is part of daily communication and interactions. The key is to harness the benefits of using open AI tools to enhance performance, synthesize information, or basically make our lives easier. However, individuals need to critically assess AI generated information and critically decide if the information is accurate and aligned to the needs of users.

## BACKGROUND

This chapter aims to guide Higher Education Institutions through the complex ethical landscape of Artificial Intelligence. Combining a conceptual analysis of AI's evolving role in colleges and universities with an extensive literature review, this research explores the necessity for a proactive approach in HEIs regarding the ethical applications of AI, particularly focusing on issues like cheating, plagiarism, academic integrity, and student privacy. The literature review synthesizes key findings, leading to a proposed practical framework designed to assist HEIs in developing, revising, or finalizing their AI ethics policies. A wide array of resources dedicated to AI governance and ethics are also provided.

Additionally, the chapter considers future research and the ethical implications of emerging generative AI technologies. A concise overview of Google's latest Large Language Model (LLM), Gemini, is presented. This AI resource is expected to be released to the public in 2024.

Furthermore, this chapter outlines the results from action research conducted with Purdue University Global and Google on a new AI tool called NotebookLM, showcasing a pioneering partnership in its pilot phase for real-world academic application. Selected for its innovation and developmental challenges, this case study was analyzed independently to emphasize its groundbreaking features. Despite limiting wider comparative insights, the study reveals promising initial results and areas for improvement in NotebookLM. Future research should validate these findings, include comparative studies to deepen understanding, and assess NotebookLM's impact and potential enhancements in educational settings.

Finally, the chapter concludes with a curated selection of additional readings and resources, enriching the reader's understanding of the topics discussed.

## Proactive Stance on AI in Higher Education

The rapid expansion of Artificial Intelligence in academia has brought to the forefront of consciousness the need for proactive engagement with its ethical implications. As teaching and learning assisted by AI shifts from theoretical discussion to implemented reality in traditional and online classrooms, college and university administrators and faculty must pivot from a reactive to a proactive mindset, prioritizing pedagogical innovation that leverages AI's strengths while safeguarding the human facets of education (Fuchs, 2023). This approach is essential to ensure that the ethical considerations of AI are thoroughly integrated into university policies, coursework, and the deployment of AI technology in academic settings. In this context, it is imperative to address the ethical implications of AI in higher education. Stakeholders need to understand the cyclical effects of AI and education, particularly in ensuring fair and unbiased pattern recognition and decision-making in educational applications (Cardona et al., 2023).

Furthermore, the integration of AI in higher education necessitates a careful examination of how it may impact equity for students, emphasizing the importance of ethical considerations in the design and deployment of AI technologies. In the complex landscape of Artificial Intelligence in higher education, a holistic approach that prioritizes ethics and human-centric efforts are essential. AI should be viewed as a tool to enhance teaching and learning, not as a replacement for college professors (Hie & Thouary, 2023). While AI offers numerous advantages in enhancing education at the university level, it also presents challenges, such as data privacy concerns, over-reliance on technology, and the need to address the lack of understanding and knowledge among faculty and students (Bailey, 2023). In the subsequent segments, we will explore the ethical considerations in university policies on AI, the advantages of using AI in higher education, and the challenges associated with its implementation. By addressing these critical aspects, we aim to provide a comprehensive understanding of the proactive stance required to navigate the ethical implications of AI in higher education.

The launch of ChatGPT in November 2022, followed by the introduction of other models such as Bard, Perplexity, and Claude, has sparked an urgent debate in higher education on how academics should respond to this abrupt onset of AI generative technology (Extance, 2023). There are major concerns on how AI can play a role in plagiarism and cheating (Bailey, 2022), as AI can now quickly generate human-like text and answers without citations. If students utilize these features improperly by passing off AI-written content as their own original work, it undercuts academic integrity. More broadly, some fear AI may hinder authentic learning, substituting rote information for genuine comprehension, critical analysis, and conceptual thinking (Torres & Mayo, 2023). However, it is important to remember that technology is value neutral. The impact of each tool depends on how users choose to apply it. Rather than reacting fearfully to AI systems, colleges and universities have an opportunity to revisit foundational questions on educational ethics. How can HEIs actively teach and model ethical decision-making for learners? What specific strategies can be used to integrate this concept into course outcomes and classroom lessons? How do generative AI and other digital technologies challenge academic integrity and foster personal responsibility for learning in students? What skills and values should higher education impart to students in an age of information abundance? How can academics maintain rigor and integrity? Institutions willing to lead this open discourse will emerge better positioned to develop forward-thinking policies around AI and academics.

Once college and university communities have established a consensus on the ethical standards that govern academic integrity, they can move forward with developing, revisiting, or revising policies related to academic integrity and the ethical use of generative AI. Any established guidelines must align with the institution's educational philosophy while addressing context-specific issues around plagiarism and cheating. They should be comprehensive yet adaptable, keeping pace with AI's ongoing evolution. Whether addressing AI or analog forms of misconduct, the goal remains the same, upholding academic integrity. Consequently, institutions of higher education should consistently orient students toward ethics and self-directed learning while also communicating clear expectations and consequences for policy violations (Torres & Mayo, 2023). With sound policies and communal buy-in, academics can harness AI responsibly by using technology to further pedagogy rather than hinder it.

As the process of creating, revising, or finalizing institution protocols on Artificial Intelligence in higher education settings move forward, a proactive policy development approach is key to ensuring academic integrity and fostering a thriving learning environment. This requires a comprehensive and collaborative approach that involves all key stakeholders, including students, faculty members, administrators, and educational technology experts. Addressing critical aspects, such as the role of AI in enhancing student engagement, critical thinking, and problem-solving skills, defining ethical behavior in teaching and assessment in the context of AI, and establishing clear guidelines on technology use and requirements for both students and faculty are essential (Chan, 2023). Without active anticipation and response to these issues, policies might become fragmented and ineffective, ultimately undermining academic integrity in the face of AI. Institutions must embrace a proactive approach to policy development, ensuring that AI is integrated into the academic landscape in a way that aligns with the values of academic integrity, equity, and innovation.

Another significant and proactive step to enhance the effectiveness of academic integrity policies surrounding Artificial Intelligence in HEIs is to ensure faculty involvement at every stage of policy development. Faculty, being on the front lines of education, play an essential role in both virtual and traditional classrooms. Their engagement in policy creation ensures that these guidelines are practical, relevant, and widely accepted. Furthermore, as AI becomes increasingly prevalent in business and industry, equipping students with AI skills is imperative for their success in a dynamic economy. Faculty members who proactively incorporate AI in their teaching not only enhance course learning objectives but also improve the AI competencies of students (Hie & Thouary, 2023). This proactive approach benefits students, faculty, and HEIs. To address the varying attitudes and skill levels regarding AI among instructors, ongoing training should be provided to ensure all professors can effectively integrate AI into their teaching.

Implementing Artificial Intelligence systems and policies in higher education involve more than just maintaining academic integrity. Ensuring student privacy and human oversight are also critical concerns. Without proper supervision, there is a risk of privacy and ethical violations that could lead to legal trouble (Rodriguez, 2020). Understanding and adhering to federal laws are critical in this context. The Family Educational Rights and Privacy Act (FERPA) safeguards students' educational records, limiting unauthorized access and disclosure of personal information. Similarly, the Individuals with Disabilities Education Act (IDEA) provides protections for students with disabilities, including in the context of digital data management. HEIs must navigate these laws alongside relevant state legislations to ensure compliance (Cardona et al., 2023). Given AI systems' ability to collect extensive personal data, it is vital for administrators to develop policies that responsibly handle data and prevent misuse (Dolan & Yasin, 2023).

To address these concerns, HEIs should engage all necessary stakeholders in the policy-making process. This engagement, as Nguyen et al. (2023) suggest, should involve open communication channels to gather diverse perspectives and concerns. Collecting divergent viewpoints fosters comprehensive policies that meet community needs and expectations while distributing ownership and responsibility. Once appropriate oversight is in place, building literacy around AI ethics, privacy, and equity helps foster principled applications (Cornell, 2023). Having fair and accountable AI adoption in higher education requires a balanced approach that prioritizes privacy, legal compliance, human oversight, and decision-making processes that include diverse perspectives.

## Creating Effective and Adaptable Policies

As previously discussed, Artificial Intelligence has become an indispensable force in HEIs. Early adopters among the student body are already actively engaging with AI, and it is anticipated that AI's utilization will continue to expand across the student population (Coffey, 2023). While AI presents a multitude of benefits, concerns surrounding plagiarism and cheating are both significant and valid (Hoffman, 2023). Consequently, HEIs must proactively develop and implement policies and protocols that effectively address these emerging challenges.

Drafting effective policies to address academic integrity is important to ensure the responsible use of AI tools and to clearly communicate the consequences of policy violations to learners. By providing clear guidance on AI's appropriate integration into academic work and ensuring that students fully comprehend these guidelines, administrators and faculty can foster a transparent and equitable learning environment (Chan, 2023).

To successfully integrate AI protocols into the fabric of higher education, a paradigm shift from a reactive to a proactive approach is essential. Throughout 2023, foundational discussions regarding the nature of AI, its influence on teaching and learning, its potential benefits and drawbacks, and the stance that educational institutions should adopt, should have already taken place. If these discussions have not occurred, they must be initiated without delay. Recognizing AI as a permanent fixture in the educational landscape, it is imperative for administrators and faculty to establish clear positions on its integration. This proactive acknowledgment paves the way for the development, modification, and finalization of academic integrity policies tailored to the AI era. Ideally, this process should be embedded within a strategic, forward-thinking plan, as discussed in section 1, that positions HEIs to effectively navigate the ever-evolving technological landscape.

Creating effective and ethical AI policies in HEIs require a tailored approach that reflects the unique contexts of each institution. This process involves critical consideration of various key topics to ensure policies are well-rounded and address the diverse aspects of AI usage in educational settings. A recommended framework for developing these policies encompassing essential topics is detailed below.

- **Defining AI and Institutional Scope:** Establish a clear and consistent understanding of AI within the institution. Define AI, its various forms, and its potential applications within the educational context. Specify permitted and restricted AI usage in research, classroom settings, and assessments, ensuring alignment across the institution (Cordona et. al, 2023).
- **AI's Role in Education:** Reaffirm the core purpose of education that centers on fostering critical thinking, original thought, and active learning. Articulate how AI can support these goals, emphasizing its ability to enhance personalized learning, deter academic dishonesty, and enrich the educational experience.
- **Guidelines for AI Usage:** Provide clear and comprehensive guidelines for acceptable and unacceptable AI applications in academic contexts. Include concrete examples to illustrate permissible and prohibited AI practices. This section should serve as a guide for students and faculty in navigating AI's role in education effectively (Chan, 2023).
- **Enhancing AI Literacy:** Commit to educating all stakeholders about AI's potential and limitations. Offer ongoing training and workshops to ensure informed and effective use of AI within the HEI. This includes providing opportunities for hands-on experience with AI tools and technologies.
- **Ensuring Equitable AI Access:** Address access disparities to AI tools. HEIs should guarantee equitable access to AI technologies for all learners, thereby promoting an inclusive learning environment (D'Agostino, 2023). This may involve providing financial support, targeted training, or access to AI-enabled learning resources.
- **Data Protection and Privacy:** Implement robust data protection strategies for AI applications, complying with privacy laws and ethical guidelines. Prioritize the safeguarding of sensitive student information, ensuring transparency in data collection, storage, and usage practices (Paykamian, 2023).
- **Confronting AI Bias:** Acknowledge and actively mitigate biases in AI systems. Provide guidelines for identifying, addressing, and reporting biases in AI-generated content (Chan, 2023). This may involve implementing bias detection tools, conducting regular audits of AI systems, and providing a mechanism for reporting and addressing biased content.
- **Policy Violation Protocols:** Define clear repercussions for policy violations. Establish a transparent process for reporting AI misuse and outline consistent, proportionate consequences for not following the rules. This ensures accountability and maintains the integrity of the AI policy (Dolan and Yasin, 2023).
- **Dynamic Policy Review:** Commit to regularly updating policies to reflect AI and technological advancements. Involve stakeholder feedback in periodic reviews to ensure the policy's relevance and effectiveness. This adaptability allows the institution to stay ahead of the curve and address emerging AI challenges (Chan, 2023).
- **Support and Training for AI Integration:** Offer comprehensive resources and support for AI utilization in educational contexts. This includes providing diverse learning materials, expert consultations, and a dedicated support system for ethical AI usage queries. This support network

ensures that educators and students have the necessary guidance to integrate AI seamlessly into their teaching and learning practices (Chan, 2023).

- **Collaborative Policy Development with AI Providers:** Build partnerships with AI technology providers to inform policy development. Collaborate with AI companies to customize AI tools to align more closely with educational objectives and values. This collaborative approach ensures that AI tools are developed and implemented in a way that aligns with the institution's educational mission and values.

By implementing these comprehensive guidelines, HEIs can effectively harness the power of AI to enhance education while safeguarding ethical principles, promoting inclusivity, and protecting student privacy.

## SOLUTIONS AND RECOMMENDATIONS

The information below presents key strategies for ethically incorporating Artificial Intelligence in higher education, stressing the importance of leadership, ethical governance, resource allocation, and expert consultation for policy development. It proposes recommendations to address ethical issues like plagiarism and privacy, emphasizing the need for ongoing policy evaluation and updates. It highlights securing funding for AI infrastructure, the benefits of partnering with government and industry for ethical AI tools, and the importance of staying updated on AI governance through continuous learning and networking. This guidance aims to help HEIs navigate AI integration by customizing these strategies to address their unique challenges effectively.

- **Leadership Commitment:** Secure strong buy-in from institutional leaders. Cultivate an ethos of ethical governance and best practices for AI within HEIs. This commitment must permeate all levels of the institution to engage all stakeholders effectively (Complete College, 2023).
- **Resource Utilization and Expert Consultation:** Utilize existing institutional resources and expertise. Engage industry experts to create ethically sound policies and practices for integrating AI in educational environments (Tasneem, 2023).
- **Ongoing Policy Assessment:** Regularly reevaluate AI policies to reflect technological advancements, address academic integrity, safeguard student privacy, and meet evolving institutional needs (Chan, 2023).
- **Funding for AI Infrastructure:** Allocate both short-term and long-term funds to build a robust infrastructure. These efforts should support AI initiatives, including continuous research and training programs (Complete College, 2023).
- **Collaboration with Government and Industry:** Establish partnerships with governmental bodies and industry leaders. These collaborations should aim to research, experiment, and develop new, ethically responsible AI tools that enhance teaching and learning (PR Newswire, 2023).
- **Continuous Learning and Networking:** Stay informed about developments in AI governance and ethics. Engage with a wide range of resources to understand the implications and opportunities of AI in the educational sector.

As we recognize the importance of HEI stakeholders staying informed and engaging with diverse resources to grasp the full scope of AI's implications and opportunities in education, it becomes essential to identify authoritative sources on said topics. In this context, Gordon (2023) steps in as a pivotal resource, offering a comprehensive list of research specifically tailored to enhance our understanding of AI governance and AI Ethics. Below is a curated list of resources aimed at guiding HEIs through the process of critically evaluating and leveraging insights to address and navigate the ethical challenges of AI within academic settings effectively.

- **Carnegie Mellon** has an interesting research paper which discusses international frameworks and highlights the variances impacting governance. It highlights that numerous codes of conduct or lists of principles for the responsible use of AI already exist. UNESCO and the OECD/G20 are the two most widely endorsed. In recent years, various institutions have been working to turn these principles into practice through domain-specific standards. For example, the European
- **Google's AI Principles**. In addition to Google's AI principles, they also include a clear description of areas they will not engage in AI use cases. Some of these areas noted are: technologies that cause or are likely to cause overall harm like in weapons or other technologies whose principal purpose or implementation is to cause or directly facilitate injury to people. Technologies that gather or use information for surveillance violating internationally accepted norms and also technologies whose purpose contravenes widely accepted principles of international law and human rights.
- **IBM** has a wealth of information on AI Governance and Ethical Guidelines.
- **IEEE** is the world's largest technical professional organization for the advancement of technology, which has created the Ethically Aligned Design business standards.
- **The AI Now Institute** focuses on the social implications of AI and policy research in responsible AI. Research areas include algorithmic accountability, antitrust concerns, biometrics, worker data rights, large-scale AI models and privacy. The report, "AI Now 2023 Landscape: Confronting Tech Power", provides a deep dive into many ethical issues that can be helpful in developing a responsible AI policy.
- **The Australian government** released the **AI Ethics Framework** that guides organizations and governments in responsibly designing, developing and implementing AI.
- **The Berkman Klein Center for Internet & Society at Harvard University** fosters research into the big questions related to the ethics and governance of AI.
- **The Canadian Ethical AI Framework and Positioning** includes Canada's AI ethics and guiding principles as well as communication on legislation developments.
- **The CEN-CENELEC Joint Technical Committee on Artificial Intelligence (JTC 21)** is an ongoing EU initiative for various responsible AI standards. The group plans to produce standards for the European market and inform EU legislation, policies and values.
- **The European Commission** proposed what would be the first legal framework for AI, which addresses the risk of AI and aims to provide AI developers, deployers and users with a clear understanding of the requirements for specific uses of AI.
- **The Montreal AI Ethics Institute**, a nonprofit organization, regularly produces "State of AI ethics reports" and helps democratize access to AI ethics knowledge.
- **The OECD AI Ethical Principles** is one of the first AI ethical frameworks developed. The OECD has undertaken empirical and policy activities on AI in support of the policy debate, starting with a Technology Foresight Forum on AI in 2016 and an international conference on AI: Intelligent Machines, Smart Policies in 2017. The organization has conducted analytical and measurement work that provides an overview of the AI technical landscape, maps economic and social impacts

of AI technologies and their applications, identifies major policy considerations, and describes AI initiatives from governments and other stakeholders at national and international levels.

- **The Singapore government,** which has been a pioneer and released the Model AI Governance Framework to provide actionable guidance for the private sector on how to address ethical and governance issues in AI deployments.
- **The Institute for Technology, Ethics and Culture (ITEC) Handbook** is a collaborative effort between Santa Clara University's Markkula Center for Applied Ethics and the Vatican to develop a practical, incremental roadmap for technology ethics. The handbook includes a five-stage maturity model, with specific measurable steps that enterprises can take at each level of maturity. It also promotes an operational approach for implementing ethics as an ongoing practice, akin to DevSecOps for ethics.
- **The ISO/IEC 23894:2023 IT-AI-Guidance on risk management** standard describes how an organization can manage risks specifically related to AI. It can help standardize the technical language characterizing underlying principles and how these principles apply to developing, provisioning or offering AI systems.
- **The NIST AI Risk Management Framework (AI RMF 1.0)** guides government agencies and the private sector on managing new AI risks and promoting responsible AI. Abhishek Gupta, founder and principal researcher at the Montreal AI Ethics Institute, pointed to the depth of the NIST framework, especially its specificity in implementing controls and policies to better govern AI systems within different organizational contexts.
- **The Nvidia/NeMo Guardrails** provides a flexible interface for defining specific behavioral rails that bots need to follow. It supports the Colang modeling language. One chief data scientist said his company uses the open source toolkit to prevent a support chatbot on a lawyer's website from providing answers that might be construed as legal advice.
- **The Stanford Institute for Human-Centered Artificial Intelligence (HAI)** provides ongoing research and guidance into best practices for human-centered AI. One early initiative in collaboration with Stanford Medicine is Responsible AI for Safe and Equitable Health, which addresses ethical and safety issues surrounding AI in health and medicine.
- **The UK AI Ethical Framework** for AI decision making. According to a recent EU survey and a British Computer Society survey in the UK, there is a distinct distrust in the regulation of advanced technology. A review by the Committee on Standards in Public Life found that the government should produce clearer guidance on using artificial intelligence ethically in the public sector. Hence, this framework is advancing the UK AI ethical governance approach.
- **The University of Turku (Finland)**, in coordination with a team of academic and industry partners, formed a consortium and created the **Artificial Intelligence Governance and Auditing (AIGA) Framework,** which illustrates a detailed and comprehensive AI governance life cycle that supports the responsible use of AI.
- **The USA Blueprint for an AI Bill of Rights** is a guide for a society that protects all people from these threats—and uses technologies in ways that reinforce our highest values.
- **The World Economic Forum's "The Presidio Recommendations on Responsible Generative AI** white paper includes 30 "action-oriented" recommendations to "navigate AI complexities and harness its potential ethically" (p. 1). It includes sections on responsible development and release of generative AI, open innovation and international collaboration, and social progress.

As new open AI sources emerge in the market, individuals and organizations need to think critically about ethical AI use, develop a framework for ethical use, and offer training to educate employees, students, faculty, and all users on ethical use of open AI. By implementing these recommendations, HEIs can proactively and responsibly integrate AI into their systems, ensuring both ethical compliance and educational advancement.

## FUTURE RESEARCH DIRECTIONS

New directions for open AI use in academics, personal use, and business contexts is unlimited but warrants critical thought when deciding on the ethical use of AI. As discussed in the previous sections, when adopting new generative AI tools, it is critical decision-makers establish an ethical framework and clear policies. AI can also hallucinate, creating information that is inaccurate or producing outcomes not desired by the user. As further explained by IMB (20240), "AI hallucination is a phenomenon wherein a large language model (LLM)—often a generative AI chatbot or computer vision tool—perceives patterns or objects that are nonexistent or imperceptible to human observers, creating outputs that are nonsensical or altogether inaccurate," (para. 1). As such, critical thought, clear ethical policies, and training for users need to be put into place.

New AI tools such as NotebookLM and Gemini are currently being refined and being implemented on an experimental basis into online classrooms and in Google Labs in an experimental capacity. Technology is constantly changing and assessing the value of new open AI tools depends on the user exercising critical thought.

### Gemini

Google Labs announced the release of an experimental open AI tool known as Gemini. Gemini is a multimodal AI tool. Gemini is Google's latest large language model (LLM). An LLM is the system that underpins the types of AI tools you have probably seen and interacted with on the internet. For example, GPT-4 powers ChatGPT Plus, OpenAI's advanced paid-for chatbot (Blake, 2023). In other words, it can deal with various forms of input and output, including text, code, audio, images, and videos. That gives it a lot of flexibility to perform a wide range of tasks. Google claims Gemini is "…the first model to outperform human experts on MMLU (Massive Multitask Language Understanding), one of the most popular methods to test the knowledge and problem-solving abilities of AI models," (Google DeepMind, 2023, p.1).

Google demonstrated the capabilities of Gemini in a "hands-on" video demonstrating the potential of the new AI tool. The "hand-on" can be viewed from this URL: https://www.youtube.com/watch?v=UIZAiXYceBI. In the video, Gemini could be seen following a paper ball hidden under a cup and understanding a user's sleight-of-hand coin trick. It could predict what a dot-to-dot puzzle showed before a single line was drawn and explain when one path on a map might lead to danger and one may lead to safety.

Since Gemini is still in the experimental stage, it is too soon to assess if Gemini will perform the way the architects intended it to work. However, Gemini should be on the watch list for emerging technologies in 2024.

## NotebookLM

NotebookLM is an experimental product designed from the ground up using the power and promise of language models. It is also a very different kind of notebook — one in which the user can use existing documents and notes to help the user to learn, create, and brainstorm. Google's new AI-powered note-taking app, NotebookLM, is still in its early stages, but it has the potential to be a powerful tool for organizing and interacting with the user's research.

NotebookLM is a chatbot that is trained on the user's own notes and documents. It can help users find information, generate summaries, and answer questions. It can also generate source guides, which are summaries of the user's documents that include suggested questions to ask. According to Pierce (2023):

Google's new AI-powered note-taking app, NotebookLM, is still in its early stages but has the potential to be a powerful tool for organizing and interacting with your notes. The app allows you to import your own documents and then generate summaries, outlines, and answers to your questions based on the content of those documents. It can also identify the most relevant information in your documents and provide citations for its answers. NotebookLM is still under development, but it has the potential to be a valuable tool for students, researchers, and anyone else who needs to organize and manage large amounts of information (p.1).

## NotebookLM Overview

NotebookLM is an AI-powered notebook that allows users to explore ideas and answer questions using the power of a language model grounded in documents that you select. Think of it as a personalized AI collaborator that has read all the materials important to the user's work and that can instantly summarize or create the information the user needs. The NotebookLM AI learns from the "sources" the user shares with it, helping the user remember important facts, make new connections, or synthesize information across multiple documents. Since NotebookLM is still in the experimental stage, it is best viewed on a desktop computer. The font may appear too small on mobile devices. NotebookLM is a standalone experiment available in Google Labs, and Google does not have plans to make it part of any other Google products currently (Google p.1).

According to Martin and Johnson (2023), "NotebookLM is an experimental product designed to use the power and promise of language models paired with your existing content to gain critical insights, faster. Think of it as a virtual research assistant that can summarize facts, explain complex ideas, and brainstorm new connections — all based on the sources you select. A key difference between NotebookLM and traditional AI chatbots is that NotebookLM lets you "ground" the language model in your notes and sources. Source-grounding effectively creates a personalized AI that's versed in the information relevant to you," (p.1). With so many documents spread across so many locations, it is hard to make the most of the information the user has curated. That is why Google researchers have started exploring how to apply the power of AI to riff on specific information that the user has selected. NotebookLM is an app that offers a space where users can collaborate with AI to brainstorm ideas, ask questions, and organize thoughts — all drawn from the user's own source documents (Google p.1). The next sections will address the benefits and challenges of NotebookLM.

## Benefits of Using NotebookLM

There are several benefits to using NotebookLM versus other AI generative tools such as ChatGPT. A big difference between NotebookLM and other generative AI tools such as ChatGPT is that information is gained from an individual's own documents. Once the user has selected the specific Google Doc, several functions can be performed:

- **Get a summary:** When a Google Doc is first loaded into NotebookLM, it will automatically generate a summary, along with key topics and questions to ask so the user gains a better understanding of the material.
- **Ask questions:** When the user is ready to go deeper into use of the tool, the user may ask questions about the documents that have been uploaded. A few examples are listed below and illustrated in Figure 2:
    - A medical student could upload a scientific article about neuroscience and tell NotebookLM to "create a glossary of key terms related to dopamine."
    - An author working on a biography could upload research notes and ask a question like: "Summarize all the times Houdini and Conan Doyle interacted" (Google NotebookLM Onboarding and FAQ, p.1).
- **Generate ideas:** NotebookLM is not just for Question and Answer. Innovators in the Google Lab have found some of its more delightful and useful capabilities are when it is able to help people come up with creative new ideas. For example:
    - A content creator could upload their ideas for new videos and ask: "Generate a script for a short video on this topic."
    - Or an entrepreneur raising money could upload their pitch and ask: "What questions would potential investors ask?"

There are additional benefits of Notebook LM. When the researcher asked NotebookLM, "What are the features of NotebookLM?" The AI tool generated responses captured in Figure 1. Screenshot of NotebookLM's response to the benefits of using the tool.

- **NotebookLM can be used to summarize facts, explain complex ideas, and brainstorm new connections.**
- It can be used to ask questions about documents and generate ideas.
- NotebookLM can be used to generate summaries or outlines of documents.
- It can be used to find commonalities across documents, identify the most important information in a document, and generate surprising insights.
- NotebookLM can be used to create a personalized AI assistant that is versed in the information relevant to you.

## Utilizing Features in NotebookLM

Information overload is a challenge for everyone—especially knowledge workers, students, and academics faced with synthesizing information from multiple sources. Making connections from curated sources

*Figure 1. Screenshot of NotebookLM's response to the benefits of using the tool*

? **NotebookLM can be used to summarize facts, explain complex ideas, and brainstorm new connections.**
? It can be used to ask questions about documents and generate ideas.
? NotebookLM can be used to generate summaries or outlines of documents.
? It can be used to find commonalities across documents, identify the most important information in a document, and generate surprising insights.
? NotebookLM can be used to create a personalized AI assistant that is versed in the information relevant to you.

*Figure 2. Screenshot of notes sample in a PR520 fall 2023 Purdue global course generated from NotebookLM*

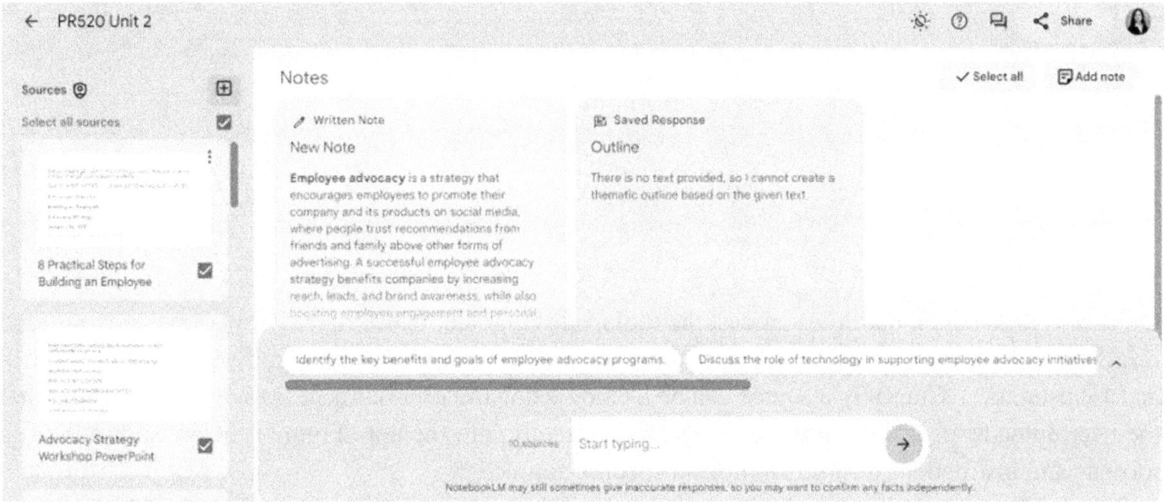


is time consuming and challenging. Researchers at Google Labs set out to address the question, "How do we go from information to insight?" (Martin & Johnson, P.1).

NotebookLM has several features such as "Notes" and the capability to create study guides based on curated resources.

Users can write down and save notes in a "Note" in each notebook. Users can also save a useful response from NotebookLM to your Note by clicking on the "send to Notes" button under a response. Users can navigate to the note by clicking on the "Open Notes" button in the bottom left corner. Figure 2 below is a screenshot of a notes sample in a PR520 Fall 2023 Purdue Global course captured in NotebookLM.

NotebookLM also has the capability to generate guides for users based on curated content by the user. Figure 3 below shows a sample guide that synthesized salient points from the content generated by the user.

## NotebookLM FAQs

While NotebookLM's source-grounding does seem to reduce the risk of model "hallucinations," it is always important to fact-check the AI's responses against the user's original source material. When

*Figure 3. Sample guide generated in NotebookLM*

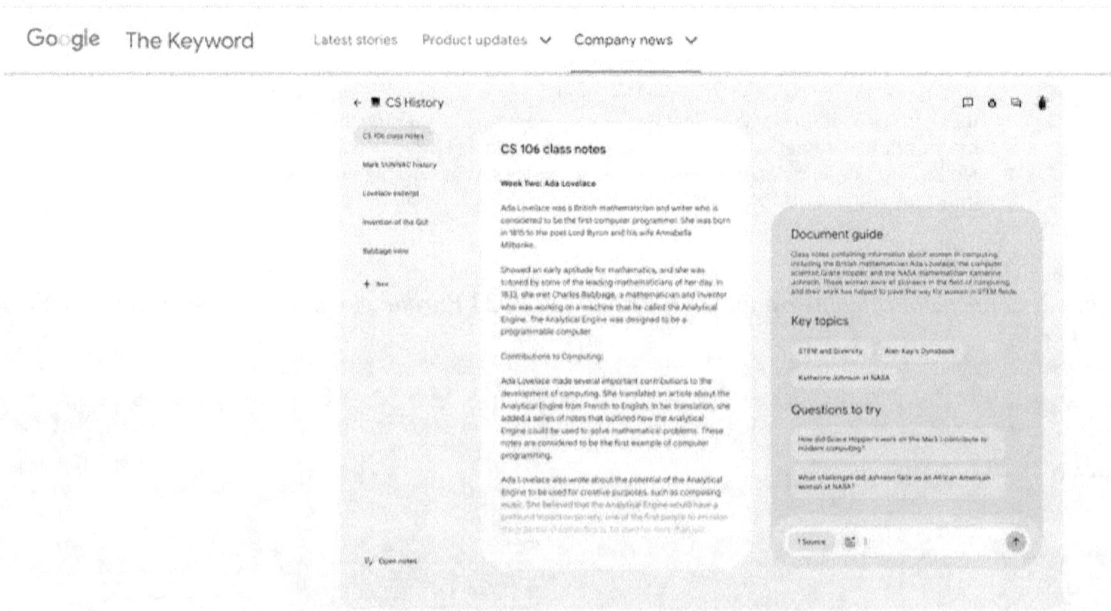


the user is drawing on multiple sources, the tool makes that fact-checking easy by accompanying each response with citations, showing the user the most relevant original quotes from your sources (Martin and Johnson, p.1). Currently a source can be a Google doc the user has access to, a local PDF file that the user uploads, or text the user can copy-paste directly into the app. Figure 4 below shows the file formats and instructions for users to upload specific sources.

Notebooks can also be shared with colleagues or classmates by clicking on the “share” icon in the tool bar in the top right corner of the app. Users can grant either *Viewer* or *Editor* access to another user by adding their email address:

- a viewer will have read-only access to all the source documents the user shared with them in the Notebook
- an editor will be able to view, add or remove sources in your shared Notebook as well as sharing it further with other users

Users can also talk to multiple notebooks at one time. Users can use the *Source* button right next to the input box to toggle between talking to a single source vs. talking to all sources in a Notebook. Citations are also available in the notebooks. A citation is a specific block of text from your source that is considered relevant to the question you input and used by NotebookLM to build the response for the user. Figure 5 below is a screenshot of a sample citation in NotebookLM.

For most questions or instructions, the user gives NotebookLM, the app will return cited passages to support the AI’s response. These are direct quotes from the source documents related to the user’s query. Seeing these passages can help users check the AI’s accuracy and find the original quote.

*Figure 4. Format for uploading sources in NotebookLM from PR520 fall term for Purdue University Global*

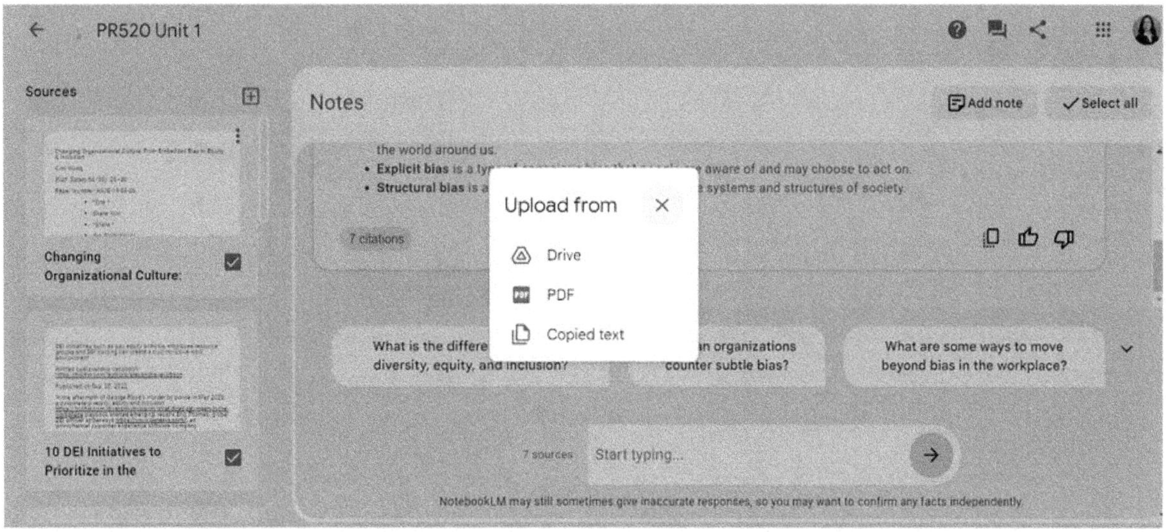


## OTHER FAQS RELATED TO EXPERIMENTAL NotebookLM

### What's a Document Guide?

When the user uploads a new source to NotebookLM, the app instantly creates a "source guide" that summarizes the document and offers Key Topics and Questions to try to help the user better understand the source material.

### Where Can I Write my Notes?

Each project contains a notes doc where you can jot down your own ideas or capture information that NotebookLM uncovers from your sources. Each response from the AI has a "save to notes" icon on the right side, and the user can also copy and paste key passages into the notes doc.

*Figure 5. Sample citation in NotebookLM*

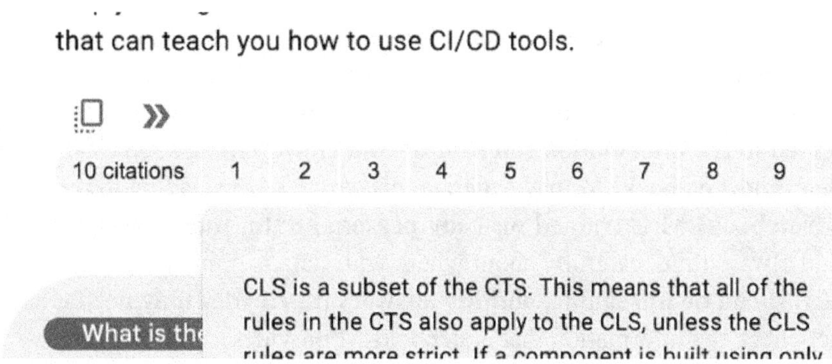

*Figure 6. NotebookLM query on limitations of using the tool*

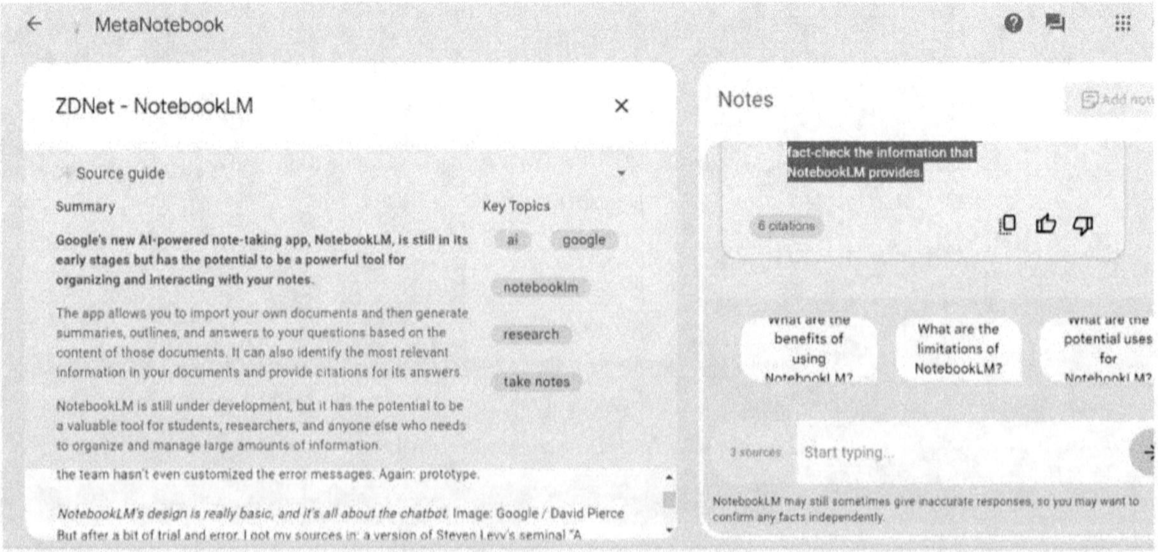


## What Type of Files do you Support?

Currently, NotebookLM supports the following source types: Google Docs, local PDF files, or manual copy-paste.

## Limitations

As with any technological tool, especially one in the experimental stage, there are limitations. Stevenson (2023) posed a question regarding, "What are the limitations of using NotebookLM?" NotebookLM generated responses captured in Figures 6 and 7 listed below.

As with any technological tool, there are limitations and ethical considerations. Clear policies also need to be developed for the instructor and student to ensure ethical use of the AI tool. What are the limitations? As the faculty member piloting NotebookLM in a fall 2023 PR520 graduate course, Stevenson (2023) NotebookLM used prompt engineering to raise the following question: What are the limitations of NotebookLM? The following responses were generated from NotebookLM:

- **NotebookLM is still in beta and has some limitations.** For example, it can only accept and import Google Docs, and each project can have up to five sources, each source can be up to 10,000 words long.
- **The sources that you provide to NotebookLM are used to create a personalized AI assistant that is versed in the information relevant to you.** However, the sources are not always accurate or complete, and NotebookLM may generate inaccurate or misleading information.
- **Because NotebookLM is trained on your personal data, there are some privacy concerns.** You should always think carefully about where your data is going and how it might be used.
- **NotebookLM can be unreliable, and the answers it provides may not be helpful.** It is important to fact-check the information that NotebookLM provides.

*Figure 7. NotebookLM generated a response to a question on limitations*

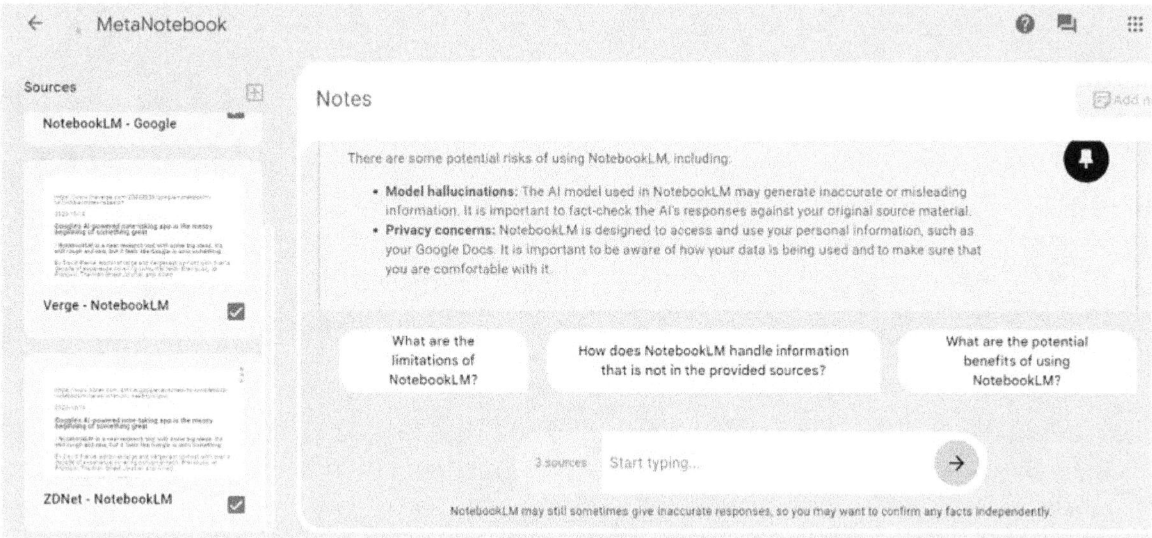


As with any technology tool, there are risks related to using NotebookLM. Privacy is a concern for most generative AI tools including NotebookLM. Stevenson (2023) posed a question related to the potential risks using NotebookLM. Figure 7 below reveals the results of the query.

## Using Experimental NotebookLM in the Classroom

As part of the Purdue University Global's (PG) Classroom of the Future initiative, the institution has been invited to be the first higher education institution to test drive Google's groundbreaking new AI technology, NotebookLM. NotebookLM is being used in selected PG courses during the November 2023 term as an optional/enrichment activity and not as a required tool. Faculty will ideate and implement the activity selected for use in their course. Examples include a hands-on activity during a seminar or an optional task for a discussion post. A Community of Practice includes ideation sessions, feedback loops, focus groups between faculty and Google researchers, and opportunities for collaboration will be part of the experience.

Faculty participating in the experimental program are required to offer insights and suggestions that will have a direct impact on making this tool more effective and user-friendly. Experience using NotebookLM includes feedback from the faculty member teaching the course and learners currently enrolled in an active fall 2023 course. Examples used in this document were generated from co-author Dr. Carolyn Stevenson's Master's level Diversity, Equity, and Inclusion course, PR520. Selected faculty interface with Google researchers to provide feedback through focus groups and surveys. Information gleaned from students was also shared with the intention of improvements for future iterations.

As with any generative AI tool, ethics, adaptable policies, and recommendations for practical use need to be considered. Leaders responsible for making AI implementation decisions need to ensure there are policies and procedures in place to safeguard against unethical use of AI. Notebook LM and Gemini are newer AI tools, and the launch of future AI will continue to grow and become integrated across the professions. It is important to have a firm foundation in ethical use and training for all us-

*Figure 8. NotebookLM generated responses to a question on potential risks*

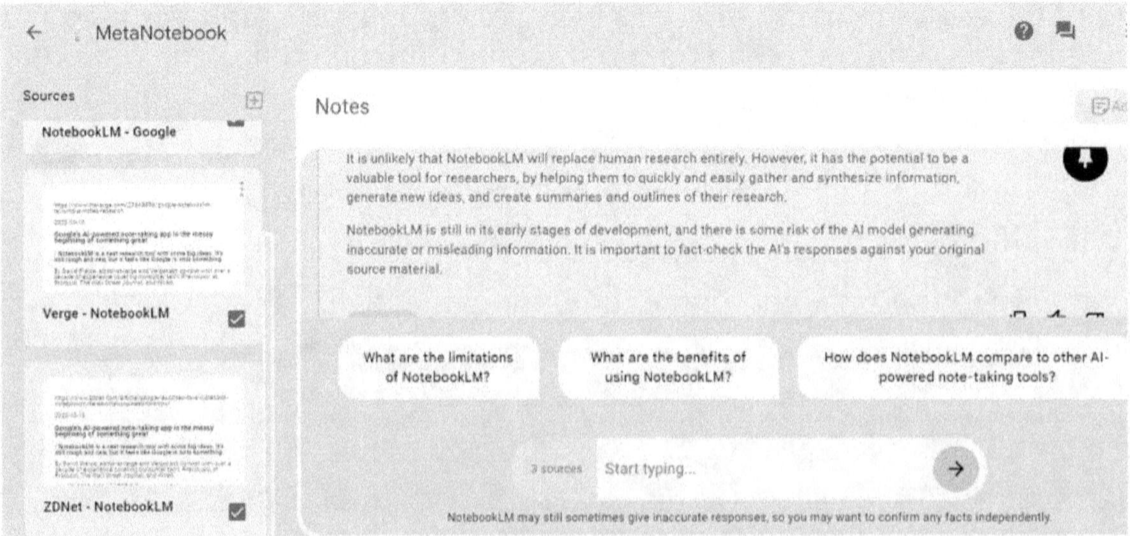


ers to ensure generative AI is used to enhance the workflow, while maintaining critical thought and human considerations.

## CONCLUSION

In conclusion, we strongly advocate for strategic AI adoption in academic settings by Higher Education Institutions. This commitment must prioritize ethical considerations and human-centric teaching approaches, ensuring that AI's integration upholds academic integrity, addresses privacy and data protection, and remains accessible to all. HEIs face the challenge of balancing AI's benefits with potential risks, such as AI-assisted plagiarism and privacy breaches.

To effectively navigate these complexities, HEIs must establish dynamic and comprehensive policies. These policies should clearly articulate AI's educational role, outline responsible usage guidelines, and promote AI literacy among both students and faculty. Equally important is the guarantee of equitable AI access, vigilant protection of data and privacy, proactive measures against AI biases, and the establishment of clear protocols for policy infringements. Collaboration with AI providers in policy development is also important.

The ultimate objective is to utilize AI as a tool, not as a substitute for human educators. This necessitates ongoing faculty training to ensure adept integration of AI resources in teaching methodologies. The focus should remain steadfast on preparing students for the ethical use of Artificial Intelligence. The goal is to foster an academic environment that champions privacy, equity, innovation, and above all, maintains the integrity of the educational process.

Additionally, AI generated tools hold potential and challenges for researchers and learners. Technology is constantly evolving and tools such as Google's NotebookLM and Gemini use the power and promise of language models paired with existing content to quickly gain critical insights. As this experimental open AI tool gets refined, it holds potential for use by students, faculty, and individuals. NotebookLM can house resources

*Figure 9. NotebookLM generated responses to a question on replacing human research*

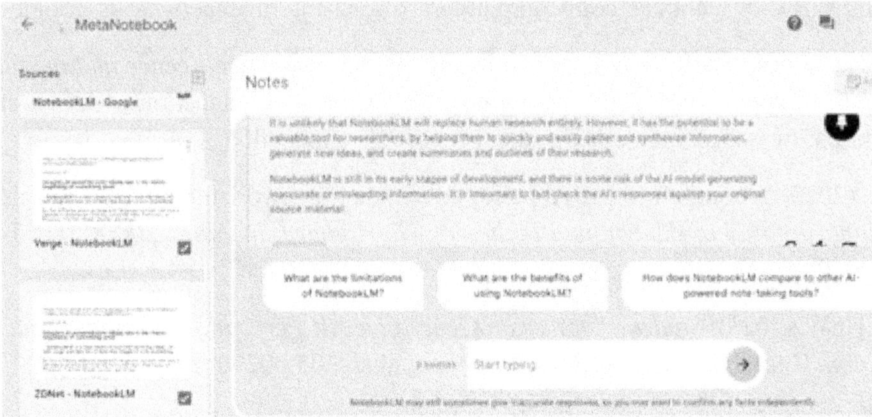


and notes as well as save students time by utilizing specific resources curated by the instructor. There is a great deal of potential for personal use in areas such as research and job searches. As with any new technology, the user must make ethical decisions on using any generative AI tool. Establishing an ethical framework for using AI is critical. Gemini holds the promise of offering an enhanced version of ChatGPT. Since both NotebookLM and Gemini are in the experimental stages, time is needed for refinement prior to use by the general population.

Will NotebookLM replace human research? Stevenson (2023) asked NotebookLM the question using NotebookLM and the response yielded a response that it is “unlikely to replace human research.” Additional information is illustrated in Figure 8 below.

To conclude, open AI tools can be used in the teaching, learning and research process. There are benefits and limitations to using generative AI tools. It is important for the user to understand the ethical implications of using generative AI. For faculty teaching at all levels, it is important to inform students of ethical guidelines, proper citation of AI generated content, and limitations of AI. Students and faculty need to be adequately trained in ways generative AI tools can enhance teaching and learning, rather than replace it. The need for application of critical thought by humans will never be replaced. Using AI tools to assist in teaching and learning is a value add, not a replacement for traditional learning and application of content. The need for application of critical thought by humans will never be replaced. Using AI tools to assist in teaching and learning is a value add, not a replacement for traditional learning and application of content. Organizations and institutions of higher learning should also clear guidelines for ethical use to ensure accuracy and quality of information gathered from using open AI.

## REFERENCES

Bailey, J. (2022). Why teachers are worried about AI. *Plagiarism Today.* https://www.plagiarismtoday.com/2022/12/07/why-teachers-are-worried-about-ai/

Bailey, J. (2023). AI in education: The leap into a new era of machine intelligence carries risks and challenges, but also plenty of promise. *Education Next*, *23*(4), 28–35. https://www.educationnext.org/a-i-in-education-leap-into-new-era-machine-intelligence-carries-risks-challenges-promises/

Blake, A. (2023). What is Google's Gemini? Everything you need to know about Google's next-gen AI. *Tech Radar*. https://www.techradar.com/computing/artificial-intelligence/what-is-google-gemini

Cardona, M. A., Rodriguez, R. J., & Ishmael, K. (2023). *Artificial Intelligence and the future of teaching and learning: Insights and recommendations*. Department of Education – Office of Educational Technology. https://www2.ed.gov/documents/ai-report/ai-report.pdf

Chan, C. K. Y. (2023). A comprehensive AI policy education framework for university teaching and learning. *International Journal of Educational Technology in Higher Education*, *20*(38), 1–25. doi:10.1186/s41239-023-00408-3

ChanC. K. Y.HuW. (2023). *Students' voices on Generative AI: Perceptions, benefits, and challenges in higher education*. Cornell University. https://arxiv.org/abs/2305.00290

Coffey, L. (2023). *Students outrunning faculty in AI use*. Inside Higher Ed. https://www.insidehighered.com/news/tech-innovation/artificial-intelligence/2023/10/31/most-students-outrunning-faculty-ai-use

Complete College America. (2023). *Attainment with AI: Making a real difference in college completion with Artificial Intelligence*. Complete College America. https://completecollege.org/wp-content/uploads/2023/11/CCA_AttainmentwithAI.pdf

Cornell University Center for Teaching Innovation. (2023). *Ethical AI for teaching and learning*. Cornell University. https://teaching.cornell.edu/generative-artificial-intelligence/ethical-ai-teaching-and-learning

D'Agostino, S. (2023). *How AI tools both help and hinder equity*. Inside Higher Ed. https://www.insidehighered.com/news/tech-innovation/artificial-intelligence/2023/06/05/how-ai-tools-both-help-and-hinder-equity

Dolan, D., & Yasin, E. (2023). *A guide to generative AI policy making*. Inside Higher Ed. https://www.insidehighered.com/views/2023/03/22/ai-policy-advice-administrators-and-faculty-opinion

Extance, A. (2023). ChatGPT has entered the classroom: How LLMs could transform education. *Nature*. https://www.nature.com/articles/d41586-023-03507-3

Fuchs, K. (2023). *Exploring the opportunities and challenges of NLP models in higher education: Is Chat GPT a blessing or a curse?* Frontiers. https://www.frontiersin.org/articles/10.3389/feduc.2023.1166682/full

Google. (2023). *Hands-on with Gemini: Interacting with multimodal AI* [Video]. YouTube. https://www.youtube.com/watch?v=UIZAiXYceBI

Google. (2023). *NotebookLM onboarding and FAQ*. Internal training document.

Google DeepMind. (2023). *Welcome to the Gemini era*. Google. https://deepmind.google/technologies/gemini/#introduction

Hie, A., & Thouary, C. (2023). *How AI is reshaping higher education*. AACSB. https://www.aacsb.edu/insights/articles/2023/10/how-ai-is-reshaping-higher-education

Hoffman, R. (2023). *AI in the classroom: The potential drawbacks and benefits of AI in education*. Greylock. https://greylock.com/greymatter/ai-in-the-classroom/

IBM. (2024). *What are AI hallucinations?* IBM. https://www.ibm.com/topics/ai-hallucinations

Marr, B. (2023). A short history of ChatGPT: How we got to where we are today. *Forbes*. https://www.forbes.com/sites/bernardmarr/2023/05/19/a-short-history-of-chatgpt-how-we-got-to-where-we-are-today/?sh=4c973d2c674f

Martin, R., & Johnson, S. (2023). *Introducing NotebookLM.* Google. https://blog.google/technology/ai/notebooklm-google-ai/

Newswire, P. R. (2023). Education and tech leaders come together to offer guidance on integrating AI safety into classrooms worldwide. *PR News Wire*. https://www.prnewswire.com/news-releases/education-and-tech-leaders-come-together-to-offer-guidance-on-integrating-ai-safely-into-classrooms-worldwide-301812478.html

Nguyen, A., Ngo, H. N., Hong, Y., Dang, B., & Thi Nguyen, B.-P. (2023). Ethical principles for Artificial Intelligence in education. *Education and Information Technologies*, *28*(4), 4221–4241. doi:10.1007/s10639-022-11316-w PMID:36254344

Paykamian, B. (2023). *Who should be regulating AI classroom tools.* Government Technology. https://www.govtech.com/education/higher-ed/who-should-be-regulating-ai-classroom-tools

Pierce, D. (2023). *Google's AI-powered note-taking app is the messy beginning of something great.* ZDNet. https://www.zdnet.com/article/google-launches-its-ai-notebook-notebooklm-heres-what-you-need-to-know/

Rodriguez, R. (2020). Legal and human rights issues of AI: Gaps, challenges, and vulnerabilities. *Journal of Responsible Technology*, *4*, 100005. doi:10.1016/j.jrt.2020.100005

Tasneem, A. (2023). *7 missteps university leaders must avoid in their AI approach.* EAB. https://eab.com/insights/blogs/strategy/missteps-university-leaders-ai-approach/

Torres, J. T., & Mayo, C. P. T. (2023). *AI eroding AI? A new era for Artificial Intelligence and academic integrity*. Faculty Focus. https://www.facultyfocus.com/articles/teaching-with-technology-articles/ai-eroding-ai-a-new-era-for-artificial-intelligence-and-academic-integrity/

## ADDITIONAL READING

Bender, E. M., Gebru, T., McMillan-Major, A., & Shmitchell, S. (2021). *FAccT '21: Proceedings of the 2021 ACM Conference on Fairness, Accountability, and Transparency*, (pp. 610–623). ACM.

Bergman, P., Kopko, E., & Rodriguez, J. E. (2021). *Using predictive analytics to track students: Evidence from a seven-college experiment.* (NBER Working Paper No. 28948).

Brynjolfsson, E. (2022). The turing trap: The promise & peril of human-like Artificial Intelligence. *Daedalus, 151(2),* 272-287. https://digitaleconomy.stanford.edu/news/the-turing-trap-the-promise-peril-of-human-like-artificial-intelligence/

Burrows, S., Gurevych, I., & Stein, B. (2015). The eras and trends of automatic short answer grading. *International Journal of Artificial Intelligence in Education*, *25*(1), 60–117. doi:10.1007/s40593-014-0026-8

Chen, G., Chen, P., Hsu, C., Lai, C., & Lee, T. (2020). *Deep learning for NLP and speech recognition*. Springer.

Chen, H., & Denoyelles, A. (2013). Mobile learning in higher education. *New Directions for Higher Education*, 164.

Conrad, K. (2023). *Sneak preview: A blueprint for an AI bill of rights for education.* Critical AI. https://criticalai.org/2023/07/17/a-blueprint-for-an-ai-bill-of-rights-for-education-kathryn-conrad/

Devlin, J., Chang, M.-W., Lee, K., & Toutanova, K. (2019). BERT: Pretraining of deep bidirectional transformers for language understanding. In *Proceedings of the 2019 Conference of the North American Chapter of the Association for Computational Linguistics: Human Language Technologies*. arXiv. https://arxiv.org/pdf/1810.04805.pdf

Eynon, R., & Erin, Y. (2020). Methodology, legend, and rhetoric: The Constructions of AI by academia, industry, and policy groups for lifelong learning. *Science, Technology & Human Values*, *46*(1), 166–191. doi:10.1177/0162243920906475

Fischer, C., Parados, Z., & Baker, R. (2020). Mining big data in education: Affordances and challenges. *Review of Research in Education*, *44*(1), 130–160. doi:10.3102/0091732X20903304

Fryer, L. K., Ainley, M., Thompson, A., Gibson, A., & Sherlock, Z. (2017). Stimulating and sustaining interest in a language course: An experimental comparison of Chatbot and Human task partners. *Computers in Human Behavior*, *75*, 461–468. doi:10.1016/j.chb.2017.05.045

Gao, J., Galley, M., & Li, L. (2018). Neural approaches to conversational AI. *Foundations and Trends in Information Retrieval*, *13*(2-3), 127–298. doi:10.1561/1500000074

García-Martín, E., Faviola Rodrigues, C., Riley, G., & Grahn, H. (2019). Estimation of energy consumption in machine learning. *Journal of Parallel and Distributed Computing*, *134*, 75–88. doi:10.1016/j.jpdc.2019.07.007

Gillani, N., Eynon, R., Chiabaut, C., & Finkel, K. (2023). Unpacking the black box of AI in education. *Journal of Educational Technology & Society*, *26*(1), 99–111.

Godwin-Jones, R. (2017). Smartphones and language learning. *Language Learning & Technology*, *21*(2), 3–17.

Goodlad, L., & Baker, S. (2023). *Now the humanities can disrupt AI*. Public Books. https://www.publicbooks.org/now-the-humanities-can-disrupt-ai/

Grgurović, M., Chapelle, C. A., & Shelley, M. C. (2013). A meta-analysis of effectiveness studies on computer technology-supported language learning. *ReCALL*, *25*(2), 165–198. doi:10.1017/S0958344013000013

Hannan, E., & Liu, S. (2023). AI: New source of competitiveness in higher education. *Competitiveness Review*, *33*(2), 265–279. doi:10.1108/CR-03-2021-0045

Hao, K. (2019). The computing power needed to train AI is now rising seven times faster than ever before. *MIT Technology Review*. https://www.technologyreview.com/2019/11/11/132004/the-computing-power-needed-to-train-ai-is-now-rising-seven-times-faster-than-ever-before/

Hasselberger, W. (2021). Can machines have common sense? *New Atlantis (Washington, D.C.)*, *65*, 94–109.

Helmus, T. C. (2022). *Artificial Intelligence, deepfakes, and disinformation: A Primer*. RAND Corporation.

Holmes, W., Porayska-Pomsta, K., Holstein, K., Sutherland, E., Baker, T., Shum, S. B., Santos, O. C., Rodrigo, M. T., Cukurova, M., Bittencourt, I. I., & Koedinger, K. R. (2021). Ethics of AI in education: Towards a community-wide framework. *International Journal of Artificial Intelligence in Education*, *32*(3), 504–526. doi:10.1007/s40593-021-00239-1

Hwang, G. J., & Chen, N. S. (2023). Exploring the potential of generative artificial intelligence in education: Applications, challenges, and future research directions. *Journal of Educational Technology & Society*, *26*(2).

Perrotta, C., & Selwyn, N. (2020). Deep learning goes to school: Toward a relational understanding of AI in education. *Learning, Media and Technology*, *45*(3), 251–269. doi:10.1080/17439884.2020.1686017

Rudolph, J., Tan, S., & Tan, S. (2023). ChatGPT: Bullshit spewer or the end of traditional Assessments in higher education? *The Journal of Applied Learning and Teaching*, *6*(1).

Scott, K. (2022). I do not think it means what you think it means: Artificial intelligence, cognitive work & scale. *Daedalus*, *151*(2), 75–84. doi:10.1162/daed_a_01901

Talan, T., & Kalınkara, Y. (2023). The role of artificial intelligence in higher education: ChatGPT assessment for anatomy course. *Uluslararası Yönetim Bilişim Sistemleri ve Bilgisayar Bilimleri Dergisi*, *7*(1), 33–40. doi:10.33461/uybisbbd.1244777

Tjeng, V., Xiao, K., & Tedrake, R. (2019). Evaluating robustness of neural networks with mixed integer programming. In *Proceedings of the International Conference on Learning Representations (ICLR)*. https://doi.org/10.48550/arXiv.1711.07356

Tyson, L. D., & Zysman, J. (2022). Automation, AI & work. *Daedalus*, *151*(2), 256–271. doi:10.1162/daed_a_01914

Yang, S. J. H., Ogata, H., Matsui, T., & Chen, N. S. (2021). Human-centered artificial intelligence in education: Seeing the invisible through the visible. *Computers and Education: Artificial Intelligence*, *2*, 100008. doi:10.1016/j.caeai.2021.100008

Zembylas, M. (2023). A decolonial approach to AI in higher education teaching and learning: Strategies for undoing the ethics of digital neocolonialism. *Learning, Media and Technology*, *1*(48), 25–37. doi:10.1080/17439884.2021.2010094

## KEY TERMS AND DEFINITIONS

**AI Ethics:** The issues that AI stakeholders such as engineers and government officials must consider ensuring that the technology is developed and used responsibly. This means adopting and implementing systems that support a safe, secure, unbiased, and environmentally friendly approach to artificial intelligence.

**Algorithm:** A sequence of rules given to an AI machine to perform a task or solve a problem. Common algorithms include classification, regression, and clustering.

**Big data:** The large data sets that can be studied reveal patterns and trends to support business decisions. It is called "big" data because organizations can now gather massive amounts of complex data using data collection tools and systems. Big data can be collected very quickly and stored in a variety of formats.

**Chatbot:** A software application that is designed to imitate human conversation through text or voice commands.

**Cognitive Computing:** Essentially cognitive computing is the same as AI. It is a computerized model that focuses on mimicking human thought processes such as pattern recognition and learning.

**Computer Vision:** Computer vision is an interdisciplinary field of science and technology that focuses on how computers can gain understanding from images and videos.

**Data Mining:** The process of sorting through large data sets to identify patterns that can improve models or solve problems.

**Data Science:** An interdisciplinary field of technology that uses algorithms and processes to gather and analyze large amounts of data to uncover patterns and insights that inform business decisions.

**Deep Learning:** A function of AI that imitates the human brain by learning from how it structures and processes information to make decisions. Instead of relying on an algorithm that can only perform one specific task, this subset of machine learning can learn from unstructured data without supervision.

**Emergent Behavior:** Emergent behavior, also called emergence, is when an AI system shows unpredictable or unintended capabilities.

**Generative AI:** A type of technology that uses AI to create content, including text, video, code, and images. A generative AI system is trained using large amounts of data, so that it can find patterns for generating new content.

**Image Recognition:** The process of identifying an object, person, place, or text in an image or video.

**Machine Learning:** A subset of AI that incorporates aspects of computer science, mathematics, and coding. Machine learning focuses on developing algorithms and models that help machines learn from data and predict trends and behaviors, without human assistance.

**Prescriptive Analytics:** A type of analytics that uses technology to analyze data for factors such as possible situations and scenarios, past and present performance, and other resources to help organizations make better strategic decisions.

**Reinforcement Learning:** A type of machine learning in which an algorithm learns by interacting with its environment and then is either rewarded or penalized based on its actions.

**Structured Data:** Data that is defined and searchable. This includes data like phone numbers, dates, and product SKUs.

**Supervised Learning:** A type of machine learning in which classified output data is used to train the machine and produce the correct algorithms. It is much more common than unsupervised learning.

**Training Data:** The information or examples given to an AI system to enable it to learn, find patterns, and create new content.

**Transfer Learning:** A machine learning system that takes existing, previously learned data and applies it to new tasks and activities.

**Turing Test:** The Turing test was created by computer scientist Alan Turing to evaluate a machine's ability to exhibit intelligence equal to humans, especially in language and behavior. When facilitating the test, a human evaluator judges conversations between a human and machine. If the evaluator cannot distinguish between responses, then the machine passes the Turing test.

**Unstructured Data:** Data that is undefined and difficult to search. This includes audio, photo, and video content. Most of the data in the world is unstructured.

**Voice Recognition:** A method of human-computer interaction in which computers listen and interpret human dictation (speech) and produce written or spoken outputs. Examples include Apple's Siri and Amazon's Alexa, devices that enable hands-free requests and tasks.

# Chapter 6
# Exploring the Transformative Potential and Generative AI's Multifaceted Impact on Diverse Sectors

**Sadhana Mishra**
*University of Hail, Saudi Arabia*

## ABSTRACT

*In the digital landscape, where innovation knows no bounds, generative artificial intelligence (Gen AI) emerges as a true virtuoso, orchestrating a seamless symphony of creative wonders. Artificial intelligence that can generate fresh and plausible images, texts, and motion graphics rapidly is called Generative AI. It possesses the remarkable ability to effortlessly transform prompts into a rich tapestry of visual art, evocative prose, dynamic videos, and an array of captivating media. This chapter embarks on an exploratory journey, delving deep into the profound influence that Generative AI wields across a diverse spectrum of industries. Through an immersive exploration of these remarkable used cases, this chapter offers a glimpse into the boundless potential and the promises that Generative AI holds, positioning itself as a versatile catalyst for progress within the expansive industrial landscape.*

## INTRODUCTION

Artificial Intelligence (AI) embodies the artistry and engineering of creating intelligent machines, infusing them with the essence of intelligence and ingenuity (McCarthy et al., 1955). Whereas, Generative AI, a captivating offspring within the realm of Artificial Intelligence, transcends mere programming and becomes a digital art of novelty. Generative artificial intelligence, also known as GenAI, is a collection of codes and algorithms designed to develop the most plausible responses to a problem at hand. According to Morgan Stanley, this emerging technology has the potential to enable tech companies to access approximately $6 trillion in offline expenditure (Morgan Stanley, 2023). Generative AI harnesses the potency of a specific branch of deep learning, known as generative adversarial networks (GANs),

DOI: 10.4018/979-8-3693-1565-1.ch006

to craft original content that springs forth from the interplay of creative forces. Among the array of generative AI platforms, notable examples encompass ChatGPT, Google Bard, DALL-E, and Musico, each a testament to the spectrum of AI-driven creativity. New forms of content such as text, images, music, audio, and videos can be created with the help of Generative AI. It can be used to generate more realistic images of people, music, and codes. Hence, Generative AI, including Large Language Models (LLMs), possesses the remarkable capability to fabricate highly realistic simulations of real-world data, extending its creative capability beyond traditional boundaries. Through the sophisticated algorithms and immense data-driven learning, Generative AI models like LLMs excel at understanding the underlying patterns and structures within diverse datasets, enabling them to recreate data that mirrors the intricacies of the world we experience. This capability finds application across numerous domains, from generating lifelike images for creative endeavors to crafting contextually rich textual content for natural language understanding tasks, fundamentally reshaping how we interact with and generate data in the digital age.

According to Tiago Bianchi (Generative AI Global Weekly Search Trends on Google 2023 | Statista, 2023), as of February 2022, global online searches related to "generative AI" had shown a consistent rise over the preceding six months. Notably, the search volume for terms associated with generative artificial intelligence surged to a popularity score of one hundred (100) index points during the week commencing February 12, 2023. This heightened interest in "Generative AI" closely correlated with the launch of ChatGPT, an AI chatbot model developed by the United States-based research company OpenAI in November 2022. Despite ChatGPT's significant global recognition since its debut, a substantial 55 percent of surveyed U.S. adults reported having no prior knowledge of this AI-powered chatbot. Another issues associated with the use of Gen AI are related with ethical, privacy, security concerns. Hackers can invade the individual privacy, security and may create harmful content with unethical use of AI. Further this technology can create serious impact of employment status, misinformation, plagiarism and copyright infringements (Lawton George, 2023). At present, knowledge about the utilization of Generative AI is confined to selective fields, resulting in a limited availability of comprehensive literature (Gozalo-Brizuela & Garrido-Merchán, 2023; Weisz et al., 2023). The primary objective of this chapter is to address the existing gap in literature concerning the applications of Generative AI. In-addition this chapter seeks to systematically explore the prominent authors and document the pertinent uses of Generative AI across various fields associated with its application. The overarching goal is to provide readers with a thorough understanding of how Generative AI is integrated into diverse domains. By elucidating the multifaceted applications of Generative AI, this chapter aims to contribute valuable insights that will not only fill the existing void in literature but also empower readers to gain a nuanced and precise comprehension of the role Generative AI plays in different fields. Through a comprehensive exploration of its relevant uses, the chapter endeavors to shed light on the intricate intersections between Generative AI and various industries, fostering a more informed and insightful perspective among readers.

## TYPES OF GENERATIVE AI

**Generates Text:** Chatbots, text generators, and AI writing tools have the remarkable ability to generate fresh text in response to a user's input, whether it's providing an answer to a query, creating summaries, translations, or paraphrases. In some instances, these chatbots seamlessly integrate with search engines, elevating the search experience to new levels of sophistication and efficiency. Examples include Chat-GPT, Quill Bot Paraphraser, Bing AI (Caulfield Jack, 2023).

*Figure 1. Global online searches over generative AI*
*(Generative AI Global Weekly Search Trends on Google 2023 | Statista, 2023)*

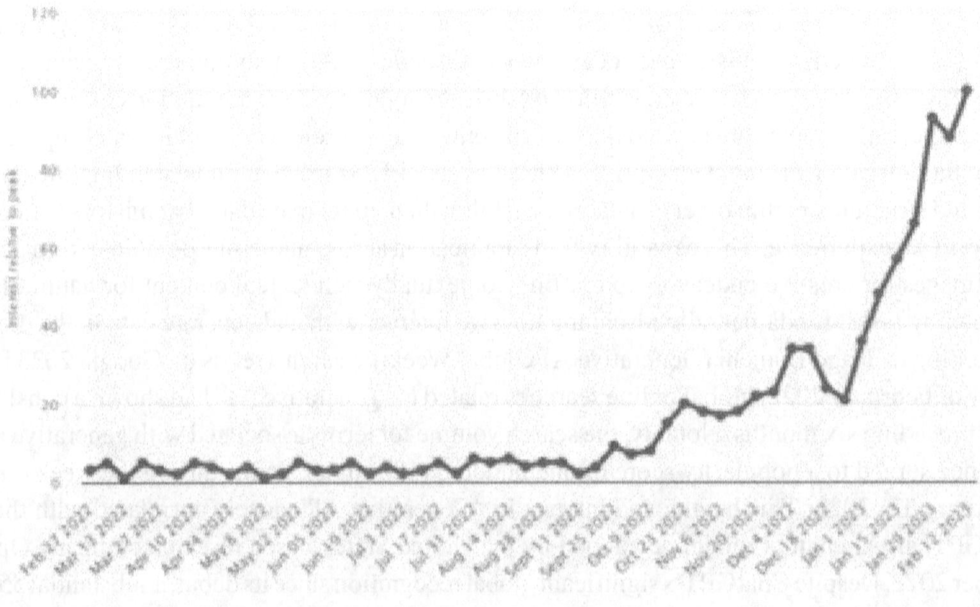

**Generates Code:** Generative AI extends its expertise into the realm of programming languages, but not limited to JavaScript. While ChatGPT highlights its versatility by handling many tasks, there are also specialized applications finely tuned to the art of code generation, offering niche expertise in this domain. Examples include OpenAI Codex, GitHub Copilot.

**Generates Images:** Language Models (LLMs) have exhibited remarkable versatility, highlighting their capacity to adapt, sometimes even beyond their core text-generation domain, delving into the realm of imagery. These applications typically receive textual prompts from users and ingeniously transform them into visual artworks, demonstrating the model's unexpected creative breadth. Examples include DALL-E, Midjourney, Stable Diffusion.

**Generates Videos:** An AI video generator represents a revolutionary tool, harnessing the immense potential of artificial intelligence to empower users in crafting distinctive videos, along with facilitating their editing endeavors. While some of these videos may originate from existing content, the majority of AI video generators operate through a captivating text-to-video process. Users input textual prompts, and like conjurers, these tools conjure entirely new videos from the ground up, breathing life into digital narratives through the wizardry of AI (Glover Ellen & Urwin Matthew, 2023). Examples include Glia Cloud, Lumen5, Pictory, Synthesia.

**Generates Audio:** Audio generative AI represents a category of artificial intelligence engineered to produce audio output from input data. This technology finds applications in crafting music and voice, employing deep learning, a machine learning approach inspired by the intricacies of human cognition (Crisara Matt, 2023). Examples include Clipchamp, Uberduck, Voicera, MusicLM, MusicGen.

*Figure 2. Primary focus areas of generative AI*
*(Gartner, 2023)*

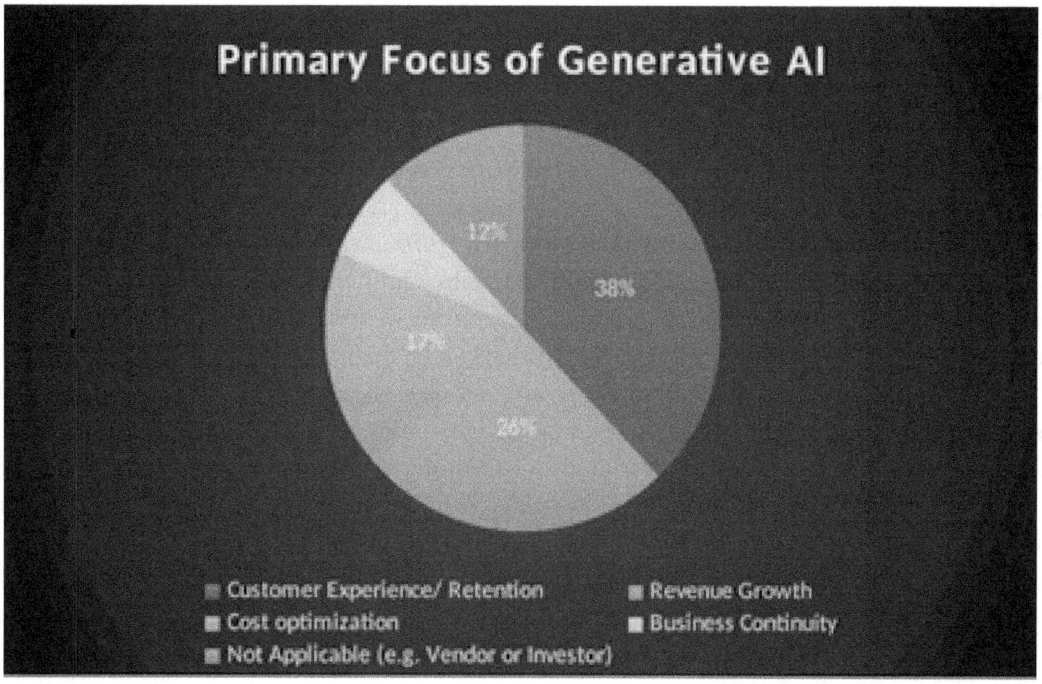


## METHODOLOGY

The methodology employed for gathering data and relevant information in this book chapter involved a qualitative approach, predominantly drawing from sources available on internet such as reports, articles, blogs, interviews of professional from leading company, trend analysis from research based firms. Qualitative data collection was considered the best method for gathering comprehensive and up-to-date information because of the ever-changing nature of Generative AI applications in many industries. There are many sources available on the Internet, for example articles, case studies, and reputable websites that provide information on topics related to Generative AI, which served as the primary sources of data for the research. These sources provided valuable insights, trends, and developments within the field, offering a comprehensive overview of the various applications and advancements in Generative AI. Various blogs written by well-known researchers contributed an additional dimension of qualitative information, offering firsthand accounts, expert opinions, and practical experience related to the application of Generative AI in a variety of fields.

The informal and conversational tone of blogs provided unique perspectives and narratives that enriched the understanding of Generative AI's real-world implications. Another important source of qualitative data was reports from respected organizations, research institutions, and industry magazines. The papers covered the use of Generative AI in many industries and provided statistical data, industry trends, and in-depth analysis. They shed light on new possibilities, difficulties, and uses for Generative AI, which is an exciting field. Overall, the qualitative methodology adopted for data collection enabled a holistic and nuanced exploration of Generative AI's relevance and integration across various fields. By leveraging diverse sources such as internet repositories, blogs, and reports, this methodology facilitated a comprehensive understanding of the multifaceted landscape of Generative AI applications.

## GENERATIVE AI IN HEALTHCARE

In the ever-evolving landscape of healthcare complexities, generative AI emerges as a guiding light of optimism. Generative AI stands as more than a fleeting trend; it represents an ever-evolving force that continually forges new horizons and drives a transformative revolution in healthcare. Within the healthcare sector, numerous hurdles loom large, including pandemics, chronic illnesses, mental health crises, and a scarcity of medical experts, Generative AI holds the promise of offering potential solutions to these pressing issues of Healthcare. According to the World Health Organization (World Health Organisation, 2016), previous studies have indicated that a significant portion, up to 50%, of medical errors within primary care settings can be attributed to administrative causes. This issue is further exacerbated by the worldwide scarcity of medical professionals. Generative AI presents a potential means to address some of these challenges, yet it demands cautious utilization due to its ongoing evolution. The precision of generative AI outputs hinges significantly on the quality of the datasets employed for their training, encompassing vital medical records, laboratory findings, and imaging studies (Sherwood Aaron, 2023).

### The Application of Generative AI in Healthcare

Clinical Decision Support: Generative AI is currently aiding healthcare practitioners, including doctors, in rendering precise and well-informed medical diagnoses. GenAI possesses the capacity to scrutinize data sourced from a patient's medical history, laboratory findings, prior therapeutic interventions, and medical imaging, including MRI scans and X-rays. It can identify potential areas of concern and propose additional examinations or treatment alternatives. Examples include "Glass AI", a system designed to facilitate medical professionals in acquiring, structuring, and curating their wealth of medical knowledge. "Regard" represents an advanced AI tool that seamlessly integrates with Electronic Health Record (EHR) systems, enhancing the efficiency and effectiveness of healthcare processes. "Kahun" is a symptom checker tool that plays a major role in the healthcare landscape by providing clinical evaluations of patients (Zhang & Boulos, 2023).

**Healthcare Management Assistance:** Advances in the automation of patient records and administrative support in healthcare settings are greatly aided by Generative AI. This helps the administration to streamline and enhance the processes associated with recording and managing patient information. Examples include "Microsoft Copilot," which incorporates the efficacy of generative AI into standard office programs like Word, PowerPoint, Teams, and more. This integration is designed to elevate productivity and streamline tasks across a spectrum of professional applications. "Nuance's advanced speech recognition technology" plays a pivotal role in enhancing clinical documentation. It empowers healthcare professionals to effortlessly dictate their notes directly into the Electronic Health Record (EHR) system, streamlining the documentation process (Zhang & Boulos, 2023).

**Patient Engagement:** Patient engagement extends far beyond mere participation; it revolves around forging a profound and meaningful bond between healthcare providers and individuals seeking care. When patients actively engage in their healthcare journey, the outcomes are profound, encompassing improved treatment adherence, lowered healthcare expenses, and heightened patient contentment (Ruiz

Josefina, 2023). Examples include "Hippocratic AI" directs its efforts toward the development of a specialized Large Language Model (LLM) tailored explicitly for healthcare. Its objective is to provide an LLM that centers around the patient's experience, placing a strong emphasis on qualities such as empathy, care, compassion, and the generation of patient-friendly responses. This approach intended to elevate patient engagement and facilitate more effective patient outreach. "Grid space" represents an enterprise-level solution driven by generative AI. It excels in automating patient outreach through its capabilities, which include managing phone calls, providing responses to inquiries, and efficiently managing administrative duties.

**Synthetic Data Generation:** The utilization of generative AI models to produce synthetic data has become an asset for healthcare researchers. Synthetic data, designed to replicate real-world data while safeguarding patient confidentiality, serves as a vital resource for the development and testing of AI models without jeopardizing the security of sensitive information (Alyoubi et al., 2020). Examples include "Syntegra Medical Mind" leverages generative AI to create authentic synthetic patient records derived from genuine healthcare data sources such as Electronic Health Records (EHRs). "DALL-E 2", an advancement by OpenAI, stands as another remarkable model designed for the generation of images from textual descriptions. Its training encompassed an extensive dataset comprising billions of text-image pairs, enabling it to master the art of crafting lifelike synthetic images.

**Improved drug discovery and development:** Generative AI is rapidly propelling advancements in drug discovery, revolutionizing the planning and execution of clinical trials, and ushering in an era of precision medicine therapies. This transformative technology is reshaping the landscape of healthcare and pharmaceuticals, offering unprecedented opportunities for innovation and tailored treatments. Examples include "Insilico Medicine", a pioneering company, has created a generative AI platform named Pharma.AI, which stands at the forefront of innovative drug design. This advanced technology has been instrumental in the development of novel therapeutic solutions for complex diseases like fibrosis and cancer, showcasing its remarkable potential to revolutionize the field of drug discovery and significantly impact the future of healthcare (*Pharma.AI*, 2023).

## GENERATIVE AI IN EDUCATION

Artificial intelligence (AI) has brought about a profound transformation in the realm of educational technology (EdTech), with Generative AI taking a leading role in driving this revolutionary shift. This technology is spearheading innovation and reshaping the educational landscape. Educational institutions are proactively incorporating Generative AI into their EdTech offerings. Through the utilization of Generative AI methodologies, these institutions are pioneering the development of inventive tools aimed at producing tailored learning resources, interactive assignments, simulations, and virtual laboratories. These resources are designed to foster interactive, hands-on learning encounters, stimulate creativity, and furnish real-time guidance through virtual tutors and mentors. The smooth integration of Generative AI into educational platforms is fundamentally transforming the learning experience, benefiting both students and educators alike (Dwivedi Deep, 2023).

## The Application of Generative AI in Education

Elevating Student Ingenuity and Involvement: Generative AI tools possess the capacity to spark and nurture students' creativity, granting them the means to craft unique and imaginative content, spanning art, music, and storytelling. Generative AI opens doors to dynamic and interactive learning experiences that cultivate curiosity and actively engage students in the educational process. One of the significant advantages of incorporating generative AI into the classroom is its capacity to accommodate a diverse spectrum of learning styles and preferences. Through the generation of content personalized to the unique needs of each student, generative AI contributes to making learning more inclusive and accessible to a broader array of learners, including those who might encounter challenges with conventional teaching approaches. This adaptability in educational content not only enhances engagement but also supports the varied ways in which students absorb and process information, ultimately fostering a more inclusive and effective learning environment (Rollins Mark, 2023). Examples include "GPT-3", a "large language model" (LLM) crafted by OpenAI, mirrors the attributes and limitations commonly found in AI-generated content, as seen in its response to our initial sentence (Davenport Thomas H. & Mittal Nitin, 2022).

**Personalized Lessons:** Generative AI aids in the development of individualized learning materials and instructional content by scrutinizing student data, encompassing learning behaviors, areas of interest, and academic performance. With the guidance of AI algorithms, the educational curriculum, resources, and activities can be dynamically adapted to cater to each student's distinct requirements and preferences. This customization amplifies student engagement, motivation, and knowledge retention by delivering content that directly aligns with their learning objectives. Examples include "Knowji" stands as an AI-powered application designed to enrich vocabulary acquisition for individuals of all age groups. "Speechify" represents a tool driven by generative AI, enabling versatile conversions between text and speech, or speech and text. It seamlessly operates on both desktop and online platforms (Karatas Gulbahar, 2023).

**Course Design:** Generative AI tools have the capability to generate educational materials like textbooks, worksheets, quizzes, and organize course materials, including syllabi, lesson plans, and assessments. Leveraging extensive databases of educational content and language processing techniques, AI models can produce resources tailored to particular subjects, grade levels, or technical objectives. Generative AI, when combined with complementary technologies like virtual reality & Augmented reality, has the capacity to craft simulations and immersive virtual environments. The fusion of generative AI with augmented reality (AR) and virtual reality (VR) empowers educators to design engaging interactive simulations, virtual laboratories, and even virtual instructional guides. Examples include "Gradescope", represents an AI-driven solution that streamlines the grading process for educators, making assessment evaluation more efficient and user-friendly (*Gradescope*, n.d.). "Synthesia" stands for an AI-powered video generator capable of transforming textual input into dynamic videos. "NOLEJ" utilizes the wealth of untapped learning resources, including textbooks, videos, and various online media, and swiftly convert them into captivating interactive content (*NOLEJ AI Keeps It Simple*, n.d.).

**Data Privacy and Security:** Generative AI draws from an extensive array of data, encompassing student, business, and potentially personal information. Safeguarding this data and upholding its confidentiality is paramount. Educational institutions and developers must adhere rigorously to stringent data protection regulations and institute robust security protocols to thwart unauthorized access, data breaches, or any form of data misappropriation (Kaur Jagreet, 2023). Synthetic data, as its name implies, is data that is artificially generated instead of originating from real-world occurrences. Examples include "Gretel", which produces synthetic datasets that emulate the key attributes of authentic data.

## GENERATIVE AI IN TOURISM AND HOSPITALITY

The hospitality and tourism sector undergoes continuous transformation, marked by the constant emergence of novel technologies and evolving trends. Among these novel technologies, Generative AI stands out as a significant game-changer poised to exert a substantial influence on the industry in the forthcoming years. Generative AI excels in swiftly delivering the most pertinent outcomes to consumers from an extensive array of choices, all while minimizing the time and effort invested in the decision-making process. Its integration into the travel industry not only promises advantages for companies but also enhances the overall experience for consumers. As of 2023, the Generative AI market boasts a size of USD 751.91 Million, and it is projected to experience substantial growth, reaching an estimated USD 3,581.95 Million by 2032, driven by a robust Compound Annual Growth Rate (CAGR) of 18.94% (Precedence research, 2023).

### The Application of Generative AI in Tourism and Hospitality

Tailored Travel Arrangements: Generative AI has brought significant transformations to the travel industry, enabling the provision of uniquely personalized travel arrangements tailored to the specific preferences and needs of customers. Generative AI bridges the gap between travelers and immersive experiences, drawing them nearer to fulfilling and captivating adventures. It furnishes travelers with recommendations derived from their search history and input, ensuring that the guidance offered is finely attuned to their preferences and needs. Examples include "Trip notes", a data-driven travel organizer that streamlines the process of trip planning, making it more effortless and efficient.

**Language translation and Customer Service:** Generative AI exhibits a remarkable capability to fluidly transition between languages, delivering real-time translation services, a crucial component of travel and hospitality chatbots. In the era of technological progress, businesses are harnessing the capabilities of generative AI-driven chatbots to provide round-the-clock customer support, particularly for inquiries and concerns related to travel. Generative AI Chatbots excels in delivering customers personalized reviews that precisely match their requirements. Examples include "Microsoft Bing", a state-of-the-art search engine that integrates advanced AI technology at its core. "ChatBot" serves as an AI-driven customer support solution that enhances service quality by simplifying processes and extending support across a wide range of channels and languages.

**Virtual Tour Companions:** Generative AI has the potential to transform into virtual tour companions, offering tourists immersive experiences of diverse destinations. This allows travelers to explore these locations without the need for physical presence. What makes these virtual tours even more captivating is their ability to craft personalized journeys tailored to individual preferences and search criteria. Examples include "P.A.D.D.Y."' AI-powered virtual tour guide created in Ireland.

**Travel Planner & Itinerary builder:** A Generative AI chatbot excels in constructing personalized itineraries by assisting travelers in a more streamlined and efficient manner, facilitating the planning of activities and excursions tailored precisely to their preferences and needs. A Generative AI Chatbot has the capability to tap into a diverse array of sources, including platforms and local guides, to offer travelers tailored and pertinent activity suggestions that align closely with their individual interests and preferences. These chatbots significantly alleviate the burden of researching transportation routes, destinations, accommodation, dining options, budget considerations, reviews, and other essential details, making the travel planning process considerably more effortless. Examples include "YaatriAI" simplifies trip planning, making it both swift and straightforward. "Wonderplan" employs a filtering mechanism to gather fundamental information about the traveler's preferences for the trip. Roam Around" allows you to input your desired destination, and within a matter of seconds, it crafts a customized travel plan for your journey.

**Predictive Analytics:** Tourism and hospitality enterprises harness the power of Generative AI to meticulously scrutinize data, uncovering valuable patterns, trends, and insights. The travel companies and tour operators can analyze vast amounts of travel-related data, such as: Booking Trends, Customer Feedback, Weather Patterns, demand for tours and activities. This analytical prowess empowers these companies and operators to optimize their operations by identifying areas that may require improvement and directing their focus towards enhancing overall efficiency and customer experiences. Generative AI possesses the capacity to analyze data sourced from social media and online reviews, providing a comprehensive perspective on the travel industry. By scrutinizing customer behaviors and preferences, this analysis equips travel companies with invaluable insights to refine their pricing structures, optimize inventory management, and tailor strategies. All these efforts converge to enhance their ability to meet their customers' needs effectively. Examples include "Bloom reach" is a cloud-based software designed for the travel industry.

## GENERATIVE AI IN MARKETING AND ADVERTISING

Generative AI holds the potential to make a significant impact in three key domains of marketing and sales: customer experience (CX), growth strategies, and productivity enhancement (Deveau Richelle et al., 2023). Generative AI offers a multitude of advantages in the realms of marketing and advertising, spanning enhanced operational efficiency, cost-effectiveness, heightened personalization, and expanded scalability. Generative AI can assist in the creation of marketing materials and deliver prompt responses to customer inquiries, streamlining marketing processes. Generative AI revolutionizes the advertising sector by showcasing its creative potential.

### The Application of Generative AI in Marketing and Advertising

**Content creation:** Generative AI boasts the ability to generate and condense various forms of content, including blog posts, social media updates, product descriptions, catchy slogans, and more. This automation not only translates into substantial time savings but also offers the advantage of continuous, round-the-clock content production at a scalable level. Content creators can employ GenAI to craft drafts, explore novel ideas, experiment with unconventional combinations, and ignite fresh avenues of inspiration for their teams, without the intent of replacing or limiting their creativity. Examples include AI video generators like Synthesia, Colossyan, Hour One, D-ID. "Jasper Campaigns", to effortlessly craft complete marketing campaigns that resonate with their brand's unique tone and voice, all through a single brief.

**Content personalization:** Generative AI possesses the capacity to customize content for individual users by analyzing their behaviors, preferences, and prior interactions through the examination of available data and search history. This personalized approach enhances the user experience and engagement significantly. Examples include Dall-E, Bard.

**Data Analysis and Sentiment Analysis:** Generative AI possesses the capabilities to delve into intricate data sets, extracting valuable insights and forecasting consumer behavior patterns. These insights, in turn, serve as a guiding compass for the formulation of highly effective marketing strategies. Generative AI can contribute to sentiment analysis by generating synthetic text data that has been annotated with various sentiments, including positive, negative, or neutral, thereby facilitating sentiment analysis model training.

**Design generation**: Generative AI extends its reach far beyond textual content, as it possesses the capacity to craft visual elements, produce remarkably lifelike images, and offer virtual try-on experiences. These capabilities serve to accelerate the design process significantly, revolutionizing the creative landscape. Heinz and nestle have both embraced Generative AI in their marketing campaigns. Examples include "Product Studio" uses AI to generate product images.

**Customer interactions:** Generative AI plays a pivotal role in enhancing customer service and support within the realm of marketing when seamlessly integrated with Chatbot and traditional AI solutions. This integration not only elevates the efficiency of customer interactions but also enriches the overall quality of service provided to customers. Generative AI finds valuable utility in customer services for three key facets: providing round-the-clock responsiveness, ensuring rapid response times, and delivering multilingual support. This technology facilitates constant accessibility for customers, expedites query resolutions, and accommodates diverse linguistic needs, ultimately bolstering the customer service experience.

## GENERATIVE AI IN MEDIA AND ENTERTAINMENT

The Generative AI in Media and Entertainment Market is projected to witness substantial growth, with an anticipated value of approximately USD 12,077 million by 2032, soaring from USD 1,202 million in 2022. This remarkable expansion is predicted to occur at a robust Compound Annual Growth Rate (CAGR) of 26.7% throughout the forecast period spanning from 2023 to 2032 (Market Research, 2023). Generative AI has instigated a profound transformation within the media and entertainment industry, reshaping the dynamics of content creation, distribution, consumption, and monetization. This innovation has ushered in new paradigms that influence the entire media and entertainment ecosystem (Garg Siddharth, 2023). Generative AI is currently in the spotlight, garnering significant attention for its potential impact on the media and entertainment sector. This technology's anticipated effects on the industry have drawn widespread interest and scrutiny.

### The Application of Generative AI in Media and Entertainment

**Content generation:** Generative AI proves its worth by automating the generation of content, spanning articles, blog posts, and social media updates. This presents a valuable time-saving asset for businesses and professionals engaged in consistent content creation, streamlining their processes and bolstering productivity. Journalists, scriptwriters, social media copywriters, bloggers, and storytellers can leverage Generative AI to swiftly craft captivating and engaging content, fostering efficiency and creative dynamism in their respective fields. Examples include "Canva" is a design platform that provides AI-powered solutions to facilitate content creation.

**Deepfakes and Face Swaps:** Generative AI serves the purpose of authentically superimposing one person's facial features onto another within video content. It also enables the creation of 'deepfakes,' which are synthetic media productions where an individual in an existing image or video can be seamlessly replaced with the likeness of another person (Kotadia Harish, 2023).

**Virtual News Anchor:** Virtual news anchors, often referred to as AI news anchors or virtual presenters, are computer-generated personas designed to deliver news content in a manner akin to human news anchors. These digital entities are a product of advanced artificial intelligence (AI) technologies and are deployed to automate and streamline the news reporting process (Okereke Frank Uche, 2023). Virtual news presenters offer several benefits: Round-the-Clock Accessibility, Expansion Capability, Cost-Effectiveness. Examples include Amazon Polly, 11labs, ChatGPT 4.

**Immersive Experience:** Generative AI possesses the ability to create immersive environments through the integration of Augmented Reality (AR) and Virtual Reality (VR), thereby facilitating and enhancing immersive experiences. Generative AI plays a pivotal role in crafting novel content, thus elevating the dynamism and engagement levels of Virtual Reality (VR) and Augmented Reality (AR) experiences. Snapchat has unveiled My AI, an AI-driven chatbot designed to respond to user queries and actively participate in conversations.

**Animation, Special Effects and Film Production**: Generative AI has the potential to transform the landscape of film and video production through the automation of diverse tasks, including video editing, color grading, and the integration of special effects. This innovation revolutionizes the traditional production process. Generative AI serves as a catalyst in expediting the creation of animations and special effects that would otherwise demand a significant amount of time to complete. This technology significantly accelerates the production process. Examples include Midjourney.

## GENERATIVE AI IN MANUFACTURING

Generative AI technology stands as a pivotal tool capable of automating and elevating diverse aspects of the manufacturing process for companies operating in the manufacturing sector. Generative AI has the potential to revolutionize conventional manufacturing procedures, unlocking fresh avenues while tackling longstanding issues. Generative AI is orchestrating a revolution within the manufacturing domain, introducing a novel paradigm that underscores the significance of open-source software and collaborative communities in the enhancement of production lines, waste reduction, and the optimization of supply chain logistics. Generative AI empowers manufacturers by guiding them towards well-informed decisions and attaining optimal design results. This not only amplifies their competitive edge but also acts as a catalyst for fostering innovation throughout the industry.

### The Application of Generative AI in Manufacturing

**Product design and development:** Generative AI stands as a formidable force in curtailing time-to-market durations, as it expedites the exploration of novel materials, drug compounds, design concepts, market insights, and more. This accelerated pace of discovery and development fuels efficiency across various industries. Generative AI extends its utility by conducting manufacturability and product performance assessments at an early developmental phase. This proactive approach allows for the early detection and resolution of potential issues, thereby leading to substantial reductions in both time and financial resources expended during this phase of product development. Examples include Wizdom.ai, Midjourney.

**Supply chain management:** Collaboration within the supply chain, particularly during the initial phases of product development, can be a complex and challenging endeavor. This complexity often arises from limited interoperability between tools and the need for extensive translation and transformation of requirement specifications and design models (Nair Harikrishnan Kesavan, 2023). Generative AI emerges as a solution to this complexity, enabling the creation of optimal supply chain models. These models consider the spectrum of supply chain operations, encompassing factors such as costs, delivery times, reliability, and more, streamlining and enhancing the supply chain collaboration process. Generative AI extends its capabilities into the realm of autonomous decision-making within supply chain management. This encompasses functions like optimizing transportation routes, overseeing inventory levels, and forecasting demand fluctuations. These AI-driven actions culminate in a bolstered supply chain, characterized by heightened resilience, responsiveness, and cost-effectiveness.

**Quality Control:** Generative AI assumes a pivotal role in elevating quality control standards within manufacturing by automating critical inspection and defect detection processes. This automation not only expedites the assessment but also enhances the precision and accuracy of quality control measures. Generative AI goes a step further in quality control optimization by meticulously scrutinizing data originating from diverse sources, such as production processes, supply chain operations, and customer feedback. This all-encompassing analysis reveals hidden tendencies and patterns, leading to better products and more satisfied buyers.

**Anticipatory Machine Maintenance:** Predictive or Anticipatory maintenance is handled by Generative AI, which analyzes large datasets collected from sensors and machines. It anticipates possible problems using pattern analysis, resulting in reduced downtime and increased production. With the help of generative AI, we can improve our use of resources and create a more sustainable world. It does this through honing production processes, reducing waste, and focusing more on environmentally friendly products and techniques.

**Production Scheduling and Stock Control:** Generative AI is reshaping manufacturing processes by automating labor-intensive jobs. This shift in thinking has the potential to not only simplify processes, but also helps to achieve unattainable levels of effectiveness and output. Generative AI emerges as a major player in demand forecasting thanks to its accurate use of customer data. In addition to assisting with efficiency and waste reduction, this data can be used to plan production and stock levels. The wasting, overproduction, and disruptive supply shortages can all be predicted by generative AI. Because of this foresight, substantial resources can be saved, and operations can be optimized.

## GENERATIVE AI IN AGRICULTURE

In terms of efficiency, production, and sustainability, generative AI has the potential to completely transform the agricultural industry. Generative AI can help farmers by providing them with data-driven insights and autonomous decision-making capabilities. This helps the farmers in crop management, predicting and managing potential issues like disease outbreaks or adverse weather conditions. These developments improve yields while decreasing resource waste, which in turn makes farming more sustainable and profitable.

## The Application of Generative AI in Agriculture

**Crop Management:** The most significant use of Generative AI in agriculture is crop management, a subfield of precision farming that makes use of state-of-the-art tools to boost productivity in the field. The pattern recognition and analysis skills of Generative AI are extremely helpful to farmers in determining when and where to plant, irrigate, and harvest. It helps farmers to improve their crop performance by spotting and interpreting trends and patterns in agricultural data (Pattam Aruna, 2023).

**Smart farming:** Generative AI is the foundation of smart agriculture, providing instantaneous observations and smart suggestions that are changing the way farmers work. This technology provides farmers with the ability to make right decisions on a micro level, which helps to ensure that resources such as water, fertilizers, and pesticides are administered accurately where and when they are required. As a result, agricultural productivity is increased, waste is reduced, and environmentally friendly practices are encouraged.

**Disease and pest management:** The use of Generative AI has opened a new era of preventative crop security by allowing for the advanced prevention and management of pests and illnesses in agriculture. Through the analysis of vast datasets encompassing environmental conditions, crop health indicators, and pest behavior, it can swiftly identify anomalies and patterns indicative of potential threats. This proactive approach allows farmers to take timely and targeted action, such as precision application of pest control measures, thereby preventing extensive crop damage.

**Crop Innovation and Sustainable Resource Allocation:** Generative AI serves as a potent ally to researchers and plant breeders in their quest to cultivate novel crop varieties that exhibit remarkable performance, heightened yield potential, and robust resistance against pests and diseases. By meticulously analyzing genetic and environmental data, generative AI identifies intricate correlations and genetic markers associated with desirable traits. Generative AI expedites the virtual screening of countless genetic combinations, significantly expediting the breeding process. This technology not only aids in the creation of crops better suited to evolving environmental challenges but also contributes to ensuring global food security and sustainable agriculture practices.

## CONCLUSION

Generative AI harbors boundless potential across diverse sectors. Its integration into various domains promises to amplify efficiency, usher in novel services, and stimulate sustainable progress. Notably, its capacity to substantially curtail time-to-market, maintain constant accessibility, operate autonomously, and empower designers positions it as a revolutionary force within industries. This, in turn, fosters the cultivation of multifaceted and imaginative designs, marking a transformative era in the industrial landscape. When looking at the capabilities of Generative AI, even product and service developments, adding new features to existing products, and creating new AI-based products are some of the areas where organizations are efficiently using this technology more often than any other. Similarly, AI is more frequently utilized for modeling the risk, measuring the performance and designing the organization. According to the recent report of Mc Kinsey, in future Generative AI will have significant impact on knowledge work. Activities involving around the decision making and collaboration will be highly impacted by the Artificial Intelligence in the future (Kinsey, 2023). In the similar vein and with the help of creative contents, content improvements, generative engineering and generative design, Generative AI is expected progress rapidly in both scientific and technological.

Like any innovation, it also has drawbacks. One of the major concerns of Generative AI is its ethical and legal use. By utilizing generative AI, cybercriminals can fabricate audio, video, and image files that violate people's privacy and disseminate false information. Another drawback is the automation of jobs. Automation driven by generative AI is a major issue right now. As it has evident in many reports that many industries such as manufacturing, marketing, healthcare, human resource extensively started using AI-backed innovations will lead to a larger unemployment rate along the road. Another big worry about Generative AI security is the possibility that hackers may obtain personal information and use it to violate people's civil and privacy rights. Market volatility is another side effect of this technology. Generative AI can create trades that can further cause market fluctuations. Another concern with Generative AI is that it might become uncontrollable if utilized anonymously, which could result in loss and unforeseen effects.

In summary, generative AI has revolutionized the way knowledge has been created and shared, with far-reaching consequences. For Generative AI to reach its full potential, serious ethical concerns must be addressed, despite the fact that it offers remarkable opportunities. In addition, to guide the development of Generative AI in an ethical way, study and cooperation among experts from other fields can be beneficial.

## REFERENCES

Aaron, S. (2023, May 12). *How will generative AI impact healthcare?* WeForum. Https://Www.Weforum.Org/Agenda/2023/05/How-Will-Generative-Ai-Impact-Healthcare/

Alyoubi, W. L., Shalash, W. M., & Abulkhair, M. F. (2020). Diabetic retinopathy detection through deep learning techniques: A review. In Informatics in Medicine Unlocked. doi:10.1016/j.imu.2020.100377

Aruna, P. (2023, May 6). Generative AI Applications: Episode 10: In Agriculture. *Medium*. Https://Medium.Com/Arunapattam/Generative-Ai-Applications-Episode-10-in-Agriculture-4ac24b6da8ea

Davenport, T. H., & Nitin, M. (2022, November 14). *How Generative AI Is Changing Creative Work*. HBR. Https://Hbr.Org/2022/11/How-Generative-Ai-Is-Changing-Creative-Work

Deep, D. (2023, July 10). *The Transformative Power of Generative AI in Education Technology*. Express Computer. Https://Www.Expresscomputer.in/Artificial-Intelligence-Ai/the-Transformative-Power-of-Generative-Ai-in-Education-Technology/100809/

Ellen, G., & Matthew, U. (2023, June 26). *15 Popular AI Video Generators*. BuiltIn. Https://Builtin.Com/Artificial-Intelligence/Ai-Video-Generator

Gartner. (2023). *Generative AI isn't just a technology or a business case*. Gartner. Https://Www.Gartner.Com/En/Topics/Generative-Ai

*Generative AI global weekly search trends on Google 2023*. (2023). Statista. Https://Www.Statista.Com/Statistics/1367868/Generative-Ai-Google-Searches-Worldwide/. https://www.statista.com/statistics/1367868/generative-ai-google-searches-worldwide/

George, L. (2023). *Generative AI Ethics: 8 Biggest Concerns and Risks*. Tech Target. https://www.techtarget.com/searchenterpriseai/tip/Generative-AI-ethics-8-biggest-concerns

Gozalo-BrizuelaR.Garrido-MerchánE. C. (2023). *A survey of Generative AI Applications*. http://arxiv.org/abs/2306.02781

Gulbahar, K. (2023, April 10). *Speech Recognition: Everything You Need to Know in 2023*. AI Multiple. Https://Research.Aimultiple.Com/Speech-Recognition/

Harish, K. (2023, July 20). *Key Use Cases for Generative AI in Media and Entertainment industry*. LinkedIn. Https://Www.Linkedin.Com/Pulse/Key-Use-Cases-Generative-Ai-Media-Entertainment-Harish-Kotadia

Jack, C. (2023, August 15). *Generative-AI*. Scribbr. Https://Www.Scribbr.Com/Ai-Tools/Generative-Ai/

Jagreet, K. (2023, July 6). *Generative AI in Education Industry | Benefits and Future Trends*. Xenon Stack. Https://Www.Xenonstack.Com/Blog/Generative-Ai-Education#:~:Text=Educators%20can%20use%20generative%20AI,%2C%20motivation%2C%20and%20academic%20achievement

Josefina, R. (2023, June 20). *Enhancing Patient Engagement with Generative AI in Healthcare*. Light IT. Https://Lightit.Io/Blog/Enhancing-Patient-Engagement-with-Generative-Ai-in-Healthcare/.

Kesavan, N. H. (2023, April 14). *Generative AI – Enterprise Use Cases for a Manufacturing Organization*. LinkedIn. Https://Www.Linkedin.Com/Pulse/Generative-Ai-Enterprise-Use-Cases-Manufacturing-Harikrishnan/

Mark, R. (2023, February 4). *Using Generative AI to Enhance Classroom Creativity and Engagement*. LinkedIn. Https://Www.Linkedin.Com/Pulse/Using-Generative-Ai-Enhance-Classroom-Creativity-Mark

Market Research. (2023). *Global Generative AI In Media And Entertainment Market*. Market Research. https://marketresearch.biz/report/generative-ai-in-media-and-entertainment-market/request-sample/

Matt, C. (2023, June 9). *AI Music Generators Make You Feel Like a Maestro. Here's How They Work*. Popular mechanics. Https://Www.Popularmechanics.Com/Technology/Audio/A44109081/Ai-Music-Generators-Explained/

McCarthy, J., Minsky, M. L., Rochester, N., & Shannon, C. E. (1955). A Proposal for the Dartmouth Summer Research Project on Artificial Intelligence: August 31, 1955 - ProQuest. *AI Magazine*.

Precedence research. (2023). *Generative AI in Travel Market*. Https://Www.Precedenceresearch.Com/Generative-Ai-in-Travel-Market

Richelle, D., Joseph, G. S., & Steve, R. (2023, May 11). *AI-powered marketing and sales reach new heights with generative AI*. McKinsey. Https://Www.Mckinsey.Com/Capabilities/Growth-Marketing-and-Sales/Our-Insights/Ai-Powered-Marketing-and-Sales-Reach-New-Heights-with-Generative-Ai

Siddharth, G. (2023). *How Generative AI Shaping the Media and Entertainment Industry*. QUY Tech. Https://Www.Quytech.Com/Blog/Generative-Ai-in-Media-and-Entertainment-Industry/. https://www.quytech.com/blog/generative-ai-in-media-and-entertainment-industry/

Stanley, M. (2023, April 18). *Tapping the $6 Trillion Opportunity in AI.* Morgan Stanley. Https://Www.Morganstanley.Com/Ideas/Generative-Ai-Growth-Opportunity

Uche, O. F. (2023, August 8). *Exploring Generative AI in Media & Entertainment: Virtual News Presenters*. LinkedIn. Https://Www.Linkedin.Com/Pulse/Exploring-Generative-Ai-Media-Entertainment-Virtual-News-Okereke

Weisz, J. D., Muller, M., He, J., & Houde, S. (2023). Toward General Design Principles for Generative AI Applications. *CEUR Workshop Proceedings*, *3359*(1), 130–144.

World Health Organisation. (2016). *Administrative Errors: Technical Series on Safer Primary Care.* WHO *Press*. https://apps.who.int/bookorders.%0Awww.who.int/patientsafety

Zhang, P., & Boulos, M. N. K. (2023). *Generative AI in Medicine and Healthcare : Promises, Opportunities and Challenges*. Research Gate..

# Chapter 7
# For Better or for Worse?
## Ethical Implications of Generative AI

**Catherine Hayes**
*University of Sunderland, UK*

## ABSTRACT

*This chapter explores how the social implications of AI are being posited, often sensationalized as a threat to humanity, rather than being framed in something humanly designed that ought to remain within the control of its maker, transparent in terms of capacity to undertake complex decision making and which most importantly is accountable for every individual action made in terms of design and programming. The aims of the chapter are threefold, namely, to consider global ethics and the impact that AI could potentially have in terms of increasing societal inequalities in terms of existing infrastructure; to provide an insight into the developmental and progressive use of AI across organizational infrastructures such as global medicine and health and the military; finally, to embed the concept of ethical AI and the potential for its praxis across all areas of its integration.*

## INTRODUCTION

*'I'm increasingly inclined to think that there should be some regulatory oversight, maybe at the national and international level, just to make sure that we don't do something very foolish. I mean with artificial intelligence we're summoning the demon.'*

(Elon Musk, MIT's AeroAstro Centennial Symposium, 2014)

This chapter will consider applied ethics in the context of pedagogical practice in relation to the potential impact of Artificial Intelligence (AI) in education and other societal infrastructures such as the military and health and medicine. The extent to which AI has been progressively integrated into society over the last three years, has exponentially increased media and scientific debates of what is and what might be, if all we are to believe about AI is realised (Bareis & Jatzenbach, 2022). As with any landmark paradigmatic shift in society's use of and access to technological advance, the widespread introduction of AI has both positive and negative aspects which can be harnessed for both human progression

DOI: 10.4018/979-8-3693-1565-1.ch007

and decimation (Mikalef et al, 2022). Initial debates surrounding AI focused predominantly on their functionalist capacities to reduce the complexity and challenge of largely physical tasks, where algorithmic decision making could be used as an adjunct support to mundane and burdensome human work (Mirbabaie et al 2022). This has now progressed to the widespread cognitive debates as to how AI can address tasks which before, seemed impossible and largely inaccessible, in terms of the decision-making processes necessary to undertake them (Madhav & Tyagi, 2022). One of the key aspects of these debates is how humanity has often designed AI in human form, to the extent technological artefacts are often perceived as robots with a high degree of sentient ambition (Owe and Baum, 2021). It is often thought AI can compete for cognitive advantage rather than simply being a design artefact used to extend the reach of humanity's applied intellect (Mele, & Russo-Spena, 2023; Yamin, et al, 2021). This has led to widespread hyperbole that AI somehow has the capacity to override human cognition and that its capacity for extended algorithmic thinking may eventually pose a huge threat to mankind, alongside offering some of the greatest technological developments of our age (Cools, Van Gorp & Opgenhaffen, 2022). Unlike other technological advances whether the choice to engage with them was always an option, AI poses a wider societal issue where that choice may no longer be possible, should the self-advancement of algorithmic decision making pose an overriding threat to humanity in terms of speed and capacity for action (Igna & Venturini, 2023). As such the ethical principles of AI are factors that all organisations now must contemplate so that the integration of ethical practice becomes a societal norm in terms of the use of AI in practice. Beyond an anthropological perspective, social ethics and the philosophies underpinning them all impact upon the capacity of organisational decision making and how AI may remain fully controllable and where the algorithms within which it operates may be constructively aligned with those affective attributes of humanity, that as a society, we would wish to promulgate, rather than any degree of negativity (Henin & Le Métayer, 2021). Whilst every organisational infrastructure operated by humans is designed on principles of altruism, equity and equality, the integration of AI poses several questions which necessitate critical reflexivity of the situated nature of their use within highly sensitive contexts such as healthcare, law, and education, all of which can have the potential for an inordinate impact on society as it is currently known (Cheng, Varshney, & Liu, 2021). Perhaps then it is rather an issue with the design of AI rather than implementation, which ought to serve the hyperbole and hypothesising that surround it (Bareis & Jatzenbach, 2022). This chapter will also explore how the social implications of AI are being posited, often sensationalised as a threat to humanity, rather than being framed in something humanly designed that ought to remain within the control of its maker, transparent in terms of capacity to undertake complex decision making and which most importantly is accountable for every individual action made in terms of design and programming (Novelli, Taddeo & Floridi, 2023).

The aims of the chapter are threefold, namely, to consider global ethics and the impact that AI could potentially have in terms of increasing societal inequalities in terms of existing infrastructure; to provide an insight into the developmental and progressive use of AI across organisational infrastructures such as global medicine and health and the military; finally, to embed the concept of ethical AI and the potential for its praxis across all areas of its integration.

## PARAMETERS OF GLOBAL INEQUALITY

Geopolitically there are several ethical debates around the role of AI (Palomeres et al, 2021). A collective sense of global responsibility necessitates ensuring not only accountability but a huge degree of reassurance that the use of AI in practice will not be of detriment to humanity in any way. In those developing countries where technological advances have not yet reached the stage of development of their Western counterparts, this poses a huge risk of adding further and widening gaps of societal inequality, where the impact of AI may still be felt but remain under the control of other nations (Haluza & Jungwirth, 2023).. Being able to functionally adapt to the use of emergent technology still predominantly remains the preserve of developed nations so from an ethical perspective the use of AI as an adjunct to human experience rather than a replacement for it is essential if we are not to lose sight of global public well-being amidst global societal interest in novel approaches to technological development (Farrow, 2021; Floridi & Cowls, 2022).

## CONTEXTS OF SOCIAL AND CIVIC RESPONSIBILITY

The irony of developing a metric framework of accountability for issues of ethical transparency, accountability, equity, and trust, is not lost on a world where the functional role of AI has far reached implications for address than those encountered in typical industrial, commercial, and educational contexts. In issues of human endeavour where predicting the future capacity and capability of AI as functionally autonomous, there are several issues of explanation which remain a challenge to justify in 21st Century developmental progress (Colaner, 2022). One of the main issues in terms of ontological and epistemological justification of AI is the lack of any apparent philosophical or theoretical underpinning for its use. Usually in the context of applied research, it is possible to provide a degree of constructive alignment, where ethical praxis is a primary and relatively predictable issue (Zhang et al, 2021). Global responsibility for a concept which has the potential to have universal impact, despite there not necessarily being a global capacity to support, sustain and ethically evaluate, is one which impacts beyond one facet of human infrastructure and pervades them all (Khogali & Mekid, 2023). Perhaps most significantly it is the degree of agency that AI occupies, which is of greatest concern to humanity and our human tendencies to frame AI as insentient threats to humanity, which have caused the greatest concerns and misunderstandings of what the ethical implications of AI might be (Hwang & Park, 2020). Beyond the technological challenges of functional AI in relation to its capability development lies the justification of man to permit its widespread inception to the digital community. Formal policy and practice initiatives have been driven by global conferences, which in their earliest years focused largely on the fundamental positive impact that AI could bring, and which largely preceded the research literature which has provided an inferential insight into what the impact of widespread AI introduction might constitute for global civic futures on both a personal and professional level (Floridi & Cowls, 2022). Leading in this arena is the UNESCO Recommendation on the Ethics of Artificial Intelligence, which using an array of Delphi methodologies have drawn on the specific expertise of specialists in the field of AI to operationally define and frame the nature of ethical frameworks and praxes in the field ought to be. In theory, human attributes and characteristics are relatively understandable choices within the consensus of those with specialist expertise within AI. However, what is more concerning is how binary decision making, without sentience or conscience can become of the preserve of man-made machines, which have no capacity for tacit discernment in relation to the binary decision-making processes that underpin the algorithms controlling them and enabling them to process a greater degree of digital information than any human has capacity for. The temporal fact that AI does not need to eat, sleep, or have any of the human 'baggage' that human capital is certain to have, ensures that progress can be made at an astonishing pace and to a level which may not have been anticipated before (Makridakis, 2017). As with any model of ethical praxis, key principles have been devised to ensure that the concepts of beneficence, non-maleficence, autonomy, justice, and explication are central to the use of ethical AI (Prem, 2023). Where the greatest challenges with this lie, though, is the fact that AI is not sentient, does not have a conscience, so in instances where automated decision making is advocated, the ethical use of AI has even greater ramifications, especially in relation to the justification of human decision making in its use (Floridi & Cowls, 2022; Mantello et al, 2023). The operational use of AI in practice is what has maintained a strategic focus on functional usage and the practicalities of system design (Peeters et al, 2021).

## SOCIO-CULTURAL PERSPECTIVES OF GOVERNANCE

Global cultural differences pose one of the greatest threats to the ethical management of AI, particularly in instances of competition (Humble, 2023). Not all use of AI will be altruistic and there are clear advantages to be gained in being able to win the race for supremacy in AI development and implementation (Qiao-Franco & Bode, 2023). Where specific ethical values are missing due to historical and local cultures then the justification of state intervention with AI may be lie beyond public questioning (Elliott, 2019). Across non-democratic countries there may even be a perceived benefit to autocratic leaders of being able to assert authority and surveillance with AI that has never been possible before (Zeng, 2020; Zuiderwijk, Chen & Salem, 2021; Young et al, 2021). It is in instances like this that human rights of privacy are destroyed, yet justified by the ideology that control of the state and state intervention is for human benefit. Any propositions to introduce global ethical frameworks of governance and accountability must first consider how best these cultural differences can be reconciled amidst such a degree of diversity, so that a universal set of principles can be agreed (Visivizi, 2021).

The whole concept of privacy is a relatively Westernized concept. Other nations value state intervention where collective control leads to individuals being responsible for their place and position in society (Döring, 2019) Not only does this link to ethics but more significantly, in countries such as China, is an integral part of morality, where moral behaviour has been an instrumental focus of attention for thousands of years prior to the introduction of AI (Nichols, 2022). Liberalist Western democracies have presented the need for individual autonomy at all costs. The rise of AI implementation has raised new questions of just how possible universal regulation of such systems can be. With such differences in state infrastructure, it is virtually impossible to see how AI ethics and governance and regulation can possibly become operational under such differences in contextual and situational specificity. Where philosophy underpins the justification of human rights, then across different global stages, this can be radically different (Zajda & Vissing, 2022).

## ACCOUNTABLE APPLIED PRACTICE IN ARTIFICIAL INTELLIGENCE

Technological emergence and machine learning continue to revolutionize societal life and will soon become existentially embedded as adjunct mechanisms of intellectual decision making, upon which we become dependent (Bareis and Katzenbach, 2022). The ethical implications of this within health and medicine cannot be overlooked, since AI, is insentient. It may be capable of making active choices via strategically designed algorithmic programming, but it has no conscience with which to couple binary or even complex decision-making processes (Islam et al, 2022). Whilst scenarios such as this are usually the preserve of Hollywood movies, the implications for humanity, lie in the degree of control and algorithmic decision making that these machines are permitted (De Bruijn, Warnier & Janssen, 2022; Nader et al, 2021). Ethical complexity is even greater when we consider that any subjective or emotive intelligence can be removed from decisions where life and death can be at stake, for instance within policing, the military or indeed medical and surgical intervention (Miller, 2022). It is where autonomous AI is introduced that the situated context of its use must be contemplated, acknowledged, and accounted for (Alter, 2022). When faced with humanoid robots, which deliver AI this becomes even more of issue, since our cognitive capacity to perceive them as 'real' instead of the machines they are, is an issue (Finkel & Krämer, 2022). We have given non sentient objects, the characteristic of humanity, yet are surprised at our reaction to just how realistic they can appear (Sweeney, 2022). Since ethical practice is based on our capacity to recognise the potential risks that AI may pose to humanity. It is the metacognitive processes of meaning making which differentiates the highly nuanced characteristics of man from the decision making itself (Ganapini et al, 2023). Searching for meaning, in this sense is central to human consciousness and arguably more importantly conscience. Being able to bind algorithms to specific parameters of focused operational practice, is straightforward in terms of ethical praxis (He, 2022). Where it becomes steadily more contentious is in instances where the intellectual capacity of AI is released within parameters of practice which also incorporate social contexts and social interaction (Chen, Sun & Wang, 2022). From an economic perspective, it is AI's potential to impact positively on areas such as enhanced productivity, the elimination of human error and the potential for more specific surgical intervention, which offer most (Goralski and Tan, 2020). It is the transparency of decision making which is imperative to ensure that man always remains in control of ongoing algorithmic processing, since if AI is permitted to iteratively improve itself without oversight and reaches decision making capacity beyond man's knowledge and scope of ethical practice, then it can be hypothesized that the disconcertment characterizing Hollywood's portrayal of AI could indeed become an unwelcome reality (Brewer et al, 2022). This would entail the AI adding algorithmic decision-making processes into originally human designed algorithms for the purposes of iterative improvement, regardless of the ongoing need for ethical praxis. It is also this potential that constitutes much of the hyperbole and media panic around just what AI might represent in terms of existential threat and how this might be best managed in the context of technological evolution (Harper, 2021).

## AGENCY AND RESPONSIBILITY IN EDUCATION

Since the progressive developmental work in AI operationalisation and implementation forms the basis for future work in the field, the need to educate and provide an insight into the agentic power necessary to manage the ethical implications of AI has never been greater (Nyholm, 2020). Beyond the functional implications of introducing emergent technologies across society, the impact and meaning making of AI on people's lives ought also to be considered so that mitigation of co-existing dangers, alongside the irrefutable benefit of AI can be accounted for and organisations held accountable for in terms of formal governance processes (Smith, 2021). Education and skills training via the embedding of strategic pedagogical intervention across curricula will become a central mechanism of ensuring due regard for the ethical implications of AI and potentially a means of harnessing strategic frameworks which safeguard aspects of professional and workforce praxis, where AI may pose complex implications It is the widespread coverage of AI implementation which means that its use is no longer a proactive choice for society, rather something which has been progressively incorporated and at the same time exploited by those who wish to harness it for reckless and illegal activity (Dauvergne, 2021). In terms of corporate responsibility though, there remains active debate over whether, the military, for example, ought to be given intellectual and operational rights to AI for autonomous use in algorithmic decision making which lacks any bias and as such any conscience in relation to the potential outcome of it, when the primary objective of its mission is simply to reach the completion of a task (Shaw et al, 2019). Issues of data protection and the use of AI algorithms to collate specific information and identify all members of society pervade all use of the technology in practice. Where ethical issues are most apparent however is that AI has the potential to override the innate epistemic bias and positionality which provides the unique decision-making skill of every individual in terms of their rights, capacities, and capabilities (Araujo et al, 2020). Justification of AI within everyday society is frequently legitimised based on the accuracy with which technology can now make complex decisions from algorithmic processing, which outweighs the capacity, capability, and speed with which a human counterpart every could. In addition to this, we see a challenge to human existentialism in that any aspect of subjectivity raised by epistemic positionality and the consequent personal bias this perpetuates can be eliminated with AI. The complex ambiguity here is that from an ethical and moral perspective, since this technology is man made, then the responsibility for what AI can do to humanity also has to be an embedded and integral part of AI design, justification, and evaluation, and the 21st Century must be educated to reflect this (Holmes et, 2021). Failing to do so could be catastrophic for the potential of what AI may progress to become and man's ability to maintain ultimate control of AI capacity and capability (Benbya, Davenport & Pachidi, 2020). As AI becomes further embedded into everyday society it is complacency which needs to be avoided as there is an educational and pedagogical responsibility to ensure that iterative generations of AI users, developers and entrepreneurs maintain absolute accountability for the emergent technology, which will inevitably be such a part of humanity's future. Ultimately, the situated and contextual nature of AI's use will determine how best algorithmic decision making can be continually monitored to ensure that whilst society is progressively developed, ethical complexities of its use can be addressed at source and with ease.

## THE EXISTENTIAL THREAT OF ARTIFICIAL GENERAL INTELLIGENCE

Artificial General Intelligence (AGI) is the term applied to a theoretical version of AI which can surpass human intellect and undertake complex decision-making processes at speeds not possible in terms of cognitive function (Helm et al, 2020). Whether AI is an existential threat to humanity remains one of contention in the media, not least because those who invented it are now calling for a measured pause in its developmental progression and have released formal statements warning that AI could potentially pose a greater threat to humanity than nuclear weapons or further global pandemics. Whilst these statements are regarded as the hyperbole of the super-rich or super invested. Whilst all scientists are aware of the need for ethical praxis in relation to technological advancement it is the routine use of AI which is causing greatest concern. Being clear evidence on the positive impact of it within fields such as medicine, in the context of drug discovery, health and education is imperative (Rajpurkar er al, 2022). Whilst AI use is purposeful and direct, the evolution of AI systems continues to astonish even those who have been instrumental in its invention (Samuel, 2023). Whether certain sectors of society should be given agency in relation to the use of AI is contentious, particularly the military, where exploitation of AI systems could pose an existential threat (Federspiel et al, 2023).

One of the commonest areas for concern, however, is the current use of AI chatbots such as ChatGTP, which are an active source of bias and the spread of potential misinformation (Najafali, 2023). Alongside this is the issue of whether human resources can be replaced by AI as a mechanism of economic savings and whether AI can produce written content, which can actively compete or surpass the capabilities of humans under any given circumstance (Alvero, 2023). In terms of explanation, it is the neural capacity of AI which demonstrates clear similarity to human neurological function. Deep learning is the process by which information is processed and learning takes place. By moving beyond the superficial mechanisms of data evaluation that initial computers could be programmed to undertake, it is the prospect of machine learning which can be developed to algorithmically perform reasoning and discernment, which is an issue, particularly in relation to the short time taken for developmental process with AI systems to be made (Zhang, Zhu & Su, 2023). Whilst presently AI is not sentient, larger debates exist around the notion of consciousness when AI develops significantly and has a far higher degree of sophistication in relation to its evolution.

## VIABLE MILITARY ETHICS AND AUTONOMOUS WEAPONRY

A fundamental discriminating feature of autonomous weaponry and its potential deployment by the military is the ethical responsibility of placing the decision to discern whether to kill or not to kill, being legitimized by a technological algorithm (Pereira, 2019). One key issue is that it cannot be guaranteed that autonomous weaponry will remain the preserve of regulated military praxis. This AI can also be harnessed by terrorists, state proponents of repression and dictators, whose agendas are far from the democratic empowerment of humanity (McKendrick, 2019). Since these progressive developments are often iterative by nature and incremental by functional design, each enhancement of their capability does not remain a topical political cause for concern, and as such ensures that informed consensual debate with the societies upon these autonomous weapons systems could impact, is not happening. An additional implication of this, is that the global markets for AI weaponry, continue to increase, with demand outstripping demand at a pace of acquisition which is incomparable with any other previous weaponry (Sauer, 2021). International security is thus threatened by the potential of error, criminal hacking, and algorithmic dysfunction, which could lead to mass loss of life and an unknown level of devastation. The introduction of international law on the regulated use of advanced weaponry across the world is one potential mechanism of mitigating the risk of this happening however this needs to occur with a high degree of expedience if timing of regulation (Werthner et al, 2023). The threat posed to democratic society by the largescale introduction of autonomous weaponry continues to grow in sync with the rate of proliferation of the systems used for their operationalization. Since the systems can operate remotely, the potential of terrorists to preserve their anonymity and consequently their arrest and detainment, means their attractiveness in terms of a weapon of choice, is growing exponentially. Terrorist capacity and capability with AI autonomous weapons also has the potential to accelerate quickly since the technology used within AI defence systems is fiscally less obtainable since it needs to discern between targeted approaches to ensure targets are engaged with expedience and safety, which necessitates more expensive functional components. Whilst this represents a true dystopia, in terms of the potential for individual and collective slaughter, in the hands of despotic, authoritarian dictators, this represents a mechanism of disempowering humanity and stripping away human rights as we currently recognise them (Fontes et al, 2022). AI can become a tool in changing sociocultural norms, where surveillance becomes a norm rather than an instance of exceptional security measure. The concept of an AI arms race poses significant questions regarding the ethical development of autonomous weaponry since the pace of development is also happening amidst the competition to be the first and be the best in terms of arms acquisition. This speed may well detract from the need to ascertain AI's safety both during military intervention and the impact on humanity if this is potentially overlooked in favour of winning an arms race (Armstrong, Bostrom & Shulman, 2016). Ultimately human intellect needs to assert authority over AI at any given time, to be able to intervene in instances where risk needs to be mitigated with immediate effect. Without globally agreed policy and practice directives the impact of AI being able to retain agency could pose a threat to global safety, regardless of their state ownership. Those countries whose cultural heritage is firmly rooted in the control of populations may be more inclined to align political strategy with the potential for thresholds of escalation in AI decision making, which bypass collective humanitarian need in favour of an ongoing means of population control (Kile, 2013). In terms of mediating risk where advanced AI weaponry could potentially act autonomously by bypassing algorithmic instruction or have some degree of systems failure where human intervention is impossible, the ethical constraints of using military artificial intelligence are significant. Since this is a possibility all military intervention always necessitates an absolute degree of human control, however the greatest dilemmas will occur if the AI becomes progressively more capable of advanced decision making prior to human's being able to keep up with their speed of action. Ethical education in the use of operationalising AI military weaponry will be essential to ensure the contextualised and situated nature of both weapons' deployment and use (Amoroso & Tamburrini, 2020). This will involve extensive simulation training and an assurance that military staff always remain in agentic control of the technological equipment, to ensure both humanitarian and personal safety (Pizzi, Romanoff & Engelhardt, 2020). Alongside the potential risks of integrating AI into military operations, it is also necessary to consider the exceptional advantages that these systems offer, in preference to potentially sacrificing troops within terrains which are hazardous due to geography, topography or sheer physical danger (Masakowski, 2020). The potential of operating machinery remotely, ensures that even in instances where multifaceted decision making needs to take place, in the most dangerous military events, forces will have the opportunity, via AI, to make decisions which are acutely responsive, timely and which no longer endanger human life.

## HEALTH AND MEDICAL ETHICAL PRAXIS

Since empathy and ethics are two of the fundamental philosophies underpinning the medical relationship between doctors and patients and that the two belie the trust within active dialogue in assessment, diagnosis and medical management, the use of AI within medical practice has implications for trust, reliability, and compassion (Hatherley, 2020). Since AI is not sentient but offers high degrees of functional reliability and validity via deep neural networks in the context of diagnosis and assessment, it is not difficult to see how AI is currently very much a valued adjunct to human medical practice, rather than as alternative or replacement of this (Salahuddin et al, 2022).

Within the context of medical and surgical practice technological development has meant an exponential increase in data collection, analysis, and reporting (Liu et al, 2022). Many of the deep learning techniques now possible because of AI deep learning capacity to not only manage data but effectively interpret it due to its capacity to recognise data based on having previously been exposed to the same or similar forms of it. This represents the potential of removing human bias and subjective decision making in the functional interpretation of results which can be undertaken with far greater efficiency and expediency than humans (Dwivedi, et al 2021). This is indeed a marked improvement for the productivity of the medical workforce but its impact on the relationships between it and the patients it serves remains, yet, largely unexplored. Within the United Kingdom, where the National Health Service (NHS) has become a victim of its own success in terms of an infinite need for resources but a finite resource pot from which to fund the assessment, diagnosis, and management of patients from the cradle to the grave, this represents an important opportunity for not only more accurate and valid medical praxis but also increased efficiency in the process (Haug & Drazen, 2023). It remains something of an irony that whilst longevity has, over the last half century increased, recent events have ensured that these statistics are now in reverse, especially with more advanced technological diagnostic support and intervention mechanisms than there have ever historically been. Acceptance of technology by the public in the context of an infrastructure where they have been acclimated to being able to have absolute trust in human compassion and empathy, is another core aspect of why being able to ensure quality and standards is imperative (Schneiderman, 2020). Historically these levels of quality standards have been rooted in the need for the validity and reliability of AI in terms of diagnostic capacity (Larson et al, 2021). In terms of the ethics of healthcare where all risk must be mitigated and accounted for. Since the algorithmic capacity of AI cannot be 100% guaranteed, but nor can the mitigation of the human risk of error, it is about ascertaining the modality which offers best levels of accuracy and ensuring that ethical practice is maintained by ensuring that regardless of how purposeful AI may be, it is certainly only an adjunct support to the qualified medical staff who make use of them in clinical settings (Aslam and Hoyle, 2022). Collective global efforts to reach agreement on the regulation of AI, in terms of its integration across both high income and developing nations is still far off where it ought to be in terms of any ethical consensus surrounding AI (Uunona & Goosen, 2023). This is further complicated by race wars which have ensued in the light of global conflict where state players are competing to outwit their enemies and civilian lives are often tokenistically protected in the name of war (Taddeo et al, 2021). The address of the sociocultural impact of AI across global societies is one which remains a core issue for the roll out of AI based decision making algorithms in practice. Also separated by heuristics and the fiscal implications of new technology use, the potential for the gap between those low-income countries who are not yet able to develop a national AI infrastructure and those who are becoming adept at it, to widen global inequalities, is a major cause for ethical concern (Nordling, 2019).

## EXPLORING ANTHROPOLOGICAL VIRTUE AND ETHICS WITHIN AI USE

Criticisms that organisations harnessing ethical praxis in relation to environmental justice within the context of AI usage are not unfounded in terms of the potential processes and outcomes being adopted at the front line of industrial, commercial, and academic institutions (Hickman & Petrin, 2021). What is necessary for address is the potential difference between the ideology of ethical responsibility and its practical implementation in work-based practice. One major concern is that if ethics is commodified and packaged as something tokenistically delivered by educators in abstraction from progressive technological emergence, that the tripartite relationship between technological advancement, power and social justice can effectively become meaningless and in some way diminished in terms of pragmatic impact and proactive conscientiousness (Burrell & Fourcade, 2021). Where knowledge comes from in terms of ongoing research of AI development and how AI ethics ought to be shaped therefore, is something stymied by standardized learning packages which are delivered (Birhane, 2021; Schiff 2021). One outcome of this is the potential debate that rather than commodifying ethics and permitting them to exist in abstraction from the use of AI, is that as an anthropological virtue, they ought to be an embedded part of professional practice and intrinsic epistemic positionality, so that tacit assumption and concerns about deviation from acceptable practice can be identified as part of an ongoing process. (Avnoon, Kotliar & Rivnai-Bahir, 2023).

## CONCLUSION

Amidst the communities of practice that constitute expertise in AI on a personal and professional level, ethics has not been an historical element of operational praxis, which is attributable to relative need, rather than the need for proactive choice. Within the 21st Century progressive development of AI, though, ethics has become central to the capacity of the discipline of computing and technology to ensure social justice and a temporal regard for the future of humanity with AI playing an ever-increasing role across so many disciplines and societal infrastructures. This chapter has considered the emergent role of AI across societies, from commerce, industry, and education to spirituality and existentialism. Whilst it is impossible to conclusively predict what the future of AI across these infrastructures will be, what can be accounted for is human responsibility for the scope of its potential reach and impact. Education of existing and forthcoming generations of society will be instrumental in ensuring that the progressive implementation of AI across societies is subject to rigorous regulation and stringent regard for accountability and responsibility of use. Whilst there will always remain the opportunity for rogue non-state jeopardy in terms of AI misuse, the core principles of ethical AI use will be shaped and adapted to the sociocultural perspectives of members states for whom the ethics of AI is a priority.

The hyperbole of whether AI will autonomously and objectively override human decision making, remains a debate which both sensationalises and highlights the once unimaginable capacity of AI within the contexts of medicine, health, and the military and also the inevitable inequity that AI will exacerbate between the West and developing nations.

How regulatory frameworks can be designed as a mechanism of consensus building across global society and its representatives remains an ongoing challenge. What is clear however is that ethical praxis must become instrumental rather than a conceptual commodity, which can be packaged to accompany education in computing, engineering, and technology over forthcoming decades. This chapter has illustrated how the progressive development from the use of AI in the manufacturing and industrial centres across the world represents an important opportunity to harness potential predictable complications and implications of AI use. To embed competence and AI literacy necessitates the active uptake of educational contexts to integrate AI literacy alongside core signature pedagogies such as Mathematics and English Language. The very possibility that AI can develop to have superhuman intelligence and far outpace humanity's intellect means embedding ethical contingency within all AI systems is imperative. When potential knowledge of the future remains entirely predictive and inferential it is these degrees of ethical contingency, which provide a haven for transparency and accountability, which ought to frame AI moral and ethical praxis.

## REFERENCES

Alter, S. (2022). Understanding artificial intelligence in the context of usage: Contributions and smartness of algorithmic capabilities in work systems. *International Journal of Information Management*, *67*, 102392. doi:10.1016/j.ijinfomgt.2021.102392

Alvero, R. (2023). ChatGPT: Rumors of human providers' demise have been greatly exaggerated. *Fertility and Sterility*, *119*(6), 930–931. doi:10.1016/j.fertnstert.2023.03.010 PMID:36921837

Amoroso, D., & Tamburrini, G. (2020). Autonomous weapons systems and meaningful human control: Ethical and legal issues. *Current Robotics Reports*, *1*(4), 187–194. doi:10.1007/s43154-020-00024-3

Araujo, T., Helberger, N., Kruikemeier, S., & De Vreese, C. H. (2020). In AI we trust? Perceptions about automated decision-making by artificial intelligence. *AI & Society*, *35*(3), 611–623. doi:10.1007/s00146-019-00931-w

Armstrong, S., Bostrom, N., & Shulman, C. (2016). Racing to the precipice: A model of artificial intelligence development. *AI & Society*, *31*(2), 201–206. doi:10.1007/s00146-015-0590-y

Avnoon, N., Kotliar, D. M., & Rivnai-Bahir, S. (2023). Contextualizing the ethics of algorithms: A socio-professional approach. *New Media & Society*, 14614448221145728. doi:10.1177/14614448221145728

Bareis, J., & Katzenbach, C. (2022). Talking AI into being: The narratives and imaginaries of national AI strategies and their performative politics. *Science, Technology & Human Values*, *47*(5), 855–881. doi:10.1177/01622439211030007

Benbya, H., Davenport, T. H., & Pachidi, S. (2020). Artificial intelligence in organizations: Current state and future opportunities. *MIS Quarterly Executive*, *19*(4).

Birhane, A. (2021). The impossibility of automating ambiguity. *Artificial Life*, *27*(1), 44–61. doi:10.1162/artl_a_00336 PMID:34529757

Burrell, J., & Fourcade, M. (2021). The society of algorithms. *Annual Review of Sociology*, *47*(1), 213–237. doi:10.1146/annurev-soc-090820-020800

Chen, J., Sun, J., & Wang, G. (2022). From unmanned systems to autonomous intelligent systems. *Engineering (Beijing)*, *12*, 16–19. doi:10.1016/j.eng.2021.10.007

Cheng, L., Varshney, K. R., & Liu, H. (2021). Socially responsible AI algorithms: Issues, purposes, and challenges. *Journal of Artificial Intelligence Research*, *71*, 1137–1181. doi:10.1613/jair.1.12814

Colaner, N. (2022). Is explainable artificial intelligence intrinsically valuable? *AI & Society*, 1–8.

Cools, H., Van Gorp, B., & Opgenhaffen, M. (2022). Where exactly between utopia and dystopia? A framing analysis of AI and automation in US newspapers. *Journalism*, •••, 14648849221122647.

Dauvergne, P. (2021). The globalization of artificial intelligence: Consequences for the politics of environmentalism. *Globalizations*, *18*(2), 285–299. doi:10.1080/14747731.2020.1785670

De Angelis, L., Baglivo, F., Arzilli, G., Privitera, G. P., Ferragina, P., Tozzi, A. E., & Rizzo, C. (2023). ChatGPT and the rise of large language models: The new AI-driven infodemic threat in public health. *Frontiers in Public Health*, *11*, 1567. doi:10.3389/fpubh.2023.1166120 PMID:37181697

De Bruijn, H., Warnier, M., & Janssen, M. (2022). The perils and pitfalls of explainable AI: Strategies for explaining algorithmic decision-making. *Government Information Quarterly*, *39*(2), 101666. doi:10.1016/j.giq.2021.101666

Döring, H. (2019). Public perceptions of the proper role of the state. In *The State in Western Europe Retreat or Redefinition?* (pp. 12–31). Routledge. doi:10.4324/9781315037479-2

Dwivedi, Y. K., Hughes, L., Ismagilova, E., Aarts, G., Coombs, C., Crick, T., Duan, Y., Dwivedi, R., Edwards, J., Eirug, A., Galanos, V., Ilavarasan, P. V., Janssen, M., Jones, P., Kar, A. K., Kizgin, H., Kronemann, B., Lal, B., Lucini, B., & Williams, M. D. (2021). Artificial Intelligence (AI): Multidisciplinary perspectives on emerging challenges, opportunities, and agenda for research, practice and policy. *International Journal of Information Management*, *57*, 101994. doi:10.1016/j.ijinfomgt.2019.08.002

Elliott, A. (2019). *The culture of AI: Everyday life and the digital revolution*. Routledge. doi:10.4324/9781315387185

Farrow, E. (2021). Mindset matters: How mindset affects the ability of staff to anticipate and adapt to Artificial Intelligence (AI) future scenarios in organisational settings. *AI & Society*, *36*(3), 895–909. doi:10.1007/s00146-020-01101-z PMID:33223620

Federspiel, F., Mitchell, R., Asokan, A., Umana, C., & McCoy, D. (2023). Threats by artificial intelligence to human health and human existence. *BMJ Global Health*, *8*(5), e010435. doi:10.1136/bmjgh-2022-010435 PMID:37160371

Finkel, M., & Krämer, N. C. (2022). Humanoid robots–artificial. Human-like. Credible? Empirical comparisons of source credibility attributions between humans, humanoid robots, and non-human-like devices. *International Journal of Social Robotics*, *14*(6), 1397–1411. doi:10.1007/s12369-022-00879-w

Floridi, L., & Cowls, J. (2022). A unified framework of five principles for AI in society. *Machine learning and the city: Applications in architecture and urban design*, 535-545

Fontes, C., Hohma, E., Corrigan, C. C., & Lütge, C. (2022). AI-powered public surveillance systems: Why we (might) need them and how we want them. *Technology in Society*, *71*, 102137. doi:10.1016/j.techsoc.2022.102137

Ganapini, M. B., Campbell, M., Fabiano, F., Horesh, L., Lenchner, J., Loreggia, A., & Venable, K. B. (2023). Thinking fast and slow in AI: The role of metacognition. In *Machine Learning, Optimization, and Data Science: 8th International Conference, LOD 2022,* (pp. 502-509). Cham: Springer Nature Switzerland.

Goralski, M. A., & Tan, T. K. (2020). Artificial intelligence and sustainable development. *International Journal of Management Education*, *18*(1), 100330. doi:10.1016/j.ijme.2019.100330

Haluza, D., & Jungwirth, D. (2023). Artificial Intelligence and Ten Societal Megatrends: An Exploratory Study Using GPT-3. *Systems*, *11*(3), 120. doi:10.3390/systems11030120

Harper, R. H. (2021). AI: The social future of intelligence. In *Routledge Handbook of Social Futures* (pp. 52–58). Routledge. doi:10.4324/9780429440717-4

Hatherley, J. J. (2020). Limits of trust in medical AI. *Journal of Medical Ethics*, *46*(7), 478–481. doi:10.1136/medethics-2019-105935 PMID:32220870

Haug, C. J., & Drazen, J. M. (2023). Artificial intelligence and machine learning in clinical medicine, 2023. *The New England Journal of Medicine*, *388*(13), 1201–1208. doi:10.1056/NEJMra2302038 PMID:36988595

He, Q., Zheng, H., Ma, X., Wang, L., Kong, H., & Zhu, Z. (2022). Artificial intelligence application in a renewable energy-driven desalination system: A critical review. *Energy and AI*, *7*, 100123. doi:10.1016/j.egyai.2021.100123

Helm, J. M., Swiergosz, A. M., Haeberle, H. S., Karnuta, J. M., Schaffer, J. L., Krebs, V. E., Spitzer, A. I., & Ramkumar, P. N. (2020). Machine learning and artificial intelligence: Definitions, applications, and future directions. *Current Reviews in Musculoskeletal Medicine*, *13*(1), 69–76. doi:10.1007/s12178-020-09600-8 PMID:31983042

Henin, C., & Le Métayer, D. (2021). Beyond explainability: Justifiability and contestability of algorithmic decision systems. *AI & Society*, 1–14.

Hickman, E., & Petrin, M. (2021). Trustworthy AI and corporate governance: The EU's ethics guidelines for trustworthy artificial intelligence from a company law perspective. *European Business Organization Law Review*, *22*(4), 593–625. doi:10.1007/s40804-021-00224-0

Holmes, W., Porayska-Pomsta, K., Holstein, K., Sutherland, E., Baker, T., Shum, S. B., ... Koedinger, K. R. (2021). Ethics of AI in education: Towards a community-wide framework. *International Journal of Artificial Intelligence in Education*, 1–23.

Humble, K. (2023). Artificial Intelligence, International Law and the Race for Killer Robots in Modern Warfare. In *Artificial Intelligence, Social Harms and Human Rights* (pp. 57–76). Springer International Publishing. doi:10.1007/978-3-031-19149-7_3

Hwang, H., & Park, M. H. (2020). The Threat of AI and Our Response: The AI Charter of Ethics in South Korea. *Asian Journal of Innovation & Policy*, *9*(1).

Igna, I., & Venturini, F. (2023). The determinants of AI innovation across European firms. *Research Policy*, *52*(2), 104661. doi:10.1016/j.respol.2022.104661

Islam, S. R., Russell, I., Eberle, W., & Dicheva, D. (2022). Instilling conscience about bias and fairness in automated decisions. *Journal of Computing Sciences in Colleges*, *37*(8), 22–31.

Khogali, H. O., & Mekid, S. (2023). The blended future of automation and AI: Examining some long-term societal and ethical impact features. *Technology in Society*, *73*, 102232. doi:10.1016/j.techsoc.2023.102232

Kile, F. (2013). Artificial intelligence and society: A furtive transformation. *AI & Society*, *28*(1), 107–115. doi:10.1007/s00146-012-0396-0

Larson, D. B., Harvey, H., Rubin, D. L., Irani, N., Justin, R. T., & Langlotz, C. P. (2021). Regulatory frameworks for development and evaluation of artificial intelligence–based diagnostic imaging algorithms: Summary and recommendations. *Journal of the American College of Radiology, 18*(3), 413–424. doi:10.1016/j.jacr.2020.09.060 PMID:33096088

Liu, H., Wang, Y., Fan, W., Liu, X., Li, Y., Jain, S., Liu, Y., Jain, A., & Tang, J. (2022). Trustworthy AI: A computational perspective. *ACM Transactions on Intelligent Systems and Technology, 14*(1), 1–59. doi:10.1145/3546872

Madhav, A. S., & Tyagi, A. K. (2022). The world with future technologies (Post-COVID-19): open issues, challenges, and the road ahead. *Intelligent Interactive Multimedia Systems for e-Healthcare Applications*, 411-452.

Makridakis, S. (2017). The forthcoming Artificial Intelligence (AI) revolution: Its impact on society and firms. *Futures, 90*, 46–60. doi:10.1016/j.futures.2017.03.006

Mantello, P., Ho, M. T., Nguyen, M. H., & Vuong, Q. H. (2023). Bosses without a heart: Socio-demographic and cross-cultural determinants of attitude toward Emotional AI in the workplace. *AI & Society, 38*(1), 97–119. doi:10.1007/s00146-021-01290-1 PMID:34776651

Masakowski, Y. R. (2020). Artificial intelligence and the future global security environment. In *Artificial Intelligence and Global Security* (pp. 1–34). Emerald Publishing Limited. doi:10.1108/978-1-78973-811-720201001

McKendrick, K. (2019). Artificial intelligence prediction and counterterrorism. London: The Royal Institute of International Affairs-Chatham House.

Mele, C., & Russo-Spena, T. (2023). Artificial Intelligence in Services. Elgar Encyclopedia of Services, 356. Elgar.

Mikalef, P., Conboy, K., Lundström, J. E., & Popovič, A. (2022). Thinking responsibly about responsible AI and 'the dark side'of AI. *European Journal of Information Systems, 31*(3), 257–268. doi:10.1080/0960085X.2022.2026621

Miller, G. J. (2022, March). Artificial Intelligence Project Success Factors—Beyond the Ethical Principles. In *Information Technology for Management: Business and Social Issues: 16th Conference, ISM 2021, and FedCSIS-AIST 2021 Track, Held as Part of FedCSIS 2021,* (pp. 65-96). Cham: Springer International Publishing.

Mirbabaie, M., Brünker, F., Möllmann, N. R., & Stieglitz, S. (2022). The rise of artificial intelligence–understanding the AI identity threat at the workplace. *Electronic Markets, 32*(1), 1–27. doi:10.1007/s12525-021-00496-x

Nader, K., Toprac, P., Scott, S., & Baker, S. (2022). Public understanding of artificial intelligence through entertainment media. *AI & Society*, 1–14. doi:10.1007/s00146-022-01427-w PMID:35400854

Nichols, R. (2022). The cultural evolution of Chinese morality, and the essential value of multi-disciplinary research in understanding it. In The Routledge International Handbook of Morality, Cognition, and Emotion in China (pp. 19-37). Routledge. doi:10.4324/9781003281566-3

Novelli, C., Taddeo, M., & Floridi, L. (2023). Accountability in artificial intelligence: What it is and how it works. *AI & Society*, 1–12. doi:10.1007/s00146-023-01635-y

Owe, A., & Baum, S. D. (2021). Moral consideration of nonhumans in the ethics of artificial intelligence. *AI and Ethics*, *1*(4), 517–528. doi:10.1007/s43681-021-00065-0

Palomares, I., Martínez-Cámara, E., Montes, R., García-Moral, P., Chiachio, M., Chiachio, J., Alonso, S., Melero, F. J., Molina, D., Fernández, B., Moral, C., Marchena, R., de Vargas, J. P., & Herrera, F. (2021). A panoramic view and swot analysis of artificial intelligence for achieving the sustainable development goals by 2030: Progress and prospects. *Applied Intelligence*, *51*(9), 6497–6527. doi:10.1007/s10489-021-02264-y PMID:34764606

Peeters, M. M., van Diggelen, J., Van Den Bosch, K., Bronkhorst, A., Neerincx, M. A., Schraagen, J. M., & Raaijmakers, S. (2021). Hybrid collective intelligence in a human–AI society. *AI & Society*, *36*(1), 217–238. doi:10.1007/s00146-020-01005-y

Pereira, L. M. (2019). Should I kill or rather not? *AI & Society*, *34*(4), 939–943. doi:10.1007/s00146-018-0850-8

Pizzi, M., Romanoff, M., & Engelhardt, T. (2020). AI for humanitarian action: Human rights and ethics. *International Review of the Red Cross*, *102*(913), 145–180. doi:10.1017/S1816383121000011

Prem, E. (2023). From ethical AI frameworks to tools: A review of approaches. *AI and Ethics*, *3*(3), 1–18. doi:10.1007/s43681-023-00258-9

Qiao-Franco, G., & Bode, I. (2023). Weaponised artificial intelligence and Chinese practices of human–machine interaction. *The Chinese Journal of International Politics*, *16*(1), 106–128. doi:10.1093/cjip/poac024

Salahuddin, Z., Woodruff, H. C., Chatterjee, A., & Lambin, P. (2022). Transparency of deep neural networks for medical image analysis: A review of interpretability methods. *Computers in Biology and Medicine*, *140*, 105111. doi:10.1016/j.compbiomed.2021.105111 PMID:34891095

Sauer, F. (2021). Lethal autonomous weapons systems. In *The Routledge Social Science Handbook of AI* (pp. 237–250). Routledge. doi:10.4324/9780429198533-17

Schiff, D. (2021). Out of the laboratory and into the classroom: The future of artificial intelligence in education. *AI & Society*, *36*(1), 331–348. doi:10.1007/s00146-020-01033-8 PMID:32836908

Shaw, J., Rudzicz, F., Jamieson, T., & Goldfarb, A. (2019). Artificial intelligence and the implementation challenge. *Journal of Medical Internet Research*, *21*(7), e13659. doi:10.2196/13659 PMID:31293245

Shneiderman, B. (2020). Bridging the gap between ethics and practice: Guidelines for reliable, safe, and trustworthy human-centered AI systems. [TiiS]. *ACM Transactions on Interactive Intelligent Systems*, *10*(4), 1–31. doi:10.1145/3419764

Smith, H. (2021). Clinical AI: Opacity, accountability, responsibility and liability. *AI & Society*, *36*(2), 535–545. doi:10.1007/s00146-020-01019-6

Sweeney, P. (2022). Trusting social robots. *AI and Ethics*, 1–8. PMID:35634257

Uunona, G. N., & Goosen, L. (2023). Leveraging Ethical Standards in Artificial Intelligence Technologies: A Guideline for Responsible Teaching and Learning Applications. In Handbook of Research on Instructional Technologies in Health Education and Allied Disciplines (pp. 310-330). IGI Global. doi:10.4018/978-1-6684-7164-7.ch014

Visvizi, A. (2021). Artificial intelligence (AI): Explaining, querying, demystifying. *Artificial Intelligence and Its Contexts: Security, Business and Governance*, 13-26.

Werthner, H., Stanger, A., Schiaffonati, V., Knees, P., Hardman, L., & Ghezzi, C. (2023). Digital Humanism: The Time Is Now. *Computer*, *56*(1), 138–142. doi:10.1109/MC.2022.3219528

Yamin, M. M., Ullah, M., Ullah, H., & Katt, B. (2021). Weaponized AI for cyber attacks. *Journal of Information Security and Applications*, *57*, 102722. doi:10.1016/j.jisa.2020.102722

Young, A. T., Amara, D., Bhattacharya, A., & Wei, M. L. (2021). Patient and general public attitudes towards clinical artificial intelligence: A mixed methods systematic review. *The Lancet. Digital Health*, *3*(9), e599–e611. doi:10.1016/S2589-7500(21)00132-1 PMID:34446266

Zajda, J., & Vissing, Y. (Eds.). (2022). *Discourses of globalisation, ideology, and human rights* (Vol. 14). Springer. doi:10.1007/978-3-030-90590-3

Zeng, J. (2020). Artificial intelligence and China's authoritarian governance. *International Affairs*, *96*(6), 1441–1459. doi:10.1093/ia/iiaa172

Zhang, B., Zhu, J., & Su, H. (2023). Toward the third generation artificial intelligence. *Science China. Information Sciences*, *66*(2), 1–19. doi:10.1007/s11432-021-3449-x

Zhang, Y., Wu, M., Tian, G. Y., Zhang, G., & Lu, J. (2021). Ethics and privacy of artificial intelligence: Understandings from bibliometrics. *Knowledge-Based Systems*, *222*, 106994. doi:10.1016/j.knosys.2021.106994

Zuiderwijk, A., Chen, Y. C., & Salem, F. (2021). Implications of the use of artificial intelligence in public governance: A systematic literature review and a research agenda. *Government Information Quarterly*, *38*(3), 101577. doi:10.1016/j.giq.2021.101577

## ADDITIONAL READING

Aung, Y. Y., Wong, D. C., & Ting, D. S. (2021). The promise of artificial intelligence: A review of the opportunities and challenges of artificial intelligence in healthcare. *British Medical Bulletin*, *139*(1), 4–15. doi:10.1093/bmb/ldab016 PMID:34405854

Bistron, M., & Piotrowski, Z. (2021). Artificial intelligence applications in military systems and their influence on sense of security of citizens. *Electronics (Basel)*, *10*(7), 871. doi:10.3390/electronics10070871

Devagiri, J. S., Paheding, S., Niyaz, Q., Yang, X., & Smith, S. (2022). Augmented Reality and Artificial Intelligence in industry: Trends, tools, and future challenges. *Expert Systems with Applications*, *207*, 118002. doi:10.1016/j.eswa.2022.118002

Haenlein, M., & Kaplan, A. (2019). A brief history of artificial intelligence: On the past, present, and future of artificial intelligence. *California Management Review*, *61*(4), 5–14. doi:10.1177/0008125619864925

Hagendorff, T. (2021). Forbidden knowledge in machine learning reflections on the limits of research and publication. *AI & Society*, *36*(3), 767–781. doi:10.1007/s00146-020-01045-4

Kaminski, M. E. (2023). Regulating the Risks of AI. *Forthcoming. Boston University Law Review. Boston University. School of Law*, 103.

Kaur, D., Uslu, S., Rittichier, K. J., & Durresi, A. (2022). Trustworthy artificial intelligence: A review. *ACM Computing Surveys*, *55*(2), 1–38. doi:10.1145/3491209

Kazansky, B., & Milan, S. (2021). "Bodies not templates": Contesting dominant algorithmic imaginaries. *New Media & Society*, *23*(2), 363–381. doi:10.1177/1461444820929316

Maibaum, A., Bischof, A., Hergesell, J., & Lipp, B. (2022). A critique of robotics in health care. *AI & Society*, *37*(2), 1–11. doi:10.1007/s00146-021-01206-z

Medaglia, R., Gil-Garcia, J. R., & Pardo, T. A. (2023). Artificial intelligence in government: Taking stock and moving forward. *Social Science Computer Review*, *41*(1), 123–140. doi:10.1177/08944393211034087

## KEY TERMS AND DEFINITIONS

**Agency:** Is the capacity for action or intervention producing a particular effect.

**AI Safety:** AI safety pertains to the interdisciplinary field which prevents the misuse, accidental or other consequences which could be the resultant outcome of an AI system.

**Algorithm:** An algorithm is a process or set of rules to be followed in decision-making or other problem-solving operations, especially by computing technology.

**Autonomous Weapons:** A classification of military systems which has the independent ability to search for, identify and engage strategic targets based upon pre-programmed constraints and restrictions.

**Epistemic Bias:** The lens of subjective interpretation which influences systematic research practice due to failing to acknowledge or detail the ideals of human impartiality and value-freedom which may potentially be influencing it.

**Existentialism:** The philosophy of the nature of human existence as determined by capacity and capability for free will and free choice.

**Hacking:** The gaining of illegal, unlawful, or unofficially unauthorised access to data in within the context of computing and technology.

**Reliability:** The extent to which a research instrument can repeatedly provide the same results in temporally separated incidences of measurement.

**Sentient/Sentience:** Sentience is the capacity of a being to experience feelings and sensations.

**Validity:** Is the state of being officially true or legally acceptable.

# Chapter 8
# Harnessing the Power of Artificial Intelligence in Law Enforcement:
## A Comprehensive Review of Opportunities and Ethical Challenges

**Akash Bag**
https://orcid.org/0000-0001-8820-171X
*Amity University, India*

**Souvik Roy**
*Adamas University, India*

**Ashutosh Pandey**
https://orcid.org/0009-0002-8273-6443
*Adamas University, India*

## ABSTRACT

*Law enforcement is joining the fast-growing artificial intelligence (AI) research field. The chapter tries to fix that. This chapter utilized a "systematic literature review." The authors gathered research papers on using algorithms and AI in police work. This was done with Scopus, a fancy academic database. They searched for papers on "law enforcement," "policing," "crime prevention," "crime reduction," and "surveillance." Combine these terms with "algorithm" or "artificial intelligence." They found that AI has great potential to aid law enforcement. It can recognize faces, forecast crimes, and track people. These AI tools usually analyze photos, behavior, language, or a combination. However, there are significant "but" ethical issues that exist. AI can cause unjust treatment, confusion about responsibility, oversurveillance, and privacy invasion. AI's benefits and cool abilities are often highlighted over its drawbacks. Another observation is that writings on the same topics agree on what AI can achieve, its potential, and what we should explore next.*

DOI: 10.4018/979-8-3693-1565-1.ch008

## INTRODUCTION

In the 1940s, the first computer in the modern sense was invented, and it wasn't long until the young academic Alan Turing began to think about the future potential of the computing machine (Anyoha, 2017). Despite the very limited performance of the computer at the time, he wondered if computers in the future would not have to think in a way that lived up to or surpassed humans. Humans use information and reason to solve problems and make decisions, so why shouldn't machines be able to do the same? Alan died in 1956, but two years later, Marvin Minsky picked up the baton and co-hosted with John McCarthy the Dartmouth Summer Research Project on Artificial Intelligence. They invited well-respected researchers from various fields for an open discussion about artificial intelligence (AI). Henceforth, artificial intelligence became an accepted term. Research continued, and the field developed, but computers' storage capacity and computing power were limiting factors, and interest in AI cooled (Anyoha, 2017). During the 80s, AI saw another boost due to increasing research funding, and combined with the rapid development of performance in the 90s, it could be concluded that AI was here to stay (Kaynak, 2021).

Today, we live in a world where AI has found its way into more and more areas of society and completely changed how we work. Politics, education, healthcare, business, and law are just some societal areas that have undergone major changes. An area that has become highly relevant recently is law enforcement, where traditional routines still dominate (Kaynak, 2021). The majority of the police work that takes place is only done after the crime has occurred, which means that large resources are required in the form of personnel, money, and time. In addition, the possibilities of solving the crime decrease as time passes. With the help of AI, the police can perform more predictive police work, i.e., to predict where, how, and when crimes will be committed to be better equipped for calling out or preventing crimes (King et al., 2020). The hope is that with AI's help, one can analyze material from video cameras in real time. Another prospect is being able to read social media to identify patterns and behaviors that may be indications that crime is about to happen. Successful implementation of this could save many lives Freeman (2020), save enormous resources, and minimize the risk of human error that can occur in today's criminal investigation (Zufferey et al., 2022). These are just some of the areas where it is suggested that AI can bring about major improvements.

Herrmann (2023) believes in its AI report that the use of AI in the judicial system must take place in such a way that it considers the general principles of human rights, democracy, justice, and the prevailing legislation. To succeed with this consideration, authorities within the legal system must work to achieve the requirements of fairness, accountability, transparency, and explain ability. In recent years, these requirements have been developed based on consensus within the AI community about what algorithms are considered to possess to instill trust. Fairness means that algorithmic decisions should not show discriminatory or unfair tendencies. The requirement calls for all AI systems to be scrutinized to ensure they comply with the right to non-discrimination. Accountability means there must be clear regulations for who is responsible for an autonomous system's decision. Transparency means that there must be clear answers about the goal of using AI in a certain context and what parts the AI consists of. It could, for example, be about which data is used. The last requirement, explain ability, is closely related to transparency but is more focused on the person affected by a certain decision being able to understand the algorithmic decision in non-technical terms (Herrmann, 2023).

In May 2017, the Indian government decided on India's digitization strategy and defined the goal that India should be the best in the world in using the possibilities of digitization (Chopra, 2017). The strategy includes five sub-goals relating to competence, security, innovation, management, and infrastructure. In the wake of the digitization strategy, the document National Direction for Artificial Intelligence was also published, with the objective that "*India shall be a leader in taking advantage of the opportunities that the use of AI can provide, to strengthen both India's welfare and Indian competitiveness*" (Chopra, 2017). The reason for the emergence of the document was motivated by the fact that AI is a rapidly developing digital field. The office also believes that for India to be the best in the world at using the possibilities of digitalization, a clear direction is required for future priorities. The Indian government office highlights AI's potential for increased economic growth and believes that the technology can solve various environmental and social challenges. The document addresses the lack of AI competence in India and encourages stronger incorporation of AI in education, research, innovation, and use, as well as frameworks and infrastructure, to take full advantage of AI's benefits. In addition, test projects are recommended for developing AI applications in the public and private sectors, as well as partnerships and collaboration with other countries regarding their use, especially within the EU.

The government assessed that India needs strong basic research and applied research in AI to ensure the supply of knowledge and skills in the field (Chopra, 2017). The report is linked to a research project in such a spirit. Assessing the risks and opportunities with AI in law enforcement is an under-researched area. Despite the opportunities created with the help of AI, the rapid development is surrounded by potential risks. It has been found, among other things, that some groups have been disadvantaged in establishing such technology (Hong & Williams, 2019). In addition, the development of AI applications by the state and companies may pose a risk of curtailing human rights if security regulations are not established (Rodrigues, 2020). After an initial scan of databases of research articles, the available literature on algorithms and AI in law enforcement appears to be quite rich. However, the literature that summarizes the risks and opportunities with these methods is limited. In that it is a growing area which, in step with technology's development, will reasonably become increasingly relevant, it is interesting to compile the risks and opportunities that today's literature highlights. This will benefit practitioners and decision-makers to conduct research and make decisions more grounded in today's research. In light of this, it was chosen to carry out a systematic literature review that accounts for current research about opportunities and risks within crime-fighting work.

The report deals with artificial intelligence, a concept still lacking a clear definition or generally accepted delimitation. Consequently, the perception of what constitutes AI has changed over time. One formulation is that AI machines are programmed to think like humans and imitate their actions (Frankenfield, 2021). Even though the description tells what AI aims for and may mean in the future, the depiction is still some way from what AI means in a scientific sense today. In a scientific context, AI is described as the property of machines that imitate human intelligence, characterized by behaviors such as cognitive ability, memory, learning, and decision-making (Chen & Wong, 2018). A discrepancy exists in how AI is used in science, where complexity varies. Considering the diffuse dividing line of what AI is, this study will include research that refers to AI but also algorithms. In the report, algorithms and AI will be used interchangeably, and references to other literature use the same terms in the original research. The remaining part of the report has been structured as follows: the next part deals with the implementation of the method and an account of the extracted data. The results of the research are then presented, which is followed by a discussion of the results. This is synthesized and then presented in the form of a conclusion. In conclusion, a reflection is given on the work's connection to sustainability and ethics.

## METHODOLOGY

The following section presents how the search was carried out, how the articles were filtered, and a compilation of the included articles.

### Implementation of Systematic Literature Review

The study has been carried out in seven steps based on the structure and approach by Angela Carrera-Rivera and others (Carrera-Rivera Et Al., 2022). As a first step, the scope of the study has been defined to clarify which fields and types of studies are relevant to the study. The area was defined as "*risks and opportunities with algorithms and artificial intelligence in law enforcement work*" based on the purpose. As a second step, the conditions for the search process were specified. The searches were chosen to be done in the database Scopus, with the words "*law enforcement*," "*policing*," "*crime prevention*," "*crime reduction*," and "*surveillance*" in combination with "*algorithm*" or "*artificial intelligence*." The words were searched for in abstracts, titles, and keywords in articles published from 2018–2022. The period was chosen based on the age of the research area. This is because of the articles that came up with our keywords from 2010. More than half were published after 2018, indicating that the area has become increasingly relevant in recent years. The field is evolving rapidly, and the articles within the selected range are highly likely to be relevant. By using articles published before our chosen time interval, the risk of including knowledge that is now outdated increases. One of the study's co-authors searched independently to ensure the process was conducted consistently. The searches were conducted during the period April 8–25, 2022.

A quality assessment was carried out as a third step, where the articles obtained from the search were assessed based on their relevance. With the justification that the research area is new and emerging, the sources were not assessed based on the number of citations since new articles naturally have few citations. On the other hand, the relevance of the sources was assessed in two stages, first through an assessment of the headings, where the headings that had no connection to the defined subject area were sorted out. In the second round, the abstracts of the articles were reviewed where they were deemed relevant only if the article was considered to address risks or opportunities with either AI or algorithms in the field of law enforcement. First, one of the study's co-authors reviewed headlines and articles. Then, the study's co-authors conducted another review independent of the previous one. This is to minimize the margin of error when selecting articles. The fourth and fifth steps involved documentation of the study's articles. The number of articles included and excluded after the relevance assessment was documented in a flow chart.

Furthermore, a form was drawn up with basic information about the selected studies, where the author, year of publication, journal, and type of study were documented. The sixth step is based on carrying out the core point of the research: the analysis. A narrative analysis was undertaken as the area researched is grounded in the use of words rather than findings in the form of numbers. A narrative analysis means the reader is guided through a story of the various findings from the studies reviewed. Compared to a meta-analysis, a narrative analysis is suitable when the research findings cannot be quantified. The narrative was chosen to be conveyed through thematic analysis, which means that the research findings are organized to identify consistent themes and patterns (Mayring, 2000). In particular, the analysis considers whether there is a summary agreement or ambiguity about the findings, followed by a discussion about what this implies. As the seventh and final step, the conclusions that can be extracted from the analysis

*Figure 1. The selection process of the literature search*

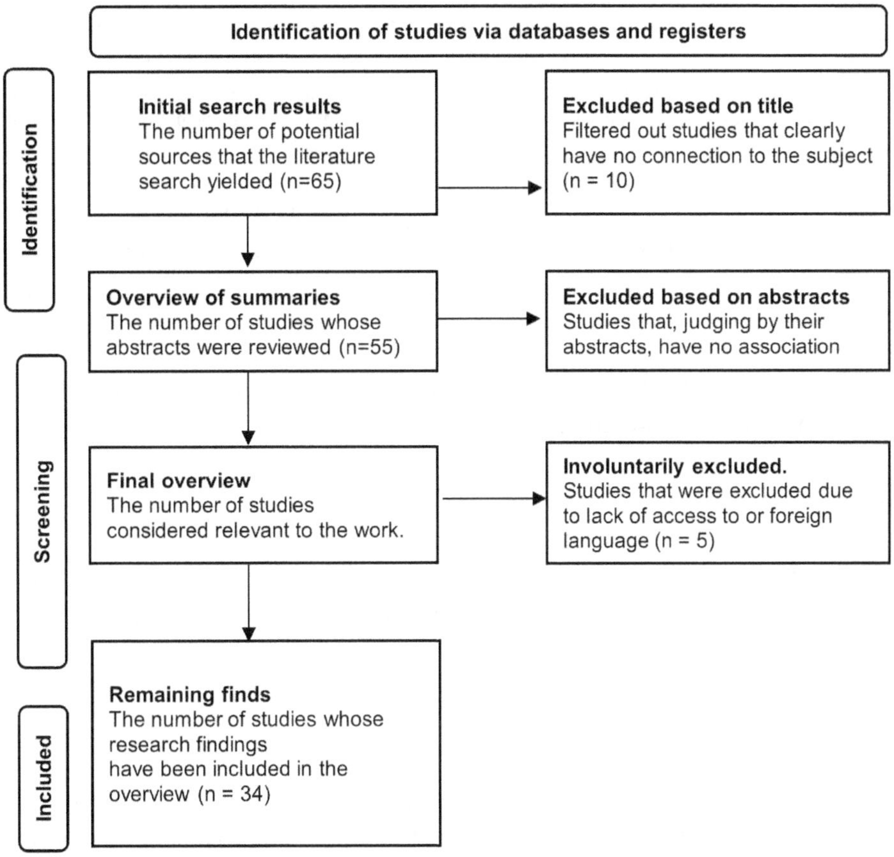


are presented coherently and unambiguously. The conclusion aims to give a clear picture of the state of the research to guide practitioners where future studies should take place.

## Selection of Articles

The literature search yielded a result of 65 articles. After carefully reviewing titles, abstracts, and content availability, 34 articles were included in the study results.

Comment: Flow chart showing which studies have been sorted out of the literature study for a given reason and which have been included.

## RESULTS

The work resulted in many opportunities and risks associated with using AI in law enforcement. To best represent the possibilities in an understandable way for the reader, they have been divided into three different areas. Some algorithms are used to predict and prevent crime from being committed, and these are reported in the Before Crime section. Others are made to assist in and detect crimes in progress, detailed in the Under Crimes section. Finally, algorithms can also serve in criminal investigations long after the

*Table 1. Compilation of identified opportunities and risks*

| | | |
|---|---|---|
| Possibilities | Before offense | Monitoring of high-risk individuals Identification of objects and movements Language analysis for behavioral forecasting Predictions and forecasts |
| | During offenses | Group behavior analysis and long-term monitoring from above Real-time traffic information and license plate identification Permanent monitoring of marine protected areas Real-time situational analysis |
| | After breakage | Drone as extended arm Intelligent identification Lie detection during interrogation Hard drive scanning Patterns and mapping |
| Risks | | Discriminatory tendencies Lack of personal responsibility Mass surveillance Erosion of privacy and integrity |

crime has been committed, which falls under the heading After crime. Finally, the risks are presented that focus on the ethical aspects presented in the studied sources.

## Possibility

This section deals with the possibilities that the various articles address. The possibilities are presented in three different subsections based on the events surrounding a crime, namely the period before the crime, the period during, and after the crime.

### Before Crime

The articles identified four types of crime prediction and crime prevention approaches: Monitoring of high-risk individuals, Identification of objects and movements, Language analysis for behavioral forecasting, and Crime predictions and forecasts.

### Monitoring of High-Risk Individuals

Several articles see great potential in using algorithms to identify and monitor persons of interest. Physical markers such as faces and fingerprints can be recognized and shared between departments and regions to monitor individuals deemed to be crime-prone (Contardo et al., 2021; Pawlicka et al., 2021; M. Smith & Miller, 2022; Wyatt, 2021). A potentially more powerful trace is digital footprints. They enable law enforcement to identify potential criminals who lack previous suspicion. For example, such tools can identify individuals who incite crime in planned demonstrations (Sánchez et al., 2022), but they can also detect people with psychological disorders dangerous to themselves and others (Bulgakova et al., 2019; Ionescu, Ghenescu, Răstoceanu, et al., 2020; Jindal & Sharma, 2018a).

### Identification of Objects and Movements

Another domain where computer image understanding comes into play is surveillance footage. With the help of computer vision, huge amounts of video material can be checked for suspicious actors (Ionescu, Ghenescu, Răstoceanu, et al., 2020; M. Smith & Miller, 2022), objects (Enriquez et al., 2019; Ionescu,

Ghenescu, Răstoceanu, et al., 2020), and actions (Rajapakshe et al., 2019). There are, for example, tools to detect and follow individuals in motion, but also tools that study the actions of groups (Ionescu, Ghenescu, Răstoceanu, et al., 2020; Rajapakshe et al., 2019). AI programs are also being developed that make it possible to detect weapons (Enriquez et al., 2019; Singh et al., 2021). Regardless of what they are designed to detect, they possess certain inherent characteristics sought in monitors. They are fast, tireless, and discreet (Wyatt, 2021).

## Language Analysis for Behavioral Forecasting

Several studies have shown that it is possible to use language analysis to predict both planned and unplanned actions before they happen (Ionescu et al., 2020; Jindal & Sharma, 2018; Ramirez et al., 2022). A popular data source is social media. Public communication and, in some cases, private calls are available there. Algorithms make it possible to analyze huge data sets (Behmer et al., 2019). The posts are fed into algorithms that roughly determine the sender's sentiment. This is done, among other things, by the algorithm identifying information that consists of combinations of names, places, organizations, and time. Users' behavior on the platform can be aggregated and compared against the activities of previous criminals. In this way, digital profiles can be constructed, which human investigators can study in more detail. These may be people suffering from severe mental illnesses (Jindal & Sharma, 2018), but the profiles can also detect groups planning atrocities (Sánchez et al., 2022). However, language analysis does not have to be limited to social media. The utility of wiretapping becomes much greater thanks to new software tools. It is now possible to connect voices to people, convert speech to text, and make out words from bad recordings (Ionescu et al., 2020). Should microphones not be available, it is even possible for computers to read lips from videos (Ionescu et al., 2020).

## Crime Predictions and Forecasts

The studies studied in the work provide several examples of methods to be used to predict the circumstances under which new crimes occur. It mentions how high-risk individuals are identified by compiling behavioral data from various data sources and analyzing this data collection (Freeman, 2020; Kaur Et Al., 2019; Smith & Miller, 2022; Zhang, 2021). With the help of algorithms trained through unsupervised learning, it is possible to note deviations from individuals' behavior patterns, at which point these can be classified into different groups based on how risky the deviant behavior, according to the algorithm, is judged to be (Jindal & Sharma, 2018b). It is also mentioned how AI adds a new dimension of accuracy in prediction; the technique is based on non-linear functions, making it possible to detect non-linear patterns (Khairuddin Et Al., 2020). However, improvements are required in the decision-making, processing, and handling of big data and a specific IT infrastructure to perform analyses when the amount of data becomes large (Bulgakova Et Al., 2019). If the infrastructure were to fall into place in the future, the police authorities are expected to have increased responsibility for handling these often private data sets (Sherer Et Al., 2018). Today, it is possible to analyze the abnormal behavior of individuals and groups to forecast whether there are risks of crime in an area by gathering information from conversations (Ionescu, Ghenescu, Rastoceanu, Et Al., 2020). However, only when the data has been analyzed can it be possible to draw adequate and sufficiently well-defined conclusions to predict where the next crime will occur (Toppireddy Et Al., 2018). In such processes, automated video surveillance becomes one of the cornerstones of predicting future crimes (Rajapakshe Et Al., 2019). A facilitating aspect in that matter is then based on the availability of data on previous crimes (Das & Das, 2019). With sufficient crime data, AI can improve police resource allocation and optimize response Chase Et Al (2017) or arrests (Jie Et Al., 2020).

## During Crime

From the articles, the following areas of use have been identified, linked to the use of AI and algorithms during the time a crime is committed: Group behavior analysis and long-term monitoring from above, Traffic information in real-time, and identification of license plates, Permanent monitoring of marine protected areas, and Situational analysis in real-time.

### Group Behavior Analysis and Long-Term Surveillance From Above

One application identified by Simpson (2021) is real-time crowd monitoring. With the help of AI, police can analyze mass behavior during peaceful and violent gatherings, where AI automatically identifies factors such as group density, speeds, and movement behavior and identifies the primary perpetrators of violence. An advantage that is highlighted with this is that it saves resources, as the police can deploy their personnel in more suitable places as a result of the better information base that the drone, in this case, has created. According to Wyatt (2021), the use of autonomous systems, such as drones, also enables monitoring for a longer period when the system largely acts on its own with limited participation from a human. The alternative is to deploy reconnaissance groups, but these cannot operate for as long, are exposed to personal risk, and are more difficult to keep hidden from counterintelligence. Drones with built-in analysis of recorded material with the help of AI are thus something that is being explored, but even less sophisticated drones that only collect material without their analysis are useful with the advent of more complex methods of video analysis with the help of AI. Therefore, The autonomous system can be used solely for information acquisition to hand over the material to other systems, which in turn carry out analyses on the large amount of data collected to find useful information (Wyatt, 2021). Even human error when reviewing video footage can be eliminated when drone surveillance is conducted with AI, according to Mac et al. (2020). Reviewing large amounts of video material to find what is needed for, for example, an investigation takes a long time, and mistakes often occur due to the monotonous and drawn-out work process. It also costs large resources in the form of human resources, which could be greatly reduced with the help of AI that filters the video material in real time without any mistakes.

### Real-Time Traffic Information and License Plate Identification

Another application area of drones highlighted is traffic monitoring (Chang et al., 2019; Du et al., 2018; Mac et al., 2020). With more and more cars in our cities, problems arise regarding how best to deal with congestion and traffic safety. Today, common surveillance cameras and infrared technology are used, among other things, to gather information about vehicle speed and traffic flow. The problem is that it is difficult to determine where these tools should be placed and that they are inflexible once placed. As a result, the necessary traffic information cannot always be obtained. A solution that Mac et al. (2020) highlight is the use of drones that, with the help of image sensors, can transmit traffic information in real time to a control station. The surveillance becomes more flexible and can cover a larger surveillance area than individual deployed cameras can.

A consequence of such systems is, for example, that information about a vehicle exceeding the speed limit is automatically received by the police, who can then act (Thoa Mac et al., 2020). Several articles highlight the possibility of deployed surveillance cameras being able to identify license plates on vehicles (Chang et al., 2019; Du et al., 2018). One possibility that is being explored is that the police should be able to give a surveillance system a registration number as input and then let the system track how the vehicle moves in real-time by using video from all cameras included in the system. By having the system predict where the vehicle is going, police can prepare an arrest in the appropriate area (Chang et al., 2019; Du et al., 2018).

## Permanent Monitoring of Marine Protected Areas

Another area of application of autonomous systems is in the marine environment (Molina-Molina et al., 2021; Rajamäki et al., 2018). Molina-Molina et al. (2021) suggest, among other things, how robots can be used to conduct permanent surveillance within marine protected areas by using AI to identify vessels that engage in suspicious activities. However, it is highlighted that it is difficult to ensure that the measured values are correct in marine environments and that the results should not be considered fully reliable. As a complement to these systems and other surveillance tools, such as helicopters, drones can also be used for information gathering (Rajamäki et al., 2018).

## Situation Analysis in Real-Time

Something that is being explored is how intelligent systems can analyze situations in real-time (Enriquez et al., 2019; Suralkar et al., 2020). Enríquez et al. (2019) explore how real-time counseling to potential victims can occur when surveillance cameras, with the help of AI, perceive potential threats on video, such as weapons. Users are advised to have an application installed on their phone that advises on how to act, depending on its position concerning the potential threat and other parameters, such as where a panicked flight occurred. This advice can be about which escape routes are most suitable or whether the person should lock themselves. The system is made possible by integrating surveillance footage with a subsystem that allows users' phones to share information of common interest with each other, also known as crowdsensing. According to the paper's authors, the challenge is to develop a method that can react quickly and correctly in an ever-changing scenario while considering multiple parameters. Suralkar et al. (2020) explore how drones with integrated AI should be able to put an event in a context. In real-time, this would mean that the drone can determine whether what is being observed is a crime in progress or a false positive result. This is in contrast to many other surveillance systems with AI, which can determine if a violent act is taking place but not if, for example, it is an innocent act due to the context (Suralkar et al., 2020).

## After Crime

From the literature, four areas have been identified where AI and algorithms create opportunities after the crime. They are Drones with extended arms, Intelligent identification of individuals and objects, lie detection during interrogations, Scanning of hard drives, and Patterns and mapping.

## Drone as an Extended Arm

According to the literature, drones have great potential to, in combination with artificial intelligence and machine learning, serve policing purposes (Contardo et el., 2021; Matthew et al., 2021; Simpson, 2021). The judiciary has recently experienced a growing problem when dealing with the dynamic nature of public gatherings are often characterized by. In addition to the fact that drones, in combination with artificial intelligence, make it possible to identify criminal activity in real-time, the application also creates valuable evidence in criminal investigations. By incorporating AI, they can independently interact with air and ground navigation systems, which plays a major role in the possibility of evidence collection even in difficult conditions (Simpson, 2021). The autonomous feature means that the drones can operate beyond lines of sight, which is a great advantage as the system can serve without direct human control (Matthew et al., 2021). In addition to documenting public gatherings, incorporating AI into video recording could mean identifying and locating criminals and assisting in rescue and search operations (Matthew et al., 2021). This is realized by taking advantage of the Internet of Things (IoT) as well as using a specific branch of artificial intelligence that is common for image analysis: convolutional neural networks (CNNs) (Contardo et al., 2021). The video analysis incorporated in drones is assumed to be of great value to countries suffering from a high presence of criminal networks (Matthew et al., 2021). Using drones and applying AI can also benefit the evaluation of police intervention in public gatherings (Simpson, 2021).

## Intelligent Identification of Individuals and Objects

Since the emergence of the camera, surveillance cameras have been an important policy tool to identify individuals who moved in a certain area around a critical time. One of the most common biometric techniques is face recognition, which has been an attractive area of research for as long as 40 years (Contardo et al., 2021). Facial recognition has the advantage that it can be used passively, which means that the person who wants to be identified does not need to be aware of the use. Previous techniques have been lacking in their effectiveness on images and video from surveillance cameras, where the conditions for the image are not ideal. AI and machine learning improve recognition technology, whereas image analysis technology based on CNN is good at recognizing faces even under difficult conditions (Contardo et al., 2021). In addition, algorithms are used to restore and improve the image quality of blurry images taken from video recordings (Sun, 2020). By analyzing cameras and networks, the technology can analyze the presence of people, which creates an opportunity to get an account of the individual's historical event pattern and behavior when searching for a person (Ionescu et al., 2020). Fingerprints are another biometric data that has been analyzed for a long time. The literature explains that traditional algorithms have performed well on fingerprints collected intentionally but performed worse on latent fingerprints, i.e., that the solution to the problem is expected to lie in developing the technology using CNNs. Other possibilities provided by AI are video analysis of license plates. By designing intelligent systems that connect data from different surveillance cameras, a search for a certain license plate number should give the police a notice when the license plate is detected by one of the cameras (Chang Et Al., 2019). Police can then get a good idea of where the suspect is headed and prepare to arrest at the appropriate location.

### Lie Detection During Interrogation

In criminal investigations, verbal information is an essential part of solving crimes. One difficulty with this type of evidence, whether it concerns testimony or interrogation of the criminal suspect, is determining the statements' integrity. The literature shows how AI-based speech analysis can assess whether the person being interrogated is lying or telling the truth (Ionescu, Ghenescu, Rastoceanu, et al., 2020; Pawlicka et al., 2021) A particular example from the literature is based on two modules linked together to detrmine the credibility (Ionescu et al., 2020). The first part is an analysis of emotional states, where data from image and audio recordings is analyzed using, among other things, deep neural networks and determines the state of mind of the interviewee. A deep neural network is a network of nodes whose structure is inspired by the human brain, which enables very efficient and fast machine learning in certain situations. The second part constitutes an analysis of the psychological state and involves real-time analysis of the heartbeat and breathing rate. The bodily metrics are only extracted from video analysis, meaning the lie detection can be done without the interviewee's awareness.

### Scan of Hard Drives

Lanagan & Choo (2021) discuss the possibility of scanning hard drives to find illegal material such as child pornography. The main area of use identified is at border controls. The article suggests looking for specific indications of child pornography, such as the hard drive containing mostly video files. This, in turn, can be the basis for a more comprehensive scan. The advantage of using AI is therefore considered to be that the hard drive can be searched efficiently, and that the personal integrity of its owner is not violated as much as if a human searched.

### Pattern and Mapping

To extract meaning from big data, the incorporation of artificial intelligence is required (Pawlicka et al., 2021). An example that is raised is how AI can solve serial crimes by discovering non-trivial connections. With the help of AI, it is also expected to become easier to reach criminal networks as the possibility of identifying individuals connected to criminal persons becomes easier (Sherer et al., 2018). It is pointed out, however, that associated individuals are not interested in themselves, but the individuals who should be studied are those connected to the criminal activity. Bulgakova et al. (2019) emphasize the importance of infrastructure to use big data in investigations.

Furthermore, Sherer et al. (2018) how judges and jurors will view the credibility of the findings that can be presented using big data as a risk. With the help of AI analysts, the work of human analysts can be streamlined, allowing humans to focus more on effective responses rather than investigation (Wyatt, 2021). AI can also potentially solve more mundane crimes, such as fraud and petty theft, more effectively than today (Pawlicka et al., 2021).

## Risks

From the literature, it was possible to distinguish four areas regarding risks associated with AI and algorithms in law enforcement. The risks identified were Discriminatory Tendencies, Lack of Personal Responsibility, Mass Surveillance, and Erosion of Privacy and Integrity.

## Discriminatory Tendencies

The literature highlights how using algorithms in policing risks perpetuating discrimination against different population groups, particularly based on ethnicity (Kitsos, 2020; Milivojevic, 2021). An example of this that Kitsos (2020) brings up is in predictive policing, where algorithms intended to coordinate police, resources are trained using previous data from police officers, who may, therefore, already have certain discriminatory tendencies. The article cites a study where it was found that young black men run a greater risk of being stopped by the police than young white men when algorithms are used to coordinate police forces precisely because of the historical data used. A method that Zhang (2021) suggests to counteract this is to let a third party weigh in on how data collected from algorithms should be interpreted and used. (Wang & Ma, (2021) also claim that AI will increase social discrimination based on ethnicity and gender, which can lead to increased crime in society.

## Lack of Personal Responsibility

The literature also discusses where the responsibility for the use of algorithms lies. Zhang (2021) points out the risk of innocent people being singled out as crime suspects simply because they fit a certain profile. In that case, an algorithm in itself forms the basis for suspicion without the police themselves being fully aware of the reasoning behind the conclusion reached. To prevent this, police officers are asked to have a critical attitude towards data and recommendations obtained from AI programs. Lanagan & Choo (2021) also argue that explainable AI (XAI) should be used in law enforcement because otherwise, a human cannot take legal responsibility for what the algorithm does. Kitsos (2020) believes that the police must create a work culture where they are responsible for their actions. Lanagan & Choo (2021) and Pasquale (2018), who believes that automation in legal matters risks undermining basic principles of responsibility if there are no people responsible for how they operate.

## Mass Surveillance

Intelligence and surveillance have always been an important tool for authorities. Following Snowden's revelations that US and European law enforcement agencies conduct mass surveillance of their citizens' electronic communications, a discussion has begun about the need for this type of surveillance and how it is carried out. The combination of the Internet and digital technologies using AI has given rise to surveillance systems that constantly monitor people's lives, resulting in an unprecedented loss of anonymity (Kitsos, 2020; Milivojevic, 2021). China's ongoing social credit rating project is made possible by extensive population monitoring and aims to give an individual or company a certain score based on how it behaves in society (Kitsos, 2020). The score determines whether you get certain privileges or not. This system exemplifies what can be made possible using mass surveillance.

## Erosion of Privacy and Integrity

When police forces use AI to predict and prevent crime, high-quality and sufficient data is required. Smith and Miller (2022) state that massive amounts of useful data are already collected, and that this information arsenal will only grow. They believe that problems arise when the data is to be accessed. The article authors want to be able to aggregate a range of data sources from different domains. These are openly available data from criminal records (Wang & Ma, 2022) and social media (Freeman, 2020), but also data from private chat conversations (Bulgakova et al., 2019), business (Freeman, 2020; Wyatt, 2021), medical records (Freeman, 2020), and tax records (Smith & Miller, 2022). Heterogeneous data collections are also problematic because they require a high level of sophistication or a lot of manual work to merge. A couple of articles, therefore, recommend national and international initiatives to create broad, common standards for data systems (Freeman, 2020; Pawlicka et al., 2021).

An issue that becomes relevant concerning these types of data sets is the issue of privacy. Crime-fighting with AI's help often involves retrieving data from several different sources, where the collected data paints a detailed picture of an individual. Often, the data used for crime forecasting consists of sensitive information such as ethnicity, skin color, political opinion, religious affiliation or other beliefs, health condition, sexual orientation, or biometric data to identify a person (Kitsos, 2020). Even less personal user data, such as paying bills and digital services, may be included (Zhang, 2021). The data approved for use on a particular platform can, together with other data taken out of context, paint a detailed picture of an individual without the individual being aware of it. A risk with the expansion of AI can thus mean an erosion of privacy and a threat to personal integrity.

From the literature, it appears that this way of using complements of data to identify individuals already exists today. It is highlighted how many Western countries strictly regulate wiretapping while recording from social media is less regulated (Rajamäki et al., 2018). Because social media is public and freely accessible, police and security officials perceive the search as justified (Rajamäki et al., 2018). The big difference between anyone and an official accessing the data is the consequences, as law enforcement scanning can have real consequences on someone's privacy and future. A case highlighted is the company Clearview AI, which, through a biometric facial recognition algorithm, used images from social media to identify perpetrators (Smith & Miller, 2022). Instead of the narrow number of images available in national databases, such as passport photos and driver's licenses, they let their product utilize the billions of images available on social media and other digital platforms. This was met with strong public criticism, legal reprisals, and a backlash from the companies providing the photographs in the form of cease-and-desist letters (Smith & Miller, 2022). They meant that Clearview AI, through its services, had violated the user agreement. It is stated that the rapid development of data availability, in combination with the increased areas of use, creates a need for new structures for consent and data security (Smith & Miller, 2022). Furthermore, Rajamäki et al. (2018) state that academic and public debates have shifted the focus from data collection to data analysis and use. A shift that, with the previous example in mind, will affect the ability to protect the individual's privacy.

## DISCUSSION

Two types of articles appear most frequently in the literature. Some advocate the implementation of digital technologies, and others propose frameworks for digital technologies. The articles are friendly to the technology covered and focus mainly on the positive aspects of an AI implementation rather than the negative ones. The articles on the same topics also agree on questions about today's capabilities, the technology's potential, and what future work should address. Since most articles express practical utility as a goal, it is remarkable how few of the presented solutions have been tested in the field. Nor does it appear in the background of the works whether there is a concrete demand for their solutions. This could indicate a dissonance between the digital tools law enforcement seeks to use and public research.

A consistent theme of the articles is the hope for what an increasingly extensive implementation of AI in law enforcement could achieve. An optimistic picture is painted regarding technological advancements, with everything from drones that can analyze real-time video footage to scan social media to detect crimes and stop them before they happen. One way to interpret the literature could be that the future of law enforcement will be characterized by extremely high-tech AI, and there is no doubt that the technology will improve.

However, the limited information about the actual implementation or realizability of the test models and frameworks that some articles propose makes it difficult to determine whether they should be seen as possibilities or whether the research is too early to make that assessment. At the same time, research at an early stage must be published, as it can illuminate new research areas. The areas of use identified in the results in the Before Crime section mainly revolve around image and text analysis. One observation is that the image and text analysis is primarily intended to identify behavior, indicating that a crime will be committed. The studies included in the study are thus based on the premise that it is possible to draw connections between a certain behavior, which is not a crime, and that an actual crime will occur. The studies do not propose how such predictions should be used, i.e., how an individual judged to have risky behavior should be treated. It would have been justified to discuss how the police should use that information since it is about correcting and pointing out people who have not yet committed a crime. Consequently, it is also possible to ask whether such an implementation would limit the framework for what is considered permissible in society since a greater number of actions (non-criminal) could still arouse the interest of the police and be grounds for suspicion.

This category is also characterized by the problem of no conclusion against which the forecast can be compared. The purpose of the technology is to predict crimes to be able to prevent them from being committed. This means there will be no real data to compare the predictions. It becomes difficult to confirm that the behavior that has been identified and corrected would lead to a crime if it had not been detected or if it was. Therefore, it can be assumed that the resource should be able to be used by the police to the same extent as other unconfirmed tips. Literature outside the conducted research highlights automation complacency as a phenomenon where people who use algorithm-based aids in decision-making become passive and do not question the technology. In the same study, a distinction is made between the term's automation complacency and automation bias, which instead describe a tendency to trust the suggestions that these aids give. A potential risk with that is thus that when setting up algorithm or AI-based technology, the human being hides behind the algorithms and thus abdicates responsibility from the consequences. On the other hand, the literature in our completed study seems to lack a discussion of whether this phenomenon may affect those responsible for using AI in law enforcement.

The areas of use presented in the section Under crime were mainly based on carrying out some image analysis, either with video recordings from stationary cameras or autonomous systems (drones), which thus adds an increased breadth to where the technology can be used. The purpose of image analysis is primarily to detect crimes in real time to stop them in the first place and assist potential crime victims with, for example, escape routes. Even when the literature lists hopeful suggestions on how image analysis can benefit law enforcement, there is a problem with how the establishment of such technology can contribute to a form of mass surveillance. An area of surveillance without equally obvious problems concerning citizen surveillance is the surveillance of marine protected areas. Deploying that type of technology offshore would potentially be perceived as less risky, as it primarily involves monitoring areas where residents' privacy does not play out. The challenge would instead be the cooperation between states, as the monitoring in such a case would have taken place in international waters (Rajamäki, 2018). Several articles deal with applications while the crime is committed test models or frameworks for a certain system without much reflection on whether it is possible to implement concerning today's technology.

An example is the article by Suralkar et al. (2020), which deals with context-aware AI. The idea that AI should be able to determine in real-time whether a physical confrontation between two people is serious or just a game may be good, but questions arise about how feasible this is. For example, it is unclear what the training data for such an AI would look like. Another example is the article written by Enríquez et al. (2019), who propose real-time advice to users in situations where a potential threat arises. The system assumes that all people involved have the same application installed for the crowdsensing system to work, and it is unclear how the system should be able to perceive the nuances among different environments that should reasonably be the basis for good advice. In the After Crime section, part of the output is the ability to see patterns and scan data to detect crimes. Another part revolves around image analysis, intended to provide clues in criminal investigations or identify behaviors that indicate a person is lying. Implementing AI or algorithm-based image analysis to provide clues in criminal investigations may be less controversial than the uses mentioned above for image analysis, as it is relatively similar to the traditional forms of work already used today. The common starting point is that there is already a suspicion of crime. The difference is that a computer goes through the video material instead of a human. Since the starting point is criminal suspicion, it can be argued that such an implementation would not be surrounded by the same degree of ethical risk as those who use image analysis before or during a crime. Common to the areas Before, During, and After the crime is that the authors were usually unable to state the accuracy of the technique. This leads to the question of how potential false positives should be handled in an actual implementation. On occasions when the police authority has made incorrect decisions, this has until now meant that the working police bear the possible criminal responsibility for the error committed. The question is whether the introduction of decision-making technology means a transfer of responsibility from human to machine. AI-developed programs also mean that the software learns itself based on training data, which means that the decisions and assumptions on which the functionality is based are not always explicitly given by a human. Milivojevic (2021) raises the question of what degree of autonomy AI programs should be treated as criminally liable.

In literature outside the results, it is reported that other sectors, such as healthcare and transport, are characterized by a similar technological development but have shifted responsibility structures. In 2011, India introduced a new patient safety law and became the first country in the world to lack individual liability of healthcare personnel for healthcare injuries . Furthermore, the automotive industry is undergoing a development towards self-driving cars, where the decisions, instead of being made by the driver, are made by algorithms. In that case, it is advocated that the car manufacturer should bear the responsibility in the event of accidents, not the driver (Z. S. Smith, 2022). Whether technology or people will be held responsible for misjudgments can also be assumed to depend on how people and society perceive technology. According to a study outside the surveyed literature by Hong & Williams (2019), "*people perceive crime-predicting AI as significantly less autonomous than crime-predicting humans*" Nevertheless, the study shows that the general perception of people is that a discriminating AI bears the same responsibility as a discriminating human at the same type of prediction. This tendency suggests that a future expansion of AI within law enforcement could mean a reduced accountability of people, due to the increased trust in technology.

The literature advocates holding individuals responsible for how AI is used, with the motivation that it places higher demands on the quality of the software, regardless of whether it concerns users or developers. If individual responsibility is demanded, the possibility of hiding behind the algorithms disappears when wrong decisions are made. This, in turn, requires that the logic behind the algorithms is comprehensible and possible for people to perceive. Personal integrity and trust in the legal process could be quickly eroded if a tip-off from an AI program that no one understood or was responsible for was enough to raise suspicion or prosecution. Holding individuals responsible thus seems like a reasonable compromise since the algorithms can then be used without the motivations behind the decisions being made being inaccessible and incomprehensible. Therefore, it is desirable to have a transparent AI, not only about the technical implementation but, above all, to understand the outcomes it generates. The interest in developing AI programs is great in other areas as well, but since the data handled is very sensitive and the consequences of mistakes are potentially enormous, it is especially important to have sound logic behind the automatic decision-making in law enforcement. In the literature, tendencies are highlighted that AI programs may act discriminatingly against certain social groups.

This is hardly something that only characterizes the fight against crime, as sources outside the literature also reported on implementations of AI in other areas that resulted in various forms of discrimination. An example is the Microsoft Twitter bot Tay, which had to be taken down after only one day of use due to its racist, anti-Semitic, and sexist statements (Taylor et al., 2018). Another example is Amazon's recruitment tool, which was found to be gender discriminatory against women (Dastin, 2022). It has only been once the AI programs have been used in practice that problems such as discrimination and injustice have emerged, indicating the importance of practical application. However, the lack of practical implementation may be because the training data required in law enforcement is often sensitive and thus surrounded by the privacy concerns raised in the literature. Even if this is the case, it, if anything, speaks even more to the need for research in the area. The literature highlights risks associated with mass surveillance and the erosion of privacy and integrity. Widespread surveillance at the community level does not come without uncertainties. Some articles in the literature admit a problem with the trade-off between surveillance and personal privacy but do not answer it convincingly. Especially not regarding articles that positively view increased surveillance of various parts of private individuals' lives. It is often defended relatively fleetingly by saying that it is for the good of the individual or that the end justifies the means.

In addition to image analysis, monitoring can be achieved through text and behavior analysis. A common place for mining such data is social media. In recent times, it has become increasingly well-known that access to social media and apps is not, in fact, free. Instead, the user pays for himself by sharing his information with the platform. From that context, the consensus has become that the use comes from the free choice to use the service, but the user may not always be sufficiently informed about the extent to which personal information is shared with the service providers. Nevertheless, there is a consensus, albeit implicit, that the user has a free choice to enter into the agreements that users and service providers enter into and that the user's consent legitimizes the provider's information collection because, as part of the value exchange, the user thus gains access to the app. However, the same consensus does not apply to the police authority's information collection, whether this is because citizens do not feel that the information, they share with the police agency pays off in enough solved crimes to sanctify the funds or the individual's fear of being under constant scrutiny from the long arm of the judiciary is uncertain.

To many, it may seem obvious that citizens' sharing of personal information should be voluntary and not forced. This is because the decision to share the information with authorities is crucial so that exercising authority is not perceived as a privacy-violating activity. At the same time, the citizens' right to not be exposed to crime is curtailed by the violation committed by the criminal. From that arises the question of which right is most important to prioritize, as well as whether most citizens in a society are prepared to share their information with police authorities to work on crime prevention proactively. The arguments about privacy violation can thus come from sides who believe that the law-abiding person has nothing to hide and should, therefore, be comfortable sharing his information.

It should be noted, however, that this applies under the assumption that the laws that exist are also passed under democratic forms and are consistent with the population's perception of what is right and wrong. It should also be noted that laws are dynamic. They change and are replaced as society develops. Social development, in turn, is often driven by popular movements and commitment. With extensive monitoring, a potential consequence could be that innovative thinking and progressive forces in society are inhibited. The issue seems to boil down to society's tolerance of certain illegal behavior from citizens. A constant collection of information that affects the whole of society for the sake of the police authority's activities can create structures where personal peculiarities are forced away, which makes society more homogeneous while general behavior becomes more law-abiding. The law regulates different crimes with different punishment scales simply because the crimes are considered to be of different seriousness. Perhaps, in the same way, there are reasons why only a certain type of more serious crime justifies the collection of personal information from the citizens of society. In the same way, in such a case, personal information should not be used to solve crimes that are not considered sufficiently serious. Ultimately, the question is about what degree of criminality justifies possible privacy-infringing information collection.

Interpol's requirements for fairness, accountability, transparency, and explain ability are relevant and can be used as guidance regarding these ethical risks. In situations where AI shows discriminatory tendencies, fairness becomes, of course, important. Accountability requires clear regulations and guidelines for how the issue of responsibility should be handled. Transparency and explain ability also have more or less to do with the question of responsibility and prevent a situation where you are forced to trust an algorithm that no one understands. No requirement directly addresses privacy concerns, but if the existing requirements are met, it can be assumed that monitoring, as well as other activities that may give rise to ethical risks, are carried out in an ethically defensible manner. In light of the results presented and the discussion held, it is recommended that more empirical research be carried out to demonstrate concrete methods and techniques that work with the technology available today. It is suggested that such projects be conducted in collaboration with law enforcement agencies to better test the technology's usefulness in practice. Future research should be conducted on more developed reasoning around the privacy issues concerning collecting the data on which much of the proposed technology depends. As the decisions of machines take on an ever-increasing role in society, more research should also be carried out into who is responsible for these decisions, as well as how it can be ensured that the machines' decisions are fair regarding human rights.

## CONCLUSION

The literature study shows that the research area of AI in law enforcement is still at an early stage. The optimism for the new digital opportunities is great, but how they should be weighed against the ethical risks accompanying them is far from obvious. The solutions proposed in the literature generally require large amounts of data in the form of text, image, or video material. The collection and handling of this material pose ethical dilemmas and coordination problems. We therefore suggest that future studies in the area pick up where today's literature ends, partly by delving into the possible consequences of the proposed technology partly by collaborating with law enforcement and testing the techniques in reality. In a world driven by rapid technological advancements, integrating Artificial Intelligence (AI) in law enforcement has emerged as a topic of great significance. This comprehensive literature review delves into the evolving landscape of AI in policing, shedding light on its potential benefits and ethical quandaries. The research identifies two prevalent articles in this domain: those that champion the adoption of digital technologies in law enforcement and others that propose frameworks for their implementation. Notably, these articles tend to emphasize the positive aspects of AI, portraying a vision of a high-tech future in policing.

However, the study raises pertinent questions about the practicality and real-world applicability of the proposed solutions. Many of these concepts lack empirical testing, leaving us to ponder whether they are genuine possibilities or still in theoretical exploration. Furthermore, it remains unclear whether there is a tangible demand for these solutions within the broader field of law enforcement. One recurring theme in the literature is the anticipation of AI's transformative potential in crime prevention, ranging from real-time video analysis to preemptive social media monitoring. Yet, the ethical implications of such extensive surveillance and predictive policing raise concerns about privacy, accountability, and the erosion of personal freedoms. The lack of concrete conclusions to validate AI-driven predictions complicates matters further. With the technology aiming to predict and prevent crimes before they occur, there is no baseline data for comparison. As a result, it becomes challenging to ascertain whether the identified behaviors would genuinely lead to crimes if left undetected. A critical issue highlighted in the study is the potential for "*automation complacency*" – a phenomenon where humans unquestioningly rely on algorithmic decision-making, shifting the responsibility away from individuals and onto machines. This shift prompts a debate on the accountability of those who utilize AI in law enforcement.

The research also indicates that similar technological shifts have occurred in other sectors, such as healthcare and transportation. These transformations have led to changes in responsibility structures, potentially indicating a broader shift towards holding technology, rather than individuals, accountable for misjudgments. Privacy, discrimination, and mass surveillance emerge as pressing concerns as AI mines vast amounts of data, particularly from social media platforms. While users may consent to share their information with private companies, the same does not apply to law enforcement agencies. Striking between preventing crime and protecting individual privacy remains a complex challenge. The study concludes by emphasizing the importance of empirical research to validate the practicality of AI in law enforcement. Collaboration between researchers and law enforcement agencies can bridge the gap between theory and real-world application. Additionally, ethical considerations, including transparency, accountability, and fairness, must guide developing and deploying AI technologies in policing. In an era where the decisions of machines are assuming greater significance, it is imperative to address the responsibility of these decisions and ensure that they align with human rights principles. As we navigate the ever-changing landscape of AI in law enforcement, these ethical considerations will remain at the forefront of discussions and policymaking.

## REFERENCES

Anyoha, R. (2017, August 28). The History of Artificial Intelligence. *Science in the News*. https://sitn.hms.harvard.edu/flash/2017/history-artificial-intelligence/

Behmer, E.-J., Chandramouli, K., Garrido, V., Mühlenberg, D., Müller, D., Müller, W., Pallmer, D., Pérez, F. J., Piatrik, T., & Vargas, C. (2019). Ontology Population Framework of MAGNETO for Instantiating Heterogeneous Forensic Data Modalities. In J. MacIntyre, I. Maglogiannis, L. Iliadis, & E. Pimenidis (Eds.), *Artificial Intelligence Applications and Innovations* (pp. 520–531). Springer International Publishing. doi:10.1007/978-3-030-19823-7_44

Bulgakova, E., Bulgakov, V., Trushchenkov, I., Vasiliev, D. V., & Kravets, E. (2019a). Big Data in Investigating and Preventing Crimes. In *Studies in Systems* (pp. 61–69). Decision and Control. doi:10.1007/978-3-030-01358-5_6

Carrera-Rivera, A., Ochoa, W., Larrinaga, F., & Lasa, G. (2022). How-to conduct a systematic literature review: A quick guide for computer science research. *MethodsX*, *9*, 101895. doi:10.1016/j.mex.2022.101895 PMID:36405369

Chang, C.-Y., Chien, L.-C., Kuo, E.-C., & Hwan, Y.-S. (2019b). Designing Intelligence system of Image Processing and Mining in Cloud-Example of New Taipei City Police Department. *IEEE International Conference on Knowledge Innovation and Invention*, (pp. 293–295). IEEE. 10.1109/ICKII46306.2019.9042641

Chase, J., Du, J., Fu, N., Le, T. V., & Lau, H. C. (2017). Law enforcement resource optimization with response time guarantees. *2017 IEEE Symposium Series on Computational Intelligence (SSCI)*, (pp. 1–7). IEEE. 10.1109/SSCI.2017.8285326

Chen, L., & Wong, G. (2018). *Transcriptome Informatics*. Science Direct. doi:10.1016/B978-0-12-809633-8.20204-5

Chopra, A. (2017, March 10). *How far has India come in its digitization journey?* Mint. https://www.livemint.com/Industry/nAIcrfPTv5G1yGLGQ54LzN/EmTech-India-2017-How-far-has-India-come-in-its-digitizatio.html

Contardo, P., Sernani, P., Falcionelli, N., & Dragoni, A. F. (2021b, July 16). Deep learning for law enforcement: A survey about three application domains. *RTA-CSIT 2021*. IEEE.

Das, P., & Das, A. (2019). *Application of Classification Techniques for Prediction and Analysis of Crime in India*., doi:10.1007/978-981-10-8055-5_18

Dastin, J. (2022). Amazon Scraps Secret AI Recruiting Tool that Showed Bias against Women. In *Ethics of Data and Analytics*. Auerbach Publications. doi:10.1201/9781003278290-44

Du, H., Xu, Z., Yan, Z., & Gao, S. (2018). *Intelligent Video Analysis Technology of Public Security Standard Sets of Data and Measurements*., doi:10.1007/978-981-10-7398-4_47

Enriquez, F., Soria Morillo, L., Alvarez-Garcia, J., Caparrini, F., Velasco-Morente, F., Deniz, O., & Vállez, N. (2019). *Vision and Crowdsensing Technology for an Optimal Response in Physical-Security*., doi:10.1007/978-3-030-22750-0_2

Frankenfield, J. (2021). *Artificial Intelligence: What It Is and How It Is Used*. Investopedia. https://www.investopedia.com/terms/a/artificial-intelligence-ai.asp

Herrmann, H. (2023). What's next for responsible artificial intelligence: A way forward through responsible innovation. *Heliyon*, *9*(3), e14379. doi:10.1016/j.heliyon.2023.e14379 PMID:36967876

Hong, J.-W., & Williams, D. (2019). Racism, Responsibility and Autonomy in HCI: Perceptions of an AI Agent. *Computers in Human Behavior*, *100*, 79–84. doi:10.1016/j.chb.2019.06.012

Ionescu, B., Ghenescu, M., Răstoceanu, F., Roman, R., & Buric, M. (2020). Artificial Intelligence Fights Crime and Terrorism at a New Level. *IEEE MultiMedia*, *27*(2), 55–61. doi:10.1109/MMUL.2020.2994403

Ionescu, B., Ghenescu, M., Rastoceanu, F., Roman, R., & Buric, M. (2020). Artificial Intelligence Fights Crime and Terrorism at a New Level. *IEEE MultiMedia, 27*(2), 55–61. doi:10.1109/MMUL.2020.2994403

Jie, Y., Liu, C. Z., Li, M., Choo, K.-K. R., Chen, L., & Guo, C. (2020). Game theoretic resource allocation model for designing effective traffic safety solution against drunk driving. *Applied Mathematics and Computation, 376*, 125142. doi:10.1016/j.amc.2020.125142

Jindal, S., & Sharma, K. (2018a). Intend to analyze Social Media feeds to detect behavioral trends of individuals to proactively act against Social Threats. *Procedia Computer Science, 132*, 218–225. doi:10.1016/j.procs.2018.05.191

Jindal, S., & Sharma, K. (2018b). Intend to analyze Social Media feeds to detect behavioral trends of individuals to proactively act against Social Threats. *Procedia Computer Science, 132*, 218–225. doi:10.1016/j.procs.2018.05.191

Kaur, B., Ahuja, L., & Kumar, V. (2019). Decision tree Model. *Predicting Sexual Offenders on the Basis of Minor and Major Victims, 193–197*, 193–197. doi:10.1109/AICAI.2019.8701276

Kaynak, O. (2021). The golden age of Artificial Intelligence. *Discover Artificial Intelligence, 1*(1), 1. doi:10.1007/s44163-021-00009-x

Khairuddin, A., Alwee, R., & Haron, H. (2020). A Comparative Analysis of Artificial Intelligence Techniques in Forecasting Violent Crime Rate. *IOP Conference Series. Materials Science and Engineering, 864*(1), 012056. doi:10.1088/1757-899X/864/1/012056

King, T. C., Aggarwal, N., Taddeo, M., & Floridi, L. (2020). Artificial Intelligence Crime: An Interdisciplinary Analysis of Foreseeable Threats and Solutions. *Science and Engineering Ethics, 26*(1), 89–120. doi:10.1007/s11948-018-00081-0 PMID:30767109

Kitsos, P. (2020, September 2). The Limits of Government Surveillance: Law Enforcement in the Age of Artificial Intelligence. *11th EETN Conference on Artificial Intelligence (SETN 2020)*. IEEE.

Lanagan, S., & Choo, K.-K. R. (2021). On the need for AI to triage encrypted data containers in U.S. law enforcement applications. *Forensic Science International Digital Investigation, 38*, 301217. doi:10.1016/j.fsidi.2021.301217

Mac, T. T., Copot, C., Lin, C.-Y., Hai, H. H., & Ionescu, C. M. (2020). Towards The Development of a Smart Drone Police: Illustration in Traffic Speed Monitoring. *Journal of Physics: Conference Series, 1487*(1), 012029. doi:10.1088/1742-6596/1487/1/012029

Matthew, U., Kazaure, J., Onyebuchi, A., Okey, O., Muhammed, I., & Okafor, N. (2021). Artificial Intelligence Autonomous Unmanned Aerial Vehicle (UAV) System for Remote Sensing in Security Surveillance. *CYBER NIGERIA, 2020*, 1–10. doi:10.1109/CYBERNIGERIA51635.2021.9428862

Mayring, P. (2000). Qualitative Content Analysis. *Forum Qualitative Sozialforschung / Forum: Qualitative. Social Research, 1*(2), 2. doi:10.17169/fqs-1.2.1089

Milivojevic, S. (2021). *Crime and Punishment in the Future Internet: Digital Frontier Technologies and Criminology in the Twenty-First Century*. Routledge. doi:10.4324/9781003031215

Molina-Molina, J. C., Salhaoui, M., Guerrero-González, A., & Arioua, M. (2021). Autonomous Marine Robot Based on AI Recognition for Permanent Surveillance in Marine Protected Areas. *Sensors (Basel), 21*(8), 8. doi:10.3390/s21082664 PMID:33920075

Pasquale, F. (2018). *A Rule of Persons, Not Machines: The Limits of Legal Automation* (SSRN Scholarly Paper 3135549). https://papers.ssrn.com/abstract=3135549

Pawlicka, A., Choraś, M., Przybyszewski, M., Belmon, L., Kozik, R., & Demestichas, K. (2021). Why Do Law Enforcement Agencies Need AI for Analyzing Big Data? *Computer Information Systems and Industrial Management: 20th International Conference, CISIM 2021.* Springer. 10.1007/978-3-030-84340-3_27

Rajamäki, J., Sarlio-Siintola, S., & Simola, J. (2018). *Ethics of Open Source Intelligence Applied by Maritime Law Enforcement Authorities.* Semantic Scholar. https://www.semanticscholar.org/paper/Ethics-of-Open-Source-Intelligence-Applied-by-Law-Rajam%C3%A4ki-Sarlio-Siintola/e76a98d86fad-d131646850bb6011ce8473469d58

Rajapakshe, C., Balasooriya, S., Dayarathna, H., Ranaweera, N., Walgampaya, N., & Pemadasa, N. (2019). Using CNNs RNNs and Machine Learning Algorithms for Real-time Crime Prediction. *2019 International Conference on Advancements in Computing (ICAC)*, (pp. 310–316). IEEE. 10.1109/ICAC49085.2019.9103425

Rodrigues, R. (2020). Legal and human rights issues of AI: Gaps, challenges and vulnerabilities. *Journal of Responsible Technology*, *4*, 100005. doi:10.1016/j.jrt.2020.100005

Sánchez, J. R., Campo-Archbold, A., Rozo, A. Z., Díaz-López, D., Pastor-Galindo, J., Mármol, F. G., & Díaz, J. A. (2022). On the Power of Social Networks to Analyze Threatening Trends. *IEEE Internet Computing*, *26*(02), 19–26. doi:10.1109/MIC.2022.3154712

Sherer, J., Sterling, N., Burger, L., Banaschik, M., & Taal, A. (2018). *An Investigator's Christmas Carol: Past.* Present, and Future Law Enforcement Agency Data Mining Practices., doi:10.1007/978-3-319-97181-0_12

Simpson, T. (2021). Real-Time Drone Surveillance System for Violent Crowd Behavior Unmanned Aircraft System (UAS) – Human Autonomy Teaming (HAT). *2021 IEEE/AIAA 40th Digital Avionics Systems Conference (DASC).* IEEE. https://www.semanticscholar.org/paper/Real-Time-Drone-Surveillance-System-for-Violent-%E2%80%93-Simpson/3d3dad177c3913d46800e286fdc0553f8b2d81d7

Singh, A., Anand, T., Sharma, S., & Singh, P. (2021). IoT Based Weapons Detection System for Surveillance and Security Using YOLOV4. *ICCES*, *2021*, 488–493. doi:10.1109/ICCES51350.2021.9489224

Smith, M., & Miller, S. (2022). The ethical application of biometric facial recognition technology. *AI & Society*, *37*(1), 167–175. doi:10.1007/s00146-021-01199-9 PMID:33867693

Smith, Z. S. (2022). Self-Driving Car Users Shouldn't Be Held Responsible For Crashes, U.K. Report Says. *Forbes*. https://www.forbes.com/sites/zacharysmith/2022/01/25/self-driving-car-users-shouldnt-be-held-responsible-for-crashes-uk-report-says/

Sun, S. (2020). Application of fuzzy image restoration in criminal investigation. *Journal of Visual Communication and Image Representation, 71*, 102704. doi:10.1016/j.jvcir.2019.102704

Suralkar, S., Gangurde, S., Chintakindi, S., & Chawla, H. (2020). An Autonomous Intelligent Ornithopter. In A. P. Pandian, R. Palanisamy, & K. Ntalianis (Eds.), *Proceeding of the International Conference on Computer Networks, Big Data and IoT (ICCBI - 2019)* (pp. 856–865). Springer International Publishing. 10.1007/978-3-030-43192-1_93

Taylor, S., Boniface, M., Pickering, B., Anderson, M., Danks, D., Følstad, A., Leese, M., Müller, V., Sorell, T., Winfield, A., & Woollard, F. (2018). *Responsible AI – Key themes, concerns & recommendations for European research and innovation*.

ToppiReddy, H. K. R., Saini, B., & Mahajan, G.ToppiReddy. (2018). Crime Prediction & Monitoring Framework Based on Spatial Analysis. *Procedia Computer Science, 132*, 696–705. doi:10.1016/j.procs.2018.05.075

Wang, H., & Ma, S. (2021). Preventing Crimes Against Public Health with Artificial Intelligence and Machine Learning Capabilities. *Socio-Economic Planning Sciences, 80*, 101043. doi:10.1016/j.seps.2021.101043

Wyatt, A. (2021). A Southeast Asian perspective on the impact of increasingly Autonomous systems on subnational relations of power. *Defence Studies, 21*(3), 271–291. doi:10.1080/14702436.2021.1908136

Zhang, R. (2021a). The AI embedding predicts the legal risks of policing and its prevention. *2021 International Conference on Computer Information Science and Artificial Intelligence (CISAI)*, 642–646. 10.1109/CISAI54367.2021.00129

Zhang, R. (2021b). *The AI embedding predicts the legal risks of policing and its prevention*. IEEE. doi:10.1109/CISAI54367.2021.00129

Zufferey, R., Tormo-Barbero, J., Feliu-Talegón, D., Nekoo, S. R., Acosta, J. Á., & Ollero, A. (2022). How ornithopters can perch autonomously on a branch. *Nature Communications, 13*(1), 7713. doi:10.1038/s41467-022-35356-5 PMID:36513661

## ADDITIONAL READINGS

Alzou, S. (2014). Artificial Intelligence in Law Enforcement, A Review. *International Journal of Advanced Information Technology, 4*(4), 1–9. doi:10.5121/ijait.2014.4401

Blauth, T. F., Gstrein, O. J., & Zwitter, A. (2022). Artificial Intelligence Crime: An Overview of Malicious Use and Abuse of AI. *IEEE Access : Practical Innovations, Open Solutions, 10*, 77110–77122. doi:10.1109/ACCESS.2022.3191790

Caldwell, M., Andrews, J. T. A., Tanay, T., & Griffin, L. D. (2020). AI-enabled future crime. *Crime Science, 9*(1), 14. doi:10.1186/s40163-020-00123-8

Jeong, D. (2020). Artificial Intelligence Security Threat, Crime, and Forensics: Taxonomy and Open Issues. *IEEE Access : Practical Innovations, Open Solutions*, *8*, 184560–184574. doi:10.1109/ACCESS.2020.3029280

Kanimozhi, N., Keerthana, N. V., Pavithra, G. S., Ranjitha, G., & Yuvarani, S. (2021). CRIME Type and Occurrence Prediction Using Machine Learning Algorithm. *2021 International Conference on Artificial Intelligence and Smart Systems (ICAIS)*, (pp. 266–273). IEEE. 10.1109/ICAIS50930.2021.9395953

King, T. C., Aggarwal, N., Taddeo, M., & Floridi, L. (2020). Artificial Intelligence Crime: An Interdisciplinary Analysis of Foreseeable Threats and Solutions. *Science and Engineering Ethics*, *26*(1), 89–120. doi:10.1007/s11948-018-00081-0 PMID:30767109

Kumar, R., Sharma, A., Kaur, M., Joshi, K., & Singh, S. (2023). Role of Artificial Intelligence to address Cyberbullying and Future Scope. *2023 International Conference on Computational Intelligence and Sustainable Engineering Solutions (CISES)*, (pp. 974–977). IEEE. 10.1109/CISES58720.2023.10182406

## KEY TERMS AND DEFINITIONS

**Artificial Intelligence (AI):** The ability of computer programs to mimic human cognitive abilities and intelligence.

**Big Data:** A term for the amount of data that has emerged in recent years and refers to data sets of such a size that they are difficult to process with traditional methods. (sw. big data)

**Computer Vision:** The ability of a computer to understand and extract content from images depending on what is being searched for. (sw. computer vision)

**Convolutional Neural Network (CNN):** A technique used in image analysis that has proven very good at recognizing faces under difficult conditions. (sw. Convolutional neural networks)

**Crowdsensing:** A technology where a large group of individuals with mobile devices capable of sensing and computing collectively share and mine information for a common interest.

**Explainable AI (XAI):** AI whose logic can be explained and is understandable to humans. (sw. explainable AI)

**Internet of Things (IoT):** The network of devices equipped with sensors, software, etc., which uses the Internet to communicate by exchanging various forms of data.

**Unsupervised Learning:** This occurs when the AI takes in data and discovers patterns without a human being involved.

## APPENDIX 1

*Table 2. List of papers included in the study*

| Author | Publication year | Journal | Type of Study |
|---|---|---|---|
| Behmer E.-J., Chandramouli K., Garrido V., Mühlenberg D., Müller D., Müller W., Pallmer D., Pérez F. J., Piatrik T., Vargas C. | 2019 | IFIP Advances in Information and Communication Technology | Conceptual article |
| Bulgakova E., Bulgakov V., Trushchenkov I., Vassilev D., Kravets E. | 2018 | Studies in Systems, Decision, and Control | Conceptual article |
| Chang C., Chien L., Kuo E., Hwan Y. | 2019 | 2nd IEEE International Conference on Knowledge Innovation and Invention 2019 | Conceptual article |
| Chase J., Du J., Fu N., Le T. V., Lau H. C. | 2018 | 2017 IEEE Symposium Series on Computational Intelligence, SSCI 2017 - Proceedings | Empirical study |
| Contardo P., Sernani P., Falcionelli N., Dragoons AF. | 2021 | CEUR Workshop Proceedings (Volume 2872) | Conceptual article |
| The P., the A. K. | 2019 | Advances in Intelligent Systems and Computing | Empirical study |
| Du H., Xu Z., Yan Z., Gao S. | 2018 | Lecture Notes in Electrical Engineering | Conceptual article |
| Enríquez F., Soria L. M., Álvarez-García J. A., Caparrini F. S., Velasco F., Deniz O., Vallez N. | 2019 | Lecture Notes in Computer Science | Conceptual article |
| Freeman S. | 2020 | Proceedings of SPIE - The International Society for Optical Engineering | Empirical study |
| Ionescu B., Ghenescu M., Rastoceanu F., Roman R., Buric M. | 2020 | IEEE Multimedia 27(2),9116069 | Conceptual article |
| They are Y., Liu C. Z., Li M., Choo K.-K. R., Chen L., Guo C. | 2020 | Applied Mathematics and Computation | Empirical study |
| Jindal S., Sharma K. | 2018 | Procedia Computer Science | Conceptual article |
| Kaur B., Ahuja L., Kumar V. | 2019 | Proceedings - 2019 Amity International Conference on Artificial Intelligence | Empirical study |
| Kitsos P. | 2020 | CEUR Workshop Proceedings | Conceptual article |
| Lanagan S., Choo K. | 2021 | Forensic Science International: Digital Investigation | Conceptual article |
| Matthew U. O., Kazaure J. S., Onyebuchi A., Daniel O. O., Muhammed I. H., Okafor | 2021 | Proceedings of the 2020 IEEE 2nd International Conference on Cyberspace | Conceptual article |
| Milivojevic S. | 2021 | Crime and punishment in the future Internet: Digital Frontier Technologies and criminology in the twenty-first century | Conceptual book |
| Molina-Molina J. C., Salhaoui M., Guerrero-González A., Arioua M. | 2021 | Intelligent Sensing Systems for Vehicle | Empirical study |
| Pawlicka A., Choraś M., Przybyszewski M., Belmon L., Kozik R., Demestichas K. | 2021 | Lecture Notes in Computer Science (including subseries Lecture Notes in Artificial Intelligence and Lecture Notes in Bioinformatics) | Conceptual article |
| Rajamäki J., Sarlio-Siintola S., Simola J. | 2018 | European Conference on Information Warfare and Security, ECCWS | Conceptual article |
| Rajapakshe C., Balasooriya S., Dayarathna H., Ranaweera N., Walgampaya N., Pemadasa N. | 2019 | 2019 International Conference on Advancements in Computing, ICAC 2019 | Empirical study |
| Ramirez J., Campo-Archbold A., Zapata A., Diaz-Lopez D., Pastor-Galindo J., Gomez Marmol F | 2022 | IEEE Internet Computing | Empirical study |
| Ridzuan Khairuddin A., Alwee R., Haron H. | 2020 | IOP Conference Series: Materials Science and Engineering | Empirical study |
| Sherer J. A., Sterling N. L., Burger L., Banaschik M., Taal A. | 2018 | Advanced Sciences and Technologies for Security Applications | Conceptual article |
| Simpson T. | 2021 | AIAA/IEEE Digital Avionics Systems Conference - Proceedings | Conceptual article |

*continued on following page*

*Table 2. Continued*

| Author | Publication year | Journal | Type of Study |
|---|---|---|---|
| Singh A., Anand T., Sharma S., Singh P. | 2021 | Proceedings of the 6th International Conference on Communication and Electronics Systems | Empirical study |
| Smith M., Miller S. | 2022 | AI and Society | Conceptual article |
| Sun S. | 2020 | Journal Of Visual Communication and Image Representation | Empirical study |
| Suralkar S., Gangurde S., Chintakindi S., Chawla H. | 2020 | Lecture Notes on Data Engineering and Communications Technologies | Conceptual article |
| Satisfactory T., Copot C., Lin C.-Y., Hong Hai H., Ionescu C. M. | 2020 | Journal of Physics: Conference Series | Conceptual article |
| Toppireddy H. K. R., Saini B., Mahajan G. | 2018 | Procedia Computer Science | Empirical study |
| Wang H., Ma S. | 2022 | Socio-Economic Planning Sciences | Empirical study |
| Wyatt A. | 2021 | Defence Studies | Conceptual article |
| Zhang R. | 2021 | Proceedings - 2021 International Conference on Computer Information Science and Artificial Intelligence, CISAI 2021 | Conceptual article |

# Chapter 9
# Lensing Legal Dynamics for Examining Responsibility and Deliberation of Generative AI-Tethered Technological Privacy Concerns:
## Infringements and Use of Personal Data by Nefarious Actors

**Bhupinder Singh**
https://orcid.org/0009-0006-4779-2553
*Sharda University, India*

## ABSTRACT

*The rapid integration of generative AI technology across various domains has brought forth a complex interplay between technological advancements, legal frameworks, and ethical considerations. In a world where generative AI has transcended its initial novelty and is now woven into the fabric of everyday life, the boundaries of human creativity and machine-generated output are becoming increasingly blurred. The paper scrutinizes existing privacy laws and regulations through the lens of generative AI, seeking to uncover gaps, challenges, and possible avenues for reform. It explores the evolution of jurisprudence in the face of technological disruption and debates the adequacy of current legal frameworks to address the dynamic complexities of AI-influenced privacy infringements. By scrutinizing cases where personal data has been exploited by nefarious actors employing generative AI for malevolent purposes, a stark reality emerges: the emergence of a new avenue for privacy breaches that tests the limits of existing legal frameworks.*

DOI: 10.4018/979-8-3693-1565-1.ch009

## INTRODUCTION

Generative AI, a cutting-edge field in artificial intelligence, has emerged as a powerful tool for creating content, such as text, images, and even entire websites that appears to be generated by humans. This technology, often exemplified by models like GPT-3 and its successors, has the ability to mimic human creativity and generate content that is often indistinguishable from what a person might produce. While generative AI has opened up exciting possibilities in various domains, it has also raised profound technological privacy concerns that demand our attention (Liu, 2022). The Generative AI operates on the principles of deep learning, utilizing massive datasets to train neural networks to understand patterns and generate content accordingly. These models have the capacity to generate human-like text, craft realistic images, and even compose music that could pass as the work of skilled artists or authors. On the surface, this may seem like a remarkable achievement, promising innovative applications across industries, from content generation to automated customer service and even creative storytelling (Huang, S., & Siddarth, D. 2023). The generative AI, driven by its insatiable appetite for data, necessitates the collection and utilization of vast amounts of personal information. The concept of responsibility takes center stage as the paper dives into the legal and ethical underpinnings of generative AI. As this technology transcends mere tools and assumes the role of a creative collaborator, the question of who bears the weight of responsibility becomes increasingly convoluted. Developers, users, and platforms converge in a complex nexus, each contributing to the ethical implications of AI-generated content and its impact on individual privacy. As generative AI operates in a domain where human intentionality merges with machine autonomy, the assignment of accountability takes on new dimensions.

Though, beneath this promising facade lie several pressing technological privacy concerns. The foremost concern is the potential misuse of generative AI. Malicious actors could employ this technology to generate convincing fake news articles, impersonate individuals, or produce fraudulent documents. Such nefarious applications have the potential to undermine trust in the digital world, fuel misinformation, and threaten the security of individuals and organizations alike (Singh, 2023). The synthesis of technological solutions, regulatory adaptations, and ethical considerations becomes paramount in shaping a future where generative AI is harnessed responsibly, and the infringements by nefarious actors are mitigated. By dissecting the evolving responsibilities, privacy challenges, and nefarious dimensions, this paper seeks not only to illuminate the complex landscape but also to provide a compass for navigating a future where AI and human values coalesce. This paper casts a penetrating gaze on the potential infringements of privacy arising from the data-hungry nature of generative AI.

There is another crucial concern pertains to privacy breaches facilitated by generative AI. These models often require access to vast amounts of data to function effectively, and as they continue to evolve, there's a risk that sensitive personal information might be used to train them. The mishandling of this data, whether intentionally or inadvertently, can result in significant privacy violations, leading to identity theft, financial fraud, or the exposure of personal details that should remain confidential (Jo, 2023). Generative AI's ability to create highly realistic deepfake content poses a substantial threat to privacy. Deepfake videos and images can convincingly depict individuals saying or doing things they never did, making it increasingly difficult to discern between reality and manipulated content. This raises concerns not only in terms of personal privacy but also for the broader societal implications, including political manipulation and character assassination.

The ethical dilemmas surrounding generative AI and privacy are also complex. As AI models generate content that mirrors human creations, it becomes challenging to determine authorship and intellectual property rights. This could lead to disputes over ownership and fair compensation for creative work, potentially diminishing the livelihoods of human creators. So, Striking the right balance between harnessing the capabilities of generative AI for innovation while safeguarding individual and societal privacy is a formidable challenge that requires careful consideration, ethical guidelines, and robust regulatory frameworks (Sharma, 2022). As we continue to navigate this evolving landscape, it is essential to address these concerns proactively to ensure that the promise of generative AI is harnessed for the benefit of humanity without compromising our fundamental right to privacy.

## Background and Context

The rapid advancement of generative artificial intelligence (AI) technologies has transformed the digital landscape, enabling machines to create and manipulate content, including text, images, and videos, that is often indistinguishable from human-generated content. While these developments bring about significant technological advancements, they also raise profound concerns regarding privacy infringements and the misuse of personal data. Nefarious actors are increasingly exploiting generative AI to engage in unauthorized data collection, impersonation, and manipulation, posing significant risks to individual privacy and societal stability.

## Research Question

The proliferation of generative artificial intelligence (AI) technologies has ushered in a new era of creativity and innovation. However, this technological advancement has also raised profound concerns related to privacy, accountability, and the potential for misuse by nefarious actors. Through an interdisciplinary approach, encompassing legal analysis, ethical considerations, and technological insights, this paper seeks to shed light on the multifaceted challenges posed by generative AI and offer pragmatic solutions that balance technological progress with the preservation of privacy and ethical values.

The research question for investigating the legal dynamics concerning the responsibility and deliberation of generative AI in the context of technological privacy concerns, particularly regarding infringements and the use of personal data by nefarious actors, could be framed as: "How can legal frameworks adapt to address the multifaceted challenges posed by generative AI technologies in safeguarding technological privacy and holding responsible parties accountable for privacy infringements, specifically in cases involving the malicious use of personal data by nefarious actors?"

## Objectives of the Paper

The objectives of the paper to encompass a comprehensive exploration of the legal aspects and challenges related to generative AI and privacy infringements to-

Analyze Current Legal Frameworks: Assess the existing legal frameworks and regulations governing AI technologies and data privacy to understand their strengths, weaknesses, and relevance in addressing the emerging challenges posed by generative AI and malicious use of personal data.

Identify Technological Privacy Concerns: Identify and categorize the specific privacy concerns associated with generative AI technologies, focusing on their potential to infringe upon individuals' rights and expose personal data to nefarious actors.

Examine Responsibility and Accountability: Investigate the notion of responsibility and accountability in the context of generative AI, considering the complex and evolving landscape of AI-generated content, and propose strategies for assigning responsibility when privacy breaches occur.

Assess Ethical Implications: Explore the ethical dilemmas surrounding generative AI and privacy, including consent, transparency, and the blurred lines between human and machine-generated content, and analyze their legal implications.

Mitigate Nefarious Use: Develop recommendations and legal mechanisms for mitigating the malicious use of generative AI, particularly in cases involving deepfake content and fraudulent activities, while safeguarding individual rights and freedom of expression.

Adaptation of Regulations: Propose adjustments and updates to existing legal frameworks to accommodate the rapid advancements in generative AI and address emerging threats, ensuring that the law remains relevant and effective.

Promote Responsible AI Deployment: Suggest guidelines and best practices for organizations and developers to promote responsible AI deployment and data handling, minimizing the risk of personal data misuse.

Consider Global Implications: Examine the global nature of AI and privacy concerns, taking into account international variations in regulations and proposing strategies for harmonizing legal approaches to better protect individuals' data on a global scale.

Recommend Policy Changes: Provide policymakers with evidence-based recommendations for shaping legal and regulatory policies that strike a balance between fostering innovation in AI technologies and preserving individuals' technological privacy.

Contribute to Academic Discourse: Contribute to the academic discourse on the intersection of generative AI, privacy, and legal dynamics by offering a comprehensive analysis of the issues, potential solutions, and future directions for research and policy development.

By pursuing these objectives, the paper aims to shed light on the intricate legal dynamics surrounding generative AI and privacy concerns, offering valuable insights for policymakers, legal professionals, researchers, and stakeholders in the field of AI ethics and regulation.

## Methodology

This research paper employs a multidisciplinary approach, drawing insights from artificial intelligence, law, and ethics. The research design includes the following components:

- Extensive literature review to understand the state of generative AI and legal frameworks.
- Examples of privacy infringements involving generative AI.
- Examination of legal documents, regulations, and policies relevant to data privacy.

## Structure of the Paper

This paper is structured as follows: Section 2 highlights the significance of research in this arena. Section 3 discusses the Growing Reliance on AI Technologies as increasing Threats to Personal data. Sections 4 specify the Nefarious Actors and Their Tactics and 5 delve into the Legal Frameworks and Regulations. Section 6 explores the Balancing Innovation and Privacy lensing Generative AI of this issue. Section 7 offers International Cooperation and Harmonization, and finally Section 8 and 9 concludes the paper and laid down the future directions and scope.

## SIGNIFICANCE OF RESEARCH

The legal dynamics surrounding the examination of responsibility and deliberation of generative AI within the context of technological privacy concerns hold immense significance in our digital age. As generative AI technologies advance, questions regarding who should be held accountable for privacy infringements and misuse of personal data become increasingly complex. These technologies can produce content that blurs the line between human and machine creation, making it challenging to attribute actions to a specific entity. Establishing legal frameworks to identify responsible parties and ensure due diligence in the use of generative AI is crucial to safeguarding individuals' privacy rights.

The threat posed by nefarious actors exploiting generative AI for personal data misuse necessitates robust legal measures. The potential for deepfake content and fraudulent activities undermines trust in digital spaces, and this erosion of trust can have profound societal and economic implications. Legal frameworks must be agile and adaptive to address emerging threats effectively, holding malicious actors accountable for their actions while simultaneously protecting innocent individuals from unwarranted consequences.

In this evolving landscape, legal authorities and policymakers face the challenge of striking a delicate balance between fostering innovation in generative AI and safeguarding privacy. The legal dynamics must account for the rapid pace of technological advancement, ensuring that regulations remain relevant and effective in deterring privacy infringements. As generative AI continues to shape our digital future, addressing these legal complexities will be essential in upholding the principles of privacy, accountability, and responsible AI deployment.

## GROWING RELIANCE ON AI TECHNOLOGIES: INCREASING THREATS TO PERSONAL DATA

The growing reliance on AI technologies and the concomitant increase in threats to personal data represent a pivotal crossroads in our digital age. As, AI becomes increasingly integrated into our daily lives, from virtual assistants and autonomous vehicles to recommendation algorithms and healthcare diagnostics, the volume and diversity of personal data collected have surged (Weisz, 2023). This data, often encompassing sensitive information like financial records, medical histories, and personal communications, have become the lifeblood of AI systems, fueling their learning and enhancing their functionality. However, this growing dependence on AI for critical tasks and services has also amplified the vulnerabilities and risks associated with personal data.

Cybersecurity threats have escalated in tandem with AI's ubiquity as malicious actors, armed with sophisticated tools and techniques, exploit weaknesses in AI systems to gain unauthorized access to personal data. This can result in devastating consequences, including identity theft, financial fraud, and the compromise of individuals' digital lives. The ability of AI to analyze vast datasets and identify patterns can also be leveraged by cybercriminals for increasingly sophisticated and targeted attacks, making the protection of personal data more challenging than ever. The generative AI technologies, such as deep learning models, enable the creation of deepfake content that can convincingly mimic real individuals (Noy, 2023). This alarming development poses a significant threat to personal privacy and the authenticity of digital content. Deepfake videos, in particular, have the potential to manipulate public opinion, defame individuals, and disrupt trust in digital media, creating a toxic environment of misinformation and mistrust. Amid this backdrop, ethical concerns loom large because the reliance on AI technologies raises critical questions about consent, transparency, and accountability. Many individuals are unaware of how their data is being used to train AI models, and they may not have given informed consent for such usage (Ahmad, W., & Dethy, E. (2019). Moreover, when AI-generated content blurs the line between human and machine creation, determining attribution and responsibility becomes a multifaceted challenge.

The legal and regulatory landscape is grappling with these evolving dynamics. Governments and organizations are tasked with creating and adapting frameworks to safeguard personal data while still fostering innovation and harnessing the benefits of AI. Striking the right balance between these objectives is a complex and ongoing process that requires continuous vigilance and adaptation. The increasing threats to personal data underscore the urgent need for comprehensive approaches to cybersecurity, ethical data usage, and responsible AI deployment. Balancing the potential of AI with the protection of individual and societal privacy rights is an imperative that will shape the future of our digital world (Singh, 2022).

Generative AI represents a watershed moment in artificial intelligence, where machines possess the capacity to produce content that closely mirrors human creativity. From text generation models like GPT-3 to image synthesis algorithms capable of crafting lifelike visuals, the innovations in generative AI are undeniably awe-inspiring (Bozkurt, 2023). However, these advancements are accompanied by significant privacy concerns. As generative AI learns from vast datasets, it often requires access to personal and sensitive information, raising questions about informed consent, data protection, and the potential for misuse. The ability of generative AI to generate deepfake content further exacerbates these privacy concerns, challenging our ability to distinguish between authentic and manipulated digital media. Balancing the potential for innovation with the imperative of safeguarding privacy rights remains a pressing challenge in the era of generative AI.

## Overview of Generative AI

Generative AI, a revolutionary branch of artificial intelligence, has redefined our understanding of machine capabilities. At its core, generative AI employs deep learning techniques to create content, whether it be text, images, music, or even videos that closely mimics human-produced material. This technology operates by training neural networks on massive datasets, enabling machines to understand patterns and generate content that is often indistinguishable from what a human could create. Generative AI models like GPT-3 and image generators like DALL-E have made headlines for their remarkable abilities (Brynjolfsson, 2023). These innovations have found applications in creative content generation, natural language understanding, healthcare diagnostics, and more. However, the rapid evolution of generative AI has also given rise to ethical and privacy concerns, challenging us to navigate the fine line between technological progress and responsible use.

## Deepfakes and the Blurring of Reality

Deepfakes, born from the capabilities of generative AI, represent a watershed moment in the convergence of technology and misinformation. These sophisticated manipulations leverage deep learning algorithms to create hyper-realistic videos, audio recordings, and images that convincingly impersonate real individuals, often seamlessly blending fact and fiction. The implications of deepfakes are far-reaching, transcending the boundaries of privacy and truth in the digital age. As these AI-generated creations blur the lines between reality and fabrication, they pose substantial threats to various facets of our society. Individuals can unwittingly become victims of character assassination, as deepfakes manipulate their words and actions (Lathrop, 2019). This technology can also undermine the credibility of digital media, eroding trust in information sources and making it increasingly challenging to discern genuine content from manipulated fabrications. The ethical and legal dilemmas surrounding deepfakes are profound, as they raise questions about consent, attribution, and accountability in an era where AI systems can replicate human behavior with alarming fidelity. As deepfake technology continues to evolve, it is imperative that we develop robust detection mechanisms, ethical guidelines, and legal frameworks to safeguard against their malicious use while preserving the integrity of information and privacy in the digital landscape.

Generative AI refers to a subset of artificial intelligence that focuses on the creation of data, often in the form of text, images, or videos (Muller, 2022). Techniques like Generative Adversarial Networks (GANs) have enabled machines to produce content that can be difficult to distinguish from content created by humans. This technological capability has raised significant privacy concerns. The Generative AI technologies are capable of:

- Generating fake content, including text, images, and videos, for malicious purposes.
- Creating deepfake videos that can impersonate individuals and spread disinformation.
- Automating the process of data collection and manipulation, including social engineering attacks.

These capabilities give rise to potential privacy infringements when used by nefarious actors. The ethical and legal implications are substantial, as they involve the manipulation of personal data and the potential for harm. Generative AI, a transformative branch of artificial intelligence, has made remarkable strides in generating human-like content, ranging from text and images to music and even video (Singh, 2022). While these advancements hold great promise in various domains, they have also given rise to profound privacy concerns that warrant careful consideration.

One of the foremost privacy concerns associated with generative AI is the potential for data misuse. These AI models require massive datasets for training, often comprising vast amounts of personal and sensitive information. The collection and utilization of such data raise ethical questions about consent and data protection. The mishandling or unauthorized access to these datasets can result in significant privacy breaches, endangering individuals' personal information and potentially leading to identity theft, financial fraud, or even the exposure of confidential medical records. Moreover, generative AI models, particularly those using deep learning techniques, have the capability to produce deepfake content, which includes convincingly realistic videos and images that can be used to impersonate individuals or fabricate events that never occurred (Jovanovic, 2022). This poses a substantial threat to personal privacy, as it becomes increasingly challenging to discern between genuine and manipulated content. Deepfakes can be exploited for various malicious purposes, from defaming individuals to undermining trust in media and public discourse.

The ethical dilemmas associated with generative AI and privacy are multifaceted. As, AI-generated content becomes indistinguishable from human creations, questions regarding authorship and intellectual property rights become complex (Qadir, 2023). Determining who is responsible for content generated by machines can be challenging, raising issues of accountability in cases of privacy infringements or content that violates ethical standards. The global nature of the internet and the borderless dissemination of content make it difficult to regulate and enforce privacy protections effectively. Privacy laws and regulations vary from one jurisdiction to another, complicating efforts to address these concerns comprehensively on a global scale. Balancing the benefits of AI with safeguarding individuals' privacy rights requires a multifaceted approach that encompasses ethical considerations, robust data protection measures, and thoughtful regulation (Foster, 2022). As generative AI continues to evolve, finding solutions to these privacy challenges becomes increasingly crucial to ensure that this technology serves as a force for good while respecting the fundamental right to privacy.

## NEFARIOUS ACTORS AND THEIR TACTICS

Nefarious actors in the digital realm represent a persistent and evolving threat to individuals, organizations, and even governments. These malicious entities employ a wide range of tactics to exploit vulnerabilities in the digital landscape for their own gain. They include cybercriminals, hacktivists, state-sponsored hackers, and more. Their tactics encompass activities such as hacking into computer systems and networks, using phishing attacks to deceive individuals into revealing sensitive information, deploying malware to compromise devices and data, and engaging in various forms of cyber-espionage. Nefarious actors can also manipulate social engineering techniques to manipulate human behavior, spreading disinformation or engaging in identity theft (Kather, 2022). The consequences of their actions can be severe, ranging from financial losses and privacy breaches to geopolitical tensions and national security threats. Combating these actors requires a multifaceted approach that involves robust cybersecurity measures, public awareness and education, international cooperation, and the development of sophisticated detection and response mechanisms to stay one step ahead of their evolving tactics.

### Types of Nefarious Actors

In the intricate landscape of digital threats, various types of nefarious actors with distinct motivations and objectives operate on the fringes of the online world. These actors encompass a wide spectrum, from financially driven cybercriminals who seek monetary gains through hacking and fraud to hacktivists who leverage their technical skills to advance political or social causes. State-sponsored hackers, often the most sophisticated, carry out espionage and cyber-attacks on behalf of governments, while script kiddies, with limited technical expertise, engage in digital mischief. Insiders within organizations may misuse their access for personal gain or sabotage. Phishers cunningly deceive individuals with the aim of extracting sensitive information, while malware authors create malicious software to compromise systems (Chui, 2023). Scammers employ various confidence tricks, and nation-state actors pursue geopolitical interests through cyber means. Organized crime groups, both online and offline, execute digital operations to further illicit activities. Recognizing these diverse categories of nefarious actors is crucial for devising comprehensive cybersecurity strategies and responses to protect against evolving digital threats and privacy breaches.

Nefarious actors in the digital realm come in various forms, each with distinct motivations and objectives. Here are some common types of nefarious actors:

Cybercriminals: These individuals or groups engage in criminal activities online with the primary goal of financial gain. They may carry out activities such as hacking into systems to steal sensitive data, launching ransomware attacks to extort money, or conducting online fraud, including identity theft and credit card fraud.

Hacktivists: Hacktivists combine hacking skills with a political or social agenda. They aim to promote a specific cause or ideology by disrupting websites, leaking sensitive information, or conducting cyberattacks against organizations or governments they oppose.

State-Sponsored Hackers: Nation-states employ highly skilled hackers to further their strategic interests. These actors engage in cyber-espionage, targeting foreign governments, organizations, or individuals to gather intelligence or disrupt critical infrastructure.

Script Kiddies: These are amateur hackers with limited technical skills who typically use pre-written scripts or tools to launch attacks. Their motivations can vary, ranging from seeking notoriety to causing mischief.

Insiders: Insiders are individuals within an organization who misuse their privileged access to commit malicious activities. They may steal sensitive data, sabotage systems, or compromise security from within.

Phishers: Phishers use social engineering techniques to deceive individuals into revealing sensitive information, such as login credentials, credit card numbers, or personal data. Phishing attacks often take the form of deceptive emails or websites.

Malware Authors: These individuals create malicious software, such as viruses, worms, Trojans, and ransomware, with the intent of infecting and compromising computer systems. Malware authors may operate independently or be part of larger criminal networks.

Scammers: Scammers employ various tactics to defraud individuals or organizations. This can include advance-fee fraud, Ponzi schemes, tech support scams, or other confidence tricks aimed at financial gain.

Nation-State Actors: Some nation-states engage in cyber-espionage or cyber-attacks to advance their geopolitical agendas, steal intellectual property, or disrupt foreign infrastructure. These operations often involve highly sophisticated and well-funded hacking teams.

Organized Crime Groups: Criminal organizations with a digital focus engage in activities such as data breaches, online fraud, and cyber-extortion. They operate similarly to traditional criminal enterprises but leverage technology for illicit gains.

Understanding the diverse motivations and tactics of these nefarious actors is essential for developing effective cybersecurity measures and strategies to protect against digital threats and privacy infringements (Porche, 2016).

## 4.2 Nefarious Actors and Privacy Infringements

The misuse of generative AI by nefarious actors is exemplified by numerous case studies and examples. For instance, deepfake videos have been used to create fake political speeches and defame individuals. Chatbots powered by AI have impersonated individuals for fraudulent purposes, leading to unauthorized data access and manipulation. Understanding these cases is crucial to assessing the legal and ethical dimensions of generative AI. Nefarious actors, often operating with malicious intent, represent a looming threat to privacy in the digital age. Their activities span a wide spectrum, from cybercriminals seeking financial gain to state-sponsored actors aiming to manipulate public opinion or engage in espionage. As technology advances, these actors have found increasingly sophisticated ways to infringe upon personal privacy, posing significant challenges for individuals, organizations, and governments (Anderson, 2023).

One of the most pervasive privacy infringements orchestrated by nefarious actors is data breaches. Cybercriminals employ a variety of tactics, including hacking, phishing, and malware, to gain unauthorized access to databases containing personal information. These breaches can expose a treasure trove of sensitive data, from credit card details and social security numbers to health records and personal emails. The consequences of such infringements can be devastating, leading to identity theft, financial losses, and long-lasting emotional distress for the victims. Another alarming manifestation of privacy infringements by nefarious actors is the creation and dissemination of deepfake content (Epstein, 2023). Using generative AI and machine learning, these actors can produce highly convincing fake videos and audio recordings, often featuring individuals saying or doing things they never did. These manipulative deepfakes can be used for various purposes, from character assassination to spreading false information, eroding trust in digital media and exacerbating the challenge of distinguishing between genuine and manipulated content.

The malicious actors engage in surveillance and espionage activities that violate personal privacy on a massive scale. State-sponsored actors, in particular, have been known to conduct cyber-espionage campaigns, targeting individuals, organizations, and even governments. They gather intelligence, trade secrets, and sensitive information, often with the goal of furthering political or economic interests. These activities not only compromise individual privacy but also have far-reaching geopolitical implications, potentially leading to diplomatic tensions and conflicts. Addressing the privacy infringements orchestrated by nefarious actors requires a multi-pronged approach. Enhanced cybersecurity measures, including robust encryption and authentication protocols, are essential to protect sensitive data from unauthorized access (Zohny, 2023). Additionally, public awareness and education play a crucial role in preventing successful phishing attacks and other forms of cybercrime. The development of AI-driven detection tools capable of identifying deepfakes and other manipulated content is also a critical step in mitigating the impact of this emerging threat. Their tactics, ranging from data breaches to deepfake creation and cyber-espionage, demand concerted efforts from individuals, organizations, and governments to safeguard personal information and protect the integrity of digital communication. Balancing the benefits of technology with the imperative of privacy protection is an ongoing challenge that underscores the importance of continuous vigilance and innovation in the realm of cybersecurity and privacy preservation.

## 4.3 Data Breaches and Unauthorized Access

The data breaches and unauthorized access pose significant privacy and security risks within the realm of generative AI. Generative AI models, like many other AI systems, often require access to extensive datasets for training and fine-tuning. These datasets can contain a plethora of personal and sensitive information, making them valuable targets for malicious actors (Sun, 2022). There are important points as how data breaches and unauthorized access manifest in the context of generative AI-

Unauthorized Access to Training Data: During the development and training of generative AI models, developers and organizations must ensure that access to training datasets is restricted to authorized personnel only. Unauthorized access, whether from internal or external actors, can result in the exposure of sensitive data. This can lead to privacy infringements and the misuse of personal information for nefarious purposes.

Data Leakage from AI Models: Trained generative AI models can inadvertently memorize and reproduce parts of the training data, including sensitive information. In some cases, this can result in data leakage through the AI-generated content, posing privacy risks to individuals whose information was part of the training dataset. Unauthorized access to the AI model's parameters or outputs could potentially lead to the extraction of sensitive data.

Privacy Violations through Content Generation: Malicious actors can leverage generative AI to create content that infringes upon privacy rights. For instance, deepfake technology can be used to produce fake videos or audio recordings that impersonate individuals, potentially leading to defamation, privacy breaches, or the dissemination of false information.

Data Harvesting and Phishing Attacks: Generative AI-generated content, such as realistic-looking emails or messages, can be used in phishing campaigns to deceive individuals into revealing sensitive information. Nefarious actors may create convincing AI-generated personas to manipulate and exploit victims.

Attacks on AI Infrastructure: The infrastructure supporting generative AI models, including servers and cloud-based resources, can be targeted for unauthorized access or disruption. A successful attack on the infrastructure can compromise the integrity and security of the AI system, potentially leading to privacy breaches.

So, addressing these risks requires a multi-pronged approach as organizations and developers working with generative AI must prioritize data protection and implement stringent access controls. They should also employ encryption and secure storage practices to safeguard training data and AI models. Detection mechanisms to identify potential data leakage and misuse are vital, as are robust cybersecurity measures to protect against unauthorized access to AI infrastructure. As generative AI continues to advance, it is essential to remain vigilant in mitigating the privacy and security risks associated with data breaches and unauthorized access (Cooper, 2023)

## 4.4 The Rise of Deepfake Manipulation

The rise of deepfake manipulation marks a critical inflection point in the evolution of digital technology, where the capabilities of generative AI intersect with the potential for widespread misinformation and privacy infringements. Deepfakes refer to artificially generated content, often in the form of hyper-realistic videos, audio recordings, or images, that seamlessly blend fabricated elements with genuine footage. These manipulations are created using deep learning algorithms that analyze and synthesize human-like features, gestures, and speech patterns, making it increasingly challenging to discern fact from fiction.

Deepfakes have emerged as a double-edged sword, offering creative possibilities in film, entertainment, and even medical research, while simultaneously presenting profound ethical and societal challenges. Malicious actors have harnessed deepfake technology for deceptive purposes, ranging from political disinformation campaigns to character assassination and privacy breaches. The implications are far-reaching, as deepfakes can be used to fabricate statements, actions, or events that never occurred, eroding trust in digital media and causing real-world consequences.

One of the most concerning aspects of deepfake manipulation is its potential to infringe upon individual privacy rights. Deepfake content can depict individuals in compromising or defamatory situations, often without their consent, leading to personal and reputational harm. Furthermore, deepfakes blur the lines of consent and attribution, as individuals can become unwitting participants in fabricated narratives. As the technology behind deepfake manipulation continues to advance, there is an urgent need for countermeasures (Peltz, 2020). These include the development of AI-driven detection tools capable of identifying deepfake content, public awareness campaigns to educate individuals about the existence and implications of deepfakes, and legal frameworks to address the malicious use of this technology.

In essence, the rise of deepfake manipulation underscores the importance of striking a delicate balance between technological innovation and responsible use. While generative AI offers exciting possibilities, its unchecked misuse poses a significant threat to privacy, trust, and the integrity of digital information. Safeguarding against the malicious use of deepfake technology is an evolving challenge that necessitates vigilance, ethical considerations, and collaborative efforts across technology, policy, and society (Longoni, 2022).

## 5. LEGAL FRAMEWORKS AND REGULATIONS

Legal frameworks and regulations play a pivotal role in addressing the complex landscape of generative AI and technological privacy concerns, particularly when it comes to infringements and the malicious use of personal data by nefarious actors. These frameworks are essential in ensuring that both the developers and users of AI technologies adhere to ethical standards, safeguard individual privacy, and are held accountable for any misuse. At the heart of this issue is the need to adapt existing legal frameworks to the rapid advancements in AI. Traditionally, legal systems have struggled to keep pace with the lightning-fast evolution of technology, and generative AI is no exception (Ahuja, 2023). To effectively examine responsibility and deliberation in cases of privacy infringements, regulations must be agile and responsive. This means establishing clear definitions and criteria for what constitutes responsible AI development, deployment, and data handling. Additionally, there must be provisions that specifically address the misuse of AI-generated content for nefarious purposes, such as deepfake creation or personal data theft.

In parallel, it is crucial to develop legal mechanisms that hold those responsible for privacy breaches accountable. This involves determining liability in cases where AI systems are involved in privacy violations (Singh, 2019). Since AI often operates autonomously, the question of responsibility becomes more complex. Legal frameworks need to establish guidelines for attributing responsibility, whether it lies with the developers, the users, or the AI systems themselves. Such legal clarity not only ensures that those affected by privacy infringements can seek justice but also serves as a deterrent against malicious actions.

Also, to effectively address privacy concerns tied to generative AI, regulators must promote transparency and ethical practices. This includes requiring developers and organizations to disclose how AI systems collect, store, and use personal data. Informed consent mechanisms should be strengthened to ensure individuals understand the implications of sharing their data with AI models. Additionally, stringent data protection regulations must be in place to govern the handling of sensitive personal information, with severe penalties for data breaches or misuse (Leiser, 2020).

The global cooperation and harmonization of regulations are also critical. The internet transcends borders, making it imperative that legal frameworks concerning generative AI and privacy are internationally consistent. Collaborative efforts among nations can establish common standards, facilitating smoother cross-border enforcement and helping prevent regulatory arbitrage.

In the face of evolving technology, legal frameworks and regulations play a critical role in protecting individual privacy. Key regulations, such as the European Union's General Data Protection Regulation (GDPR) and the California Consumer Privacy Act (CCPA), aim to safeguard personal data and provide individuals with certain rights concerning their data.

Legal frameworks and regulations pertaining to the examination of responsibility and deliberation of generative AI tethered technological privacy concerns, especially in cases of infringements and the use of personal data by nefarious actors, constitute a critical aspect of addressing the evolving challenges posed by AI technology in the digital age (Lim, 2023). While these frameworks may vary from one jurisdiction to another, several overarching principles and regulations are instrumental in shaping the legal dynamics surrounding these issues-

Data Protection Regulations: Data protection laws such as the European Union's General Data Protection Regulation (GDPR) set stringent standards for the collection, storage, and processing of personal data. These regulations require organizations to obtain informed consent, implement data security measures, and provide individuals with the right to access and control their personal information.

Intellectual Property Laws: Intellectual property frameworks play a pivotal role in addressing ownership and attribution concerns associated with generative AI. Copyright laws may need to adapt to determine authorship in cases where AI-generated content blurs the lines between human and machine creation.

Privacy Laws: Privacy laws, often intertwined with data protection regulations, establish the parameters for how personal information should be handled. They may define what constitutes a privacy violation and outline the legal consequences for breaches.

Cybersecurity Regulations: Laws related to cybersecurity oblige organizations to implement robust security measures to protect data from unauthorized access and breaches. These regulations may also stipulate reporting requirements in the event of a data breach.

Consumer Protection Laws: Consumer protection regulations safeguard individuals against deceptive practices, fraud, and the misuse of personal data. These laws can be instrumental in holding organizations accountable for unethical AI-driven activities.

Anti-Hacking and Cybercrime Laws: Legal frameworks targeting hacking, cybercrime, and unauthorized access are essential in deterring malicious actors who seek to exploit generative AI for illicit purposes.

Ethical AI Guidelines: While not legally binding, ethical guidelines and principles for AI development and deployment, such as those proposed by organizations like the IEEE or AI ethics boards, provide valuable ethical considerations that can inform the legal landscape.

International Agreements: Given the global nature of the internet and digital technologies, international agreements and treaties may be necessary to ensure consistency and cooperation in addressing AI-related privacy concerns and cyber threats.

It's important to recognize that the legal frameworks surrounding generative AI and privacy are still evolving and face ongoing challenges as technology advances. The development of comprehensive and adaptable legal mechanisms is crucial for balancing the benefits of generative AI with the protection of privacy and ethical considerations (Rivera, 2014). Legal professionals, policymakers, and technologists must collaborate to create regulatory environments that can effectively address the responsibilities and dilemmas arising from the intersection of AI and privacy concerns.

## 5.1 Challenges in Legal Adaptation

There are multifarious challenges in legal adaptation to the rapid advancements in generative AI and technological privacy concerns are manifold. The dynamic nature of AI technology often outpaces the ability of legal frameworks to keep up. Some of the primary challenges include the need to define and allocate responsibility in cases of AI-generated content, especially when it blurs the lines between human and machine authorship. Moreover, ensuring transparency and explainability in AI systems, which is crucial for accountability, can be difficult when dealing with complex, deep learning models. Privacy laws must grapple with the vast amounts of data required to train AI models and the potential for misuse or data breaches. Additionally, achieving international harmonization of AI regulations presents challenges, as legal systems vary widely across regions, making cross-border enforcement and compliance complex. The legal community faces the formidable task of striking a balance between fostering innovation and safeguarding individual rights and societal values in an era increasingly influenced by AI and generative technologies. Addressing these challenges requires ongoing collaboration among legal experts, policymakers, technologists, and the broader public to shape effective and adaptive legal frameworks (Gabison, 2016).

## 5.2 Accountability and Responsibility in AI-Generated Content

The accountability and responsibility in the realm of AI-generated content are complex and multifaceted issues. As generative AI technology advances, determining who is accountable for content created by machines becomes increasingly challenging. Traditional legal frameworks often rely on human actors as responsible parties, but when AI systems autonomously generate content, questions arise regarding the allocation of responsibility (Kenneally, 2012). The ethical considerations also come into play, as the lack of human intent in AI-generated content blurs the lines of moral responsibility. In cases of privacy infringements, defamation, or even misinformation, determining culpability becomes intricate. The legal frameworks should evolve to encompass AI-specific accountability provisions, defining clear lines of responsibility while considering the role of human oversight and ethical guidelines. The ultimate goal is to strike a balance that encourages responsible AI development and usage, mitigates potential harm, and preserves accountability in an era where machines play an increasingly prominent role in content creation and decision-making.

## 5.3 Data Protection and Privacy Regulations

The data protection and privacy regulations play a crucial role in governing the use of generative AI, especially concerning the handling of personal data. Here are some key aspects of data protection and privacy regulations in the context of generative AI-

Informed Consent: Regulations often require organizations and developers using generative AI to obtain informed consent from individuals whose data is being used for training or other purposes. This consent should be explicit and transparent, ensuring that individuals understand how their data will be used.

Data Minimization: Privacy regulations emphasize the principle of data minimization, which means that only the minimum amount of personal data necessary for a specific purpose should be collected and processed. Generative AI developers should apply this principle when curating datasets.

Data Security: Generative AI systems must adhere to stringent data security measures to protect personal information from breaches and unauthorized access. Encryption, access controls, and secure storage practices are essential components of data protection.

User Rights: Privacy regulations often grant individuals certain rights over their data, including the right to access, correct, or delete their personal information. Organizations deploying generative AI systems must facilitate these rights.

Purpose Limitation: Data collected for generative AI purposes should only be used for the intended purposes disclosed to individuals during data collection. Repurposing data for unrelated uses may violate privacy regulations.

Accountability and Transparency: Organizations must be accountable for how they use generative AI and should be transparent about their data practices. Privacy policies and disclosures should detail how AI-generated content is created and how personal data is handled.

Cross-Border Data Transfer: When data is transferred across borders, it must comply with relevant regulations. Some jurisdictions have specific requirements for cross-border data transfer, which generative AI developers need to adhere to.

Impact Assessments: Some regulations may require organizations to conduct privacy impact assessments before deploying generative AI systems to evaluate the potential risks to individuals' privacy and take steps to mitigate them.

Data Retention Policies: Organizations should establish clear data retention policies, specifying how long personal data will be stored and when it will be deleted to minimize privacy risks.

Data Breach Reporting: Regulations often mandate the prompt reporting of data breaches to authorities and affected individuals. Generative AI developers should have mechanisms in place to detect and respond to data breaches.

Children's Data: Special provisions exist in many regulations to protect the privacy of children's data. Generative AI applications targeting or involving children require heightened safeguards and parental consent.

The compliance with data protection and privacy regulations is paramount for organizations and developers working with generative AI (Phillips, 2017). The failure to adhere to these regulations can result in significant legal and financial consequences, as well as damage to reputation. As generative AI continues to advance, staying current with evolving privacy laws and best practices is essential to navigate this complex regulatory landscape effectively.

## 6. BALANCING INNOVATION AND PRIVACY: GENERATIVE AI

The balancing innovation and privacy in the realm of generative AI presents one of the most pressing challenges of our digital age. The transformative potential of generative AI is undeniable, from revolutionizing creative industries to enhancing healthcare diagnostics and facilitating personalized content recommendations. However, this very potential is intricately intertwined with the need to protect individual privacy and guard against the misuse of personal data. As generative AI models become increasingly sophisticated and data-hungry, they raise concerns about informed consent, data security, and the potential for deep privacy infringements (Gerlick, 2020).

The finding of equilibrium requires a multifaceted approach as on one hand, it necessitates the development and adherence to robust data protection regulations that prioritize individuals' rights to control their personal information. Such regulations must evolve to encompass the unique challenges posed by generative AI, addressing issues of consent for data usage, the ethical implications of AI-generated content, and the clear definition of data ownership in an AI-driven ecosystem. On the other hand, fostering innovation requires a supportive environment that encourages responsible AI development and usage (Dul, 2022). This entails creating mechanisms for transparent AI development practices, developing ethical guidelines that inform AI design, and ensuring that AI systems are equipped with safeguards to prevent malicious use. Striking this balance is crucial to ensure that generative AI continues to drive progress and innovation while preserving the fundamental principles of privacy, ethics, and individual autonomy in the digital landscape. It calls for collaboration among policymakers, technologists, ethicists, and the public to forge a path that navigates the exciting frontiers of AI while safeguarding the values that underpin a just and equitable society.

## 6.1 AI-Driven Detection of Deepfakes

AI-driven detection of deepfakes represents a critical frontier in the ongoing battle against the malicious use of synthetic media. Deepfake technology has become increasingly sophisticated, making it challenging to discern authentic content from manipulated or fabricated material. In response, AI-powered detection tools leverage machine learning and deep neural networks to scrutinize digital media for subtle inconsistencies, anomalies, and artifacts that are indicative of deepfake manipulation. These detection systems hold promise in identifying fake videos, audio recordings, and images, helping to mitigate the potential harm caused by the spread of deceptive or harmful content (Hoofnagle, 2019). However, the cat-and-mouse game between deepfake creators and detection algorithms persists, pushing the boundaries of AI-driven detection further. Ensuring the effectiveness and accessibility of these tools, while continually advancing their capabilities, remains essential in safeguarding the integrity of digital media and preserving trust in the information age.

## 6.2 Secure Data Handling Practices

Generative AI's ability to create content that closely mimics human creativity is a testament to its transformative potential. However, this power comes with significant responsibility, particularly in the realm of secure data handling. Generative AI models, like other AI systems, often require extensive datasets for training and fine-tuning. These datasets can contain sensitive and personal information, and their misuse can result in privacy breaches and ethical dilemmas. Therefore, adopting secure data handling practices is paramount.

To harness the benefits of generative AI while safeguarding data, organizations and developers must implement robust data protection measures. This includes stringent access controls, encryption, and anonymization techniques to protect sensitive information (Hlávka, 2020). Furthermore, data anonymization and minimization principles should be applied to ensure that only necessary data is used, and individuals' privacy rights are respected. The transparency in data usage and adherence to relevant data protection regulations, such as the GDPR, are imperative. Clear consent mechanisms and privacy policies that inform users about how their data will be used in AI training and content generation are essential. As, balancing innovation with data security is a pivotal challenge in the age of generative AI. By prioritizing secure data handling practices, we can unlock the potential of this technology while upholding individuals' rights and privacy in an increasingly data-driven world.

## 6.3 Ensuring Data Privacy in AI Training Sets

Ensuring data privacy in AI training sets is an imperative task as AI systems, including generative AI models, often rely on large and diverse datasets to learn and perform effectively. The process begins with the collection, curation, and preparation of these datasets, and it is here that privacy safeguards must be rigorously implemented. The first crucial step is data anonymization, where personally identifiable information is removed or obfuscated to prevent the identification of individuals. Furthermore, organizations must adhere to data minimization principles, ensuring that only the minimum amount of data necessary for the AI's intended purpose is collected and utilized, thereby reducing the risk of privacy breaches (Almeida, 2023).

The robust data security measures, including encryption and access controls, should be in place to protect these datasets from unauthorized access or breaches. It's essential to establish stringent access protocols, ensuring that only authorized personnel can access and use the data for AI training. Moreover, data retention policies should be defined, specifying how long the data will be retained and when it should be securely deleted to mitigate privacy risks.

In terms of transparency and compliance, organizations handling training data must be clear and transparent about their data practices. Individuals should be informed about how their data will be used, and consent mechanisms should be in place to obtain explicit consent for data usage in AI training. Compliance with relevant data protection regulations, such as the GDPR, is essential, as these regulations set forth strict requirements for data handling, security, and privacy rights (Khan, 2008).

As the AI landscape evolves, so must the approach to data privacy in training sets. Striking a balance between innovation and data protection is crucial, and it necessitates a collaborative effort between AI developers, policymakers, and privacy advocates to ensure that AI technologies can thrive while upholding the fundamental principles of data privacy and individual rights.

## 7. INTERNATIONAL COOPERATION AND HARMONIZATION

International cooperation and harmonization are indispensable in the legal dynamics for examining responsibility and deliberation of generative AI tethered technological privacy concerns, especially in cases involving infringements and the use of personal data by nefarious actors. The borderless nature of the digital realm, where AI technologies operate and data traverses, underscores the necessity for collaborative efforts among nations. A unified global approach is critical to establish consistent legal standards, share threat intelligence, and coordinate responses to cross-border privacy infringements. This cooperation can take various forms, including bilateral agreements, international treaties, and the alignment of existing legal frameworks to address emerging challenges posed by AI and data privacy (Hurel, 2018).

The international harmonization should strive to strike a delicate balance between fostering innovation and safeguarding individual rights. By harmonizing legal dynamics, nations can collectively define responsibilities, obligations, and ethical boundaries for AI developers, users, and malicious actors alike. It can lead to the establishment of common principles for data protection, cross-border data sharing, and incident reporting, strengthening the global community's ability to combat AI-related privacy threats. But achieving international cooperation and harmonization is not without its challenges. Differences in legal traditions, cultural norms, and geopolitical interests may hinder the harmonization process (Myrzashova, 2023). Moreover, the pace of technological advancement often outstrips the ability of international bodies to adapt and regulate effectively. Nevertheless, the pressing need to protect privacy in the face of evolving AI technologies necessitates sustained efforts in international cooperation, promoting responsible AI development, deterring malicious activities, and safeguarding the digital privacy rights of individuals worldwide.

## 8. CONCLUSION

The examination of legal dynamics surrounding generative AI, technological privacy concerns, and the potential for infringements and misuse of personal data by nefarious actors is a complex and dynamic field. The rise of generative AI technologies has ushered in a new era of innovation and creativity, but it has also brought forth ethical, legal, and privacy challenges that demand urgent attention. Striking a balance between fostering technological progress and safeguarding individual rights and data privacy remains a formidable task. To address these challenges effectively, stakeholders must prioritize responsible AI development, transparency, and robust legal frameworks that adapt to the evolving landscape of technology (Boddington, 2021). Moreover, international collaboration and harmonization are essential to ensure a consistent and cohesive approach to the responsible use of generative AI. In this dynamic environment, the lensing of legal dynamics provides valuable insights into the complexities surrounding AI ethics, privacy, and accountability. As the field continues to evolve, it is imperative that we remain vigilant, adaptive, and committed to upholding the principles of ethics, privacy, and the responsible use of generative AI for the benefit of society as a whole.

The imperative of responsible AI development is paramount in an era where artificial intelligence is rapidly transforming our world. While AI offers immense potential for innovation, efficiency, and improved decision-making, it also brings forth ethical, social, and legal considerations that demand careful attention. Responsible AI development encompasses transparency, fairness, accountability, and bias mitigation. It entails ensuring that AI systems are designed to respect individual privacy, human rights, and societal values. It also necessitates collaboration between developers, policymakers, and the public to establish ethical guidelines, regulatory frameworks, and best practices that promote the responsible use of AI. The consequences of neglecting responsible AI development can be profound, ranging from algorithmic bias and discrimination to the erosion of privacy and security. Therefore, embracing ethical AI practices is not just an option but an imperative to harness the full potential of AI technology while safeguarding our collective well-being and values in an AI-driven world (Mount, 2020).

So, balancing innovation, privacy, and ethics in the realm of generative AI represents a formidable yet essential task. The potential for AI to drive innovation across industries, from art and entertainment to healthcare and beyond, is undeniable. However, this progress must be guided by a steadfast commitment to protecting privacy and upholding ethical principles. Generative AI's capacity to create lifelike content and make autonomous decisions requires a thorough examination of the ethical considerations it presents. Questions about data privacy, algorithmic bias, transparency, and accountability must be addressed proactively. In doing so, it can foster a climate where innovation thrives without compromising individual rights and societal values. Ethical AI practices, coupled with robust privacy regulations, can help ensure that AI technologies are developed and deployed responsibly. Collaborative efforts among AI developers, policymakers, and the broader society are paramount to strike this delicate balance, promoting technological advancement while safeguarding privacy and ethics in an increasingly AI-driven world. Ultimately, it is the judicious integration of innovation, privacy protection, and ethical considerations that will enable us to harness the full potential of generative AI for the betterment of humanity.

Generative artificial intelligence (AI) represents a remarkable leap in technological innovation, enabling machines to generate content that closely resembles human creations. From realistic text generation to the creation of convincing deepfake videos, generative AI has immense potential across various domains. However, this technological progress has brought forth significant privacy concerns and ethical dilemmas, exacerbated by the potential for misuse by nefarious actors.

## 9. FUTURE DIRECTIONS

The future directions and challenges in the lensing of legal dynamics concerning generative AI, technological privacy concerns, and the potential for infringements by nefarious actors are undoubtedly multifaceted. As generative AI technologies continue to advance, our legal and ethical frameworks must evolve in tandem to keep pace with these developments. Future directions will likely involve refining existing regulations and creating new ones specifically tailored to generative AI. Stricter data protection and privacy laws, clearer guidelines on data ownership and consent, and enhanced mechanisms for AI accountability will be essential components of this evolution. However, challenges persist in this journey. The rapidly evolving nature of AI technology often outpaces regulatory efforts, making it challenging to enforce and adapt laws effectively. Moreover, the global nature of the internet and AI's borderless impact necessitate international collaboration and harmonization—a process that can be complex given differing legal traditions and cultural norms.

The nefarious actors are continually devising new strategies to exploit AI for malicious purposes, making it crucial to stay ahead in the battle against AI-enabled cyber threats. The lensing of legal dynamics in the context of generative AI and privacy concerns is an ongoing endeavor that requires adaptability, international cooperation, and an unwavering commitment to safeguarding individual rights. As we move forward, addressing these challenges will be crucial to ensuring the responsible development and deployment of generative AI for the benefit of society while minimizing the potential for privacy infringements by malicious actors. As AI research continues to progress, we can expect generative models to become even more sophisticated, capable of producing content that is increasingly indistinguishable from human-created work. This evolution will fuel innovation across industries, from revolutionizing content creation in media and entertainment to enhancing medical imaging and diagnostics. Generative AI's potential for personalization will enable tailored experiences in everything from education to marketing, improving user engagement and satisfaction. Moreover, as ethical concerns surrounding AI continue to gain attention, we can anticipate advancements in AI ethics, such as better tools for bias detection and mitigation, and the development of more transparent and interpretable AI systems. However, these advancements also come with challenges, particularly in the realm of responsible AI usage and potential ethical dilemmas. As generative AI technologies evolve, a critical focus on ethical development and regulation will be vital to harness their benefits while mitigating potential risks.

## REFERENCES

Ahmad, W., & Dethy, E. (2019). Preventing surveillance cities: Developing a set of fundamental privacy provisions. *Journal of Science Policy & Governance, 15*(1), 1–11.

Ahuja, K., Hada, R., Ochieng, M., Jain, P., Diddee, H., Maina, S., & Sitaram, S. (2023). Mega: Multilingual evaluation of generative ai. *arXiv preprint arXiv:2303.12528*. doi:10.18653/v1/2023.emnlp-main.258

Almeida, F. (2023). Prospects of Cybersecurity in Smart Cities. *Future Internet, 15*(9), 285. doi:10.3390/fi15090285

Anderson, L. B., Kanneganti, D., Houk, M. B., Holm, R. H., & Smith, T. (2023). Generative AI as a tool for environmental health research translation. *GeoHealth, 7*(7), e2023GH000875.

Boddington, G. (2021). The Internet of Bodies—alive, connected and collective: The virtual physical future of our bodies and our senses. *AI & Society*, 1–17. PMID:33584018

Bozkurt, A. (2023). Generative artificial intelligence (AI) powered conversational educational agents: The inevitable paradigm shift. *Asian Journal of Distance Education*, *18*(1).

Brynjolfsson, E., Li, D., & Raymond, L. R. (2023). Generative AI at work (No. w31161). National Bureau of Economic Research.

Chui, M., Hazan, E., Roberts, R., Singla, A., & Smaje, K. (2023). *The economic potential of generative AI.*

Cooper, G. (2023). Examining science education in chatgpt: An exploratory study of generative artificial intelligence. *Journal of Science Education and Technology*, *32*(3), 444–452. doi:10.1007/s10956-023-10039-y

Dul, C. (2022). Facial Recognition Technology vs Privacy: The Case of Clearview AI. *QMLJ*, *1*.

Epstein, Z., Hertzmann, A., Akten, M., Farid, H., Fjeld, J., Frank, M. R., Groh, M., Herman, L., Leach, N., Mahari, R., Pentland, A. S., Russakovsky, O., Schroeder, H., & Smith, A. (2023). Art and the science of generative AI. *Science*, *380*(6650), 1110–1111. doi:10.1126/science.adh4451 PMID:37319193

Foster, D. (2022). *Generative deep learning*. O'Reilly Media, Inc.

Gabison, G. (2016). Policy considerations for the blockchain technology public and private applications. *SMU Sci. & Tech. L. Rev.*, *19*, 327.

Gerlick, J. A., & Liozu, S. M. (2020). Ethical and legal considerations of artificial intelligence and algorithmic decision-making in personalized pricing. *Journal of Revenue and Pricing Management*, *19*(2), 85–98. doi:10.1057/s41272-019-00225-2

Gipson Rankin, S. M. (2021). Technological tethereds: Potential impact of untrustworthy artificial intelligence in criminal justice risk assessment instruments. *Washington and Lee Law Review*, *78*, 647.

Hlávka, J. P. (2020). Security, privacy, and information-sharing aspects of healthcare artificial intelligence. In *Artificial Intelligence in Healthcare* (pp. 235–270). Academic Press. doi:10.1016/B978-0-12-818438-7.00010-1

Hoofnagle, C. J., Kesari, A., & Perzanowski, A. (2019). The Tethered Economy. *Geo. Wash. L. Rev.*, *87*, 783.

Huang, S., & Siddarth, D. (2023). Generative AI and the digital commons. *arXiv preprint arXiv:2303.11074.*

Hurel, L. M. (2018). *Architectures of security and power: IoT platforms as technologies of government.* [MSc diss., London School of Economics and Political Science].

Jo, A. (2023). The promise and peril of generative AI. *Nature*, *614*(1), 214–216.

Jovanovic, M., & Campbell, M. (2022). Generative artificial intelligence: Trends and prospects. *Computer*, *55*(10), 107–112. doi:10.1109/MC.2022.3192720

Kather, J. N., Ghaffari Laleh, N., Foersch, S., & Truhn, D. (2022). Medical domain knowledge in domain-agnostic generative AI. *NPJ Digital Medicine*, *5*(1), 90. doi:10.1038/s41746-022-00634-5 PMID:35817798

KenneallyE.DittrichD. (2012). The menlo report: Ethical principles guiding information and communication technology research. SSRN 2445102. doi:10.2139/ssrn.2445102

Khan, S. M. (2008). Copyright, Data Protection, and Privacy with Digital Rights Management and Trusted Systems: Negotiating a Compromise between Proprietors and Users. *ISJLP*, *5*, 603.

Lathrop, B. (2019). The Inadequacies of the Cybersecurity Information Sharing Act of 2015 in the Age of Artificial Intelligence. *The Hastings Law Journal*, *71*, 501.

Leiser, M. R., & Dechesne, F. (2020). Governing machine-learning models: challenging the personal data presumption. *International data privacy law*, *10*(3), 187-200.

Lim, W. M., Gunasekara, A., Pallant, J. L., Pallant, J. I., & Pechenkina, E. (2023). Generative AI and the future of education: Ragnarök or reformation? A paradoxical perspective from management educators. *International Journal of Management Education*, *21*(2), 100790. doi:10.1016/j.ijme.2023.100790

Liu, D., Nanayakkara, P., Sakha, S. A., Abuhamad, G., Blodgett, S. L., Diakopoulos, N., & Eliassi-Rad, T. (2022, July). Examining Responsibility and Deliberation in AI Impact Statements and Ethics Reviews. In *Proceedings of the 2022 AAAI/ACM Conference on AI, Ethics, and Society* (pp. 424-435). AAAI. 10.1145/3514094.3534155

Longoni, C., Fradkin, A., Cian, L., & Pennycook, G. (2022, June). News from generative artificial intelligence is believed less. In *Proceedings of the 2022 ACM Conference on Fairness, Accountability, and Transparency* (pp. 97-106). ACM. 10.1145/3531146.3533077

Mount, M., Round, H., & Pitsis, T. S. (2020). Design thinking inspired crowdsourcing: Toward a generative model of complex problem solving. *California Management Review*, *62*(3), 103–120. doi:10.1177/0008125620918626

Muller, M., Chilton, L. B., Kantosalo, A., Martin, C. P., & Walsh, G. (2022, April). GenAICHI: generative AI and HCI. In CHI conference on human factors in computing systems extended abstracts (pp. 1-7). ACM. doi:10.1145/3491101.3503719

Myrzashova, R., Alsamhi, S. H., Shvetsov, A. V., Hawbani, A., & Wei, X. (2023). Blockchain meets federated learning in healthcare: A systematic review with challenges and opportunities. *IEEE Internet of Things Journal*, *10*(16), 14418–14437. doi:10.1109/JIOT.2023.3263598

NoyS.ZhangW. (2023). Experimental evidence on the productivity effects of generative artificial intelligence. *Available at* SSRN 4375283.

Peltz, J., & Street, A. C. (2020). Artificial intelligence and ethical dilemmas involving privacy. In Artificial Intelligence and Global Security: Future Trends, Threats and Considerations (pp. 95-120). Emerald Publishing Limited. doi:10.1108/978-1-78973-811-720201006

Phillips, M., Dove, E. S., & Knoppers, B. M. (2017). Criminal prohibition of wrongful re-identification: Legal solution or minefield for big data? *Journal of Bioethical Inquiry*, *14*(4), 527–539. doi:10.1007/s11673-017-9806-9 PMID:28913771

Porche, I. (2016). *Emerging cyber threats and implications*. RAND. doi:10.7249/CT453

Porsdam Mann, S., Earp, B. D., Nyholm, S., Danaher, J., Møller, N., Bowman-Smart, H., & Savulescu, J. (2023). Generative AI entails a credit–blame asymmetry. *Nature Machine Intelligence*, 1–4.

Qadir, J. (2023, May). Engineering education in the era of ChatGPT: Promise and pitfalls of generative AI for education. In *2023 IEEE Global Engineering Education Conference (EDUCON)* (pp. 1-9). IEEE. 10.1109/EDUCON54358.2023.10125121

Rivera, J., & Hare, F. (2014, June). The deployment of attribution agnostic cyberdefense constructs and internally based cyberthreat countermeasures. In *2014 6th international conference on cyber conflict (CyCon 2014)* (pp. 99-116). IEEE. 10.1109/CYCON.2014.6916398

Sharma, A., & Singh, B. (2022). Measuring Impact of E-commerce on Small Scale Business: A Systematic Review. *Journal of Corporate Governance and International Business Law*, *5*(1).

Singh, B. (2019). Profiling Public Healthcare: A Comparative Analysis Based on the Multidimensional Healthcare Management and Legal Approach. *Indian Journal of Health and Medical Law*, *2*(2), 1–5.

Singh, B. (2020). GLOBAL SCIENCE AND JURISPRUDENTIAL APPROACH CONCERNING HEALTHCARE AND ILLNESS. *Indian Journal of Health and Medical Law*, *3*(1), 7–13.

Singh, B. (2022). Understanding Legal Frameworks Concerning Transgender Healthcare in the Age of Dynamism. *ELECTRONIC JOURNAL OF SOCIAL AND STRATEGIC STUDIES*, *3*(1), 56–65. doi:10.47362/EJSSS.2022.3104

Singh, B. (2022). Relevance of Agriculture-Nutrition Linkage for Human Healthcare: A Conceptual Legal Framework of Implication and Pathways. *Justice and Law Bulletin*, *1*(1), 44–49.

Singh, B. (2022). COVID-19 Pandemic and Public Healthcare: Endless Downward Spiral or Solution via Rapid Legal and Health Services Implementation with Patient Monitoring Program. *Justice and Law Bulletin*, *1*(1), 1–7.

Singh, B. (2023). Blockchain Technology in Renovating Healthcare: Legal and Future Perspectives. In Revolutionizing Healthcare Through Artificial Intelligence and Internet of Things Applications (pp. 177-186). IGI Global.

Sun, J., Liao, Q. V., Muller, M., Agarwal, M., Houde, S., Talamadupula, K., & Weisz, J. D. (2022, March). Investigating explainability of generative AI for code through scenario-based design. In *27th International Conference on Intelligent User Interfaces* (pp. 212-228). ACM. 10.1145/3490099.3511119

van der Zant, T., Kouw, M., & Schomaker, L. (2013). *Generative artificial intelligence*. Springer Berlin Heidelberg.

Weisz, J. D., Muller, M., He, J., & Houde, S. (2023). Toward general design principles for generative AI applications. *arXiv preprint arXiv:2301.05578*.

Zohny, H., McMillan, J., & King, M. (2023). Ethics of generative AI. *Journal of Medical Ethics*, *49*(2), 79–80. doi:10.1136/jme-2023-108909 PMID:36693706

# Chapter 10

# Navigating the Legal and Ethical Framework for Generative AI: Fostering Responsible Global Governance

**Anuttama Ghose**
https://orcid.org/0000-0002-7210-4074
*Dr. Vishwanath Karad MIT-World Peace University, India*

**S. M. Aamir Ali**
https://orcid.org/0000-0002-8686-0217
*Symbiosis Law School, Symbiosis International University, India*

**Sachin Deshmukh**
https://orcid.org/0009-0004-3034-1946
*Dr. Vishwanath Karad MIT-World Peace University, India*

## ABSTRACT

*Generative AI systems have given incredible ability to independently produce a wide variety of content types, including textual, visual, and more. Complex issues with copyright protection and intellectual property rights have arisen as a result of this change. With a focus on fostering responsible global governance, this research delves into the complex legal and ethical considerations underlying Generative AI. The goal of this chapter is to take a look at the complicated legal issues that come up because of Generative AI's ability to generate material on its own. This chapter analyzes the current legal documents, legislation, and international treaties, focusing on ethical concerns. Ultimately, the authors want to have a positive impact on efforts to build responsible and efficient international frameworks for regulating Generative AI. This study provides an exhaustive case for the implementation of legal frameworks that can efficiently tackle the intricate legal and ethical quandaries posed by Generative AI, while simultaneously encouraging the progress of innovation and creativity.*

DOI: 10.4018/979-8-3693-1565-1.ch010

## INTRODUCTION

The advent of Generative AI represents a significant epoch of intellectual prowess and originality, encompassing a diverse range of applications that extend from the generation of content to the manifestation of creative endeavors. Nevertheless, this groundbreaking technology also introduces a plethora of complex legal matters that necessitate meticulous scrutiny. The present study delves into the intricate legal matters pertaining to Generative AI in the specified context, with the objective of offering valuable perspectives on the intricate legal structure necessary for its conscientious deployment.

Generative artificial intelligence (AI) systems exhibit the capacity to autonomously generate textual content, visual imagery, and various other forms of creative material, thereby introducing intricate considerations within the realm of intellectual property rights and copyright protection (Engelke, 2020). The present chapter delves into the intricate legal dimensions pertaining to the ownership and safeguarding of compositions generated by artificial intelligence (AI) systems. The present inquiry delves into matters pertaining to authorship, attribution, and the potential necessity for reassessment and modification of extant copyright statutes. The preservation of data privacy and security assumes paramount significance within the domain of artificial intelligence (Haugh, et al., 2018). The present study aims to examine the methodologies and approaches employed by Generative AI in the processing and exploitation of data, with specific attention directed towards scenarios involving the presence of sensitive or personal data. The legal facets encompass compliance with data protection norms, securing requisite consent, and addressing accountability in the event of data breaches. Furthermore, it is evident that the rise of liability concerns becomes conspicuous in instances where AI-generated content gives rise to deleterious outcomes or the propagation of erroneous data. The imperative for the evolution of legal frameworks arises from the need to delineate the precise boundaries of culpability, accountability, and the entitlements of individuals affected by the information generated by artificial intelligence. This encompasses the legal aspects pertaining to both civil and criminal affairs. The salience of transnational data transfers and the establishment of international standards for the governance of artificial intelligence is manifest within the global milieu. Legal issues encompass the imperative of harmonizing international regulations, resolving disputes concerning jurisdictional matters, and ensuring the appropriate application of artificial intelligence within the context of transnational borders (Somaya & Varshney, 2020).

The principal aim of this study is to investigate a range of legal and ethical considerations pertaining to Generative AI technology. The aforementioned concerns encompass a multitude of domains, including but not limited to intellectual property, data privacy, liability, and international governance. The proliferation of content generated by artificial intelligence (AI) engenders complex inquiries pertaining to authorship, ownership, and attribution, thereby posing challenges to the existing paradigms of intellectual property and copyright. The paramount importance of data privacy and security arises when Generative AI systems gradually interact with personal and sensitive data. A comprehensive examination is imperative in order to effectively tackle the legal dimensions surrounding the matters of consent, data protection legislation, and accountability in the event of data breaches. Furthermore, the potentiality of AI-generated content to cause harm, propagate misinformation, or blur the line between truth and falsehood gives rise to concerns regarding legal accountability and culpability (Cath, 2018).

Furthermore, this scholarly study undertakes a critical analysis of the existing global governance structures that are specifically concerned with Generative Artificial Intelligence (AI). This research assesses the relevance and effectiveness of global norms and standards within the domain, recognizing that Generative AI operates within a borderless digital realm that frequently exceeds national legal frameworks. The primary objective of this study is to investigate the progression and execution of co-operative global governance frameworks with the intention of fostering conscientious and principled utilization of Generative AI. In parallel, the study seeks to foster the cultivation of inventive and imaginative practices. The primary aim of this study is to offer a comprehensive and meticulous analysis of the intricate legal and ethical dimensions intertwined with Generative AI. The primary objective of this endeavor is to cultivate conscientious frameworks and offer informed recommendations that can effectively navigate the intricate landscape of legal and ethical considerations inherent to this particular domain (Daly et al., 2021). The overarching goal is to establish a global framework for governance that ensures the ethical and responsible deployment of Generative AI, thereby safeguarding the rights and well-being of individuals and communities, while simultaneously facilitating the continued progress of this transformative technology.

## UNDERSTANDING GENERATIVE AI THROUGH LEGAL LENS

Generative AI encompasses deep-learning models capable of producing superior text, graphics, and other forms of content by using the data they were trained on. The field of artificial intelligence has had several rounds of exaggerated promotion, but even those who doubt its potential acknowledge that the introduction of ChatGPT represents a significant shift (Choudhary & Ali, 2023). OpenAI's chatbot, with its most recent expansive language model, has the ability to compose poetry, share jokes, and produce essays that closely resemble human creations. By providing ChatGPT with a little input, it may generate love poetry like Yelp reviews or music lyrics in the distinctive manner of Nick Cave, which provokes us to explore the legal dimensions of such creativities (Ghose & Ali, 2023).

### Defining Generative AI: Legal Classification

The phrase "Artificial Intelligence" is not conducive to our public conversation. Artificial Intelligence lacks true intelligence. The phrase is very broad, without a clear definition, and hence cannot be debated with precision. But it is important for policy-makers to understand what they are encouraging or prohibiting. Passing a law to "restrict artificial intelligence" is a dangerous exercise under current definitions. Different functions of artificial intelligence create different problems for law and society. Generative AI creates not only new text, code, audio or video, but problems with deepfakes, plagiarism and falsehoods presented as convincing facts. AI that predicts whether a prisoner is likely to commit future crimes raises issues of bias, fairness and transparency. AI operating multi-ton vehicles on the road creates physical risks to human bodies. AI that masters the game the chess may not raise any societal issues at all. So why would politicians and courts treat them the same? Thus, in this study, the author attempts to categorically understand Generative AI through its functional ability in order to examine its accountability (Atlantic Council, 2023).

Generative AI utilizes predictive algorithms to generate outputs based on a given set of prompts. These AI tools may create very realistic imitations of human-like works, whether it be in the form of text or visual content, by meticulously reproducing each word or pixel. This may result in the creation of operational software code and functional websites, artistic works in the style of Jan van Eyck, scholarly articles on the use of symbols in The Scarlett Letter, or legal arguments in a contract dispute. These tools give rise to concerns around intellectual property and plagiarism, since they may be taught using unlawfully acquired content (Margetts, 2022). This technology has the capability to generate deep-fakes that are almost impossible to differentiate from concrete evidence. When it functions inadequately, it has the potential to produce utter gibberish portrayed as factual information. Generative artificial intelligence, including sophisticated language models like ChatGPT and image-generation software like Stable Diffusion, provide potent resources for people and enterprises. Additionally, they provoke significant and innovative inquiries about the utilization of data in AI models and the legal implications surrounding the outcomes of such models, such as a textual passage or a computer-generated picture (Zittrain, 2006).

## Literature Review

This chapter delves into the different legal and ethical considerations surrounding Generative AI technology. The mentioned concerns encompass a broad spectrum of areas, including intellectual property, data privacy, liability, and international governance. The exponential growth in content generated by artificial intelligence (AI) raises complex inquiries regarding the identification of the author, proprietor, and deserving recipient of recognition. These challenges pose significant obstacles to the existing frameworks of intellectual property and copyright. The research conducted by *Somaya* and *Varshey* highlights the importance of reassessing and potentially revising current copyright laws to effectively deal with issues related to the attribution, acknowledgment, and safeguarding of content produced by artificial intelligence (Somaya & Varshney, 2020). In his paper titled "AI, Society, and Governance: An Introduction," *Engelke* examines the significance and efficacy of global norms and standards in the field of Generative AI (Engelke, 2020). He acknowledges the borderless nature of AI operations, which often go beyond national legal frameworks. *Cath* also highlights the importance of advancing and implementing cooperative global governance frameworks that promote responsible and ethical use of Generative AI (Cath, 2018). The goal is to encourage innovative and creative approaches while ensuring ethical implementation of AI. The main objective of their research is to create an international framework for governing the ethical and responsible use of Generative AI. This framework aims to protect the rights and well-being of individuals and communities while supporting the ongoing advancement of AI technology. The ethical dilemma being discussed centres on the issue of bias and the concept of fairness, which poses a significant intellectual challenge. This topic is extensively explored by *Gupta*, *Parra*, and *Dennehy*, who delve into the debate surrounding the significance of cultural values and raise questions about racial and gender bias in AI creations (Gupta et al., 2022). The study also highlights the research conducted by *Haugh*, and explores the legal consequences of AI-generated content that can cause harm, spread misinformation, or blur the distinction between truth and falsehood (Haugh, et al., 2018). Clear frameworks are essential in determining accountability and culpability in such scenarios. *Barbanel* examines the proposed Algorithmic Accountability Act of 2019 to determine the accountability and liability related to AI in this context (Barbanel, 2019). *Clelia Casciola*'s work also emphasizes the importance of accountability in the context of healthcare (Casciola, 2022). The chapter explores the significant role of data privacy and security in the context of artificial intelligence, with a specific emphasis on Generative AI technology. In his paper, *Zimmer* highlights the importance of ensuring that Generative AI systems adhere to data protection regulations in order to protect sensitive and personal information (Zimmer, 2018). Katyal emphasizes the importance of obtaining consent when processing personal data within the context of Generative AI operations (Katyal, 2022). He emphasizes the necessity of holding individuals accountable in case of data breaches, highlighting the significance of implementing measures to handle and resolve violations of data privacy and security. In this context, *Dixit* also highlights the potential risks and emphasizes the importance of holding individuals accountable for data breaches (Dixit, 2023). *Katyal* and *Dixit* in their work also emphasize the need for establishing mechanisms to address and rectify breaches of data privacy and security (Dixit, 2023).

## Theoretical Groundwork on Generative AI: Exploring the Impact on Legal Practice and Jurisprudence

The emergence of Generative Artificial Intelligence (AI) has brought about significant transformations, warranting a comprehensive examination of its impact on the fundamental principles that underlie Intellectual Property Rights. Specifically, the domains of labor and personality theories are particularly relevant in this context. The conventional interpretations of the labor theory of intellectual property rights, which posit a correlation between ownership and human effort, encounter certain challenges when applied to generative artificial intelligence (AI) systems. The advent of artificial intelligence (AI) systems that autonomously produce innovative creations has engendered a conflation of the conventional correlation between human exertion and proprietorship of intellectual assets. The aforementioned assertion prompts a pivotal re-evaluation of legal frameworks to aptly integrate the collaborative dynamics intrinsic to the creative process, encompassing the participation of both human developers and autonomous AI systems. As a result, this necessitates the careful analysis and evaluation of the distribution of recognition and privileges. In a parallel vein, the theory of personality, which asserts that intellectual creations serve as extensions of an individual's identity, presents notable ethical quandaries within the contemporary era of generative artificial intelligence. The discernible attribute of content generated by artificial intelligence, disassociated from the idiosyncrasies of an individual human creator, necessitates a thorough reassessment of the ethical aspects pertaining to authorship and accountability. The complex and nuanced interaction between legal and ethical considerations assumes a prominent role, giving rise to fundamental inquiries regarding the entitlement of intellectual property rights to AI systems and, if such entitlement exists, the specific characteristics and extent of these rights. The imperative to ensure a judicious and equitable adjudication of the evolving landscape of intellectual property rights necessitates the reconciliation of legal frameworks with ethical imperatives as generative AI continues to advance (Katyal, 2022).

The cultural sciences have provided evidence to support the notion that human literary and artistic creations hold significant value for society at large. Crafted by individuals who possess a tangible existence, artistic creations serve as significant catalysts for societal and political transformation. These works exemplify innovative methodologies, thereby expanding the boundaries of societal progress and fostering novel prospects for the advancement of human civilization. The manifestation of human literary and artistic expression has the capacity to reflect deficiencies within the prevailing societal framework, unveil imperfections inherent in extant social and political circumstances, and facilitate the societal readiness for the transition towards a more optimal paradigm (Tallman et al., 1993). One could contend that AI-generated creations within the realms of literature and art lack the capacity to engender comparable stimuli for the advancement of societal circumstances. It is plausible that an artificial intelligence system could successfully emulate human creativity and produce literary and artistic works of comparable quality. However, its capacity to delve beneath the superficial veneer of a human-created masterpiece, transcending its mere aesthetic qualities, and critically evaluating its inherent message and significance within the context of contemporary societal circumstances is lacking. AI systems lack the capacity to perceive and experience social and political conditions in the same manner as humans. Humans and non-human entities exhibit differential responses to societal conditions, with the latter group demonstrating a distinct lack of susceptibility to the same influences as their human counterparts. Given the inherent limitations of an AI system's capacity to comprehend and empathise with the intricacies of contemporary societal conditions in a manner akin to human experience, it is inevitable that said system will fall short in eliciting transformative visions that align with the prevailing ethical norms that resonate with individuals' present aspirations (Tallman et al., 1993).

The subject matter at hand encompasses a more expansive socio-political framework. The advent of automated authorship and its subsequent impact on the literary and artistic market necessitates the implementation of appropriate remedial strategies and financial allocations. In the event of authors experiencing termination of their employment, it becomes imperative to ensure the provision of adequate financial assistance. The imperative for investment in training endeavors shall prove indispensable in facilitating the acquisition of novel proficiencies and credentials, thereby enabling individuals to alter their trajectory. The implementation of novel production projects would afford authors the opportunity to venture into previously unexplored domains of endeavor. In the present circumstance, the implementation of a remuneration framework that encompasses monetary provisions for the purpose of extending financial assistance, facilitating training endeavors, and fostering novel literary and artistic initiatives represents a pivotal and commendable measure (Thomas et al., 2015).

It is worth noting that human literary and artistic endeavors possess inherent societal value. According to the scholarly perspective put forth by Barton Beebe, the significance of engaging in aesthetic practice and aesthetic play holds particular relevance for individuals in their daily lives. The inherent worth of aesthetic play lies in its active engagement with the assimilation, appropriation, and innovative recombination of aesthetic expression (Beebe, 2017). The aforementioned phenomenon serves as a reservoir of gratification, ethical and governmental refinement, imaginative autonomy, and personal fulfillment. The deprivation of opportunities for human well-being, moral and political development, imaginative liberty, and self-actualization occurs as a consequence of the increasing reliance on machines for the execution of aesthetic play. The displacement of human authors from the literary and artistic domain by machines not only results in a loss of societal role models for human aesthetic engagement but also engenders a deprivation of such models within society. The utilization of generative AI systems affords human users the means to engage in artistic exploration by manipulating various styles and motifs (Thomas et al., 2015). However, it is crucial to distinguish the process of formulating and inputting prompts into an AI system from the concept of aesthetic play. The process of creation, which serves as the focal point of aesthetic involvement, is subsequently executed not by the human user, but rather by the AI system. The aforementioned phenomenon elicits concerns regarding its potential societal ramifications. In a hypothetical scenario where machines assume a predominant role in the realm of literary and artistic production, individuals may find themselves bereft of incentives to cultivate their own aesthetic praxis or engage in diverse modes of self-expression (Weinhardt, 2020).

## LEGAL AND ETHICAL CONSIDERATIONS SURROUNDING GENERATIVE AI

The advent of Generative Artificial Intelligence (AI) has given rise to a plethora of complex ethical and legal quandaries that permeate diverse domains. A salient ethical consideration revolves around the issue of accountability in the production of AI-generated content. The inherent autonomy exhibited by Generative AI systems engenders inquiries pertaining to the ethical ascription of creative endeavors. The complexity surrounding the definition of authorship arises when the collaborative efforts of human developers and the AI system become intertwined, thereby giving rise to uncertainties pertaining to the rightful attribution of credit and recognition for a particular creation (Henley, 1990). The aforementioned statement necessitates a reassessment of conventional conceptions pertaining to authorship, creativity, and individual agency within the framework of AI-generated content.

The ethical quandary at hand pertains to the presence of bias and the concept of fairness, which together pose a significant intellectual conundrum. Generative artificial intelligence (AI) systems acquire knowledge from extensive datasets, which may inadvertently incorporate biases that exist within the real-world context (Weinhardt, 2020). When the perpetuation of biases occurs within AI-generated content, it has the capacity to reinforce and intensify societal prejudices, thereby engendering ethical inquiries regarding the responsible deployment of AI and the potential amplification of pre-existing inequalities. It is a well-known fact that AI acquaints itself to the training data. If the training data contains social inequities and skewed decisions, the response will also be as ascertained by the data, or knowledge derived from it. This would be a deja vu of the particulars of the occurrence of the year 2016, which witnessed an AI-based conversational chatbot released by Microsoft sending acutely racially derogatory messages (Dixit, 2023). On further examination, it was discovered that it was inundated with racist language as a built-in learning feature. Because face recognition technologies are also under investigation, this is only the beginning of the problems. Blueprinted natural tendency is 70% male and 80% white, a glaring bias on human racial characteristics, according to a dataset that is used as a standard for bias analysis. When a white mask is used, software that has been "trained" not to identify faces of color may sometimes be circumvented (Gupta et al., 2022). The formidable ethical task at hand involves the delicate equilibrium between fostering innovation and upholding the ethical imperative to eradicate biases within artificial intelligence (AI) systems.

The examination of prevailing legal frameworks is imperative in light of the legal aspects pertaining to the ownership and safeguarding of intellectual property engendered by artificial intelligence. The identification and attribution of authorship and ownership rights pertaining to works generated by artificial intelligence present notable complexities, given that conventional legal frameworks predominantly perceive intellectual creations as products of human agency.

Getty Images sued Stability AI at London's High Court of Justice earlier in January 2023 (Tobin, 2023). Getty Photos maintains, in a manner consistent with the accusations made in the US case, that Stability AI violated Getty Images' copyright by training its AI on Getty Images' copyrighted photos and works. It is likely that it will have a major influence on the illegal use of copyrighted content in UK AI systems, much as its lawsuit in the US (Brittain, 2023). The issue at hand pertains to the necessity for legal systems to confront the inquiry as to whether artificial intelligence (AI) entities possess the capacity to be acknowledged as creators, or if prevailing intellectual property laws necessitate modification in order to duly recognize the collaborative involvement of both human agents and machines. On January 2023, esteemed visual artists Sarah Andersen, Kelly McKernan, and Karla Ortiz initiated a collective legal action by means of a formal complaint in the United States District Court Northern District of California San Francisco Division. The defendants named in this litigation are Stability AI Ltd. and Stability AI, Inc., Midjourney, Inc., and DeviantArt, Inc. The plaintiffs assert that their artistic creations were utilized without obtaining proper authorization as input components in the training and advancement of diverse artificial intelligence-based image generation systems, namely Stable Diffusion (developed by Stability AI), DreamStudio (created by Stability), the Midjourney Product (developed by Midjourney), and DreamUp (produced by DeviantArt). The plaintiffs further contend that Stability AI has produced reconstructed replicas of the works belonging to the plaintiffs, which they posit meet the criteria for unapproved derivative works (Loving, 2023).

The emergence of privacy concerns is a direct consequence of the extensive datasets handled by Generative AI, which possess the potential to encompass confidential personal data. The ethical and legal considerations pertaining to data privacy are magnified when artificial intelligence algorithms extract valuable information from these datasets, thereby prompting inquiries regarding consent, safeguarding of data, and the possibility of unauthorized entry. In domains such as healthcare and finance, wherein the utilization of AI-powered decision-making is progressively pervasive, ethical quandaries arise concerning the conscientious application of AI-derived insights and the potential ramifications for personal privacy. An additional ethical consideration pertains to the potential for deepfakes to engage in acts of harassment or defamation, whereby fabricated visual content is generated and disseminated to portray individuals in a derogatory or humiliating manner. Based on the findings reported by the United States government, it has been revealed that Sensity AI, a prominent company in the field, has indicated that a significant proportion, ranging from 90% to 95%, of the deepfake videos that have been circulating since the year 2018 have been generated specifically from non-consensual pornography (Gil et al., 2023).

## POLICY CONSIDERATIONS FOR NURTURING RESPONSIBLE GOVERNANCE IN THE ERA OF GENERATIVE AI

As observed above, it is imperative to adopt a comprehensive and flexible methodology in order to address the ethical and legal complexities presented by Generative AI. The establishment of regulatory frameworks that effectively navigate the intricacies of authorship, effectively address biases, adequately protect intellectual property, and steadfastly uphold privacy rights is of utmost importance in order to foster responsible innovation and guarantee that the advantages of Generative AI are ethically and sustainably harnessed.

### Policy Dimensions Surrounding Ethical Framework and Normative Guidelines

Numerous entities have generated guidelines pertaining to the ethical considerations and norms governing artificial intelligence (AI) in order to establish parameters surrounding the development and utilization of AI programs in relation to tangible occurrences in the real world (West, 2018). The numerical value provided by the user is insufficient to generate a meaningful response. The majority of these endeavors has been established within the recent years.

Governments have frequently adopted the strategy of establishing artificial intelligence commissions as a means of addressing pertinent concerns in this domain. The endeavors undertaken by the European Union (EU) are particularly remarkable. The Ethics Guidelines for Trustworthy Artificial Intelligence were released by the European Union in April 2019 (European Commission, 2019). These guidelines were formulated by a distinguished high-level expert group specializing in the field of artificial intelligence. The aforementioned document articulates a comprehensive compilation of "fundamental rights" pertaining to the domain of "trustworthy AI," thereby encompassing a wide range of commonly held principles that serve as the foundation for European establishments (European Commission, 2019). The aforementioned rights encompass a range of fundamental principles, namely individual autonomy, the preservation of human worth, the tenets of democratic governance and legal order, as well as the pursuit of equitable treatment.

The expert group of the European Union has derived seven guidelines pertaining to artificial intelligence (AI) from the aforementioned norms. AI systems ought to

- It is imperative that AI systems are subjected to human agency and oversight, acknowledging the need for human involvement in their operation and regulation.
- The technical robustness and safety of AI systems must be prioritized, emphasizing the necessity for their reliability and security.
- The preservation of privacy should be a fundamental consideration in the development and implementation of AI systems, safeguarding individuals' personal information.
- AI systems should exhibit transparency, ensuring that individuals are informed when they are engaging with an artificial system, thereby promoting honesty and clarity in human-AI interactions.
- The promotion of diversity, non-discrimination, and fairness should be integral to AI systems, fostering inclusivity and equality in their design and deployment.
- AI systems should be designed to serve the betterment of society and the environment, recognizing their potential impact on societal and environmental well-being.
- The notion of accountability is crucial in the realm of AI, necessitating that AI systems are answerable for their actions and decisions, including external scrutiny and oversight (European Commission, 2019).

One concern is if the EU would try to turn such ideas into regulations, as it did with the digital economy with the General Data Protection Regulation (herein after, GDPR). Companies doing business in Europe must comply with the GDPR, which took effect in 2018. Europe's AI ethical principles may be a first step toward worldwide AI rules like GDPR. The said act had worldwide ramifications. Japan harmonised its data privacy laws with European standards, and California passed the California Consumer Privacy Act (CCPA) in 2018 and is considering aligning it with the GDPR. Both countries wanted a GDPR "adequacy determination" in part. If a country is deemed adequate, the EU will enable its enterprises to move data from Europe to its home country (or state, in California's case). Critics say the GDPR's data protection and transparency regulations are hurting European tech businesses' AI investment. GDPR mandates corporations to provide people the right to a human review of an AI system's judgment, which boosts expenses. Without revision, the GDPR would reduce European AI investment and move it to China and the US, critics say.

Various governmental bodies and multilateral institutions have formulated AI ethics guidelines that bear resemblance to those put forth by the European Union (Yeung, 2020). The Organisation for Economic Co-operation and Development (hereinafter, OECD) initiatives stand out as prominent and contemporary examples. In the month of May in the year 2019, OECD unveiled a set of five principles that are deemed to be complementary in nature and centred around values, with the aim of fostering responsible Artificial Intelligence practices. In the month of June in the year 2019, the Group of Twenty (G20) convened and proceeded to endorse a distinct set of principles, which were exclusively derived from the principles established OECD (Yeung, 2020).

## Policy Considerations Surrounding Justice and Equity

The primary focal points revolve around the extent to which AI systems manifest, replicate, and potentially exacerbate societal issues. The ongoing discourse surrounding this issue is characterized by its dynamic nature and the presence of heightened emotional responses. The discourse at hand is particularly contentious when considering the decision-making processes of artificial intelligence systems in relation to human attributes such as gender, socioeconomic status, sexual orientation, and ethnic, racial, and religious identities. These systems can determine access to public and private resources and services, state surveillance subjects, employment screening subjects, credit scores assigned to consumers, and law enforcement and judicial interpretation and enforcement. Each of these decision-making processes evaluates and categorizes persons, including or excluding them or giving them higher or lower places, resulting in both positive and negative outcomes. Bias also refers to AI technologies' systemic flaws in prediction, especially for certain demographic groups. AI bias may occur during framing, data gathering, and data preparation. During the AI system's development, designers may intentionally or unintentionally introduce biases into the algorithm and/or training data, which may affect the analyses (Hao, 2019).

In recent times, a number of prominent instances have come to light pertaining to a phenomenon commonly denoted as "algorithmic bias" or "machine learning bias." Both Google and Amazon, as prominent technology companies, have encountered instances where their image search and hiring algorithms have been subject to scrutiny due to the presence of biases (Simonite, 2018; Dastin, 2018). These revelations have been deemed embarrassing, as they have raised concerns regarding the fairness and impartiality of these algorithms. The image search system developed by Google exhibited limitations in accurately identifying individuals belonging to ethnic minority groups. In the context of Amazon, it has been observed that the hiring algorithm employed by the organization exhibited a systematic tendency to assign higher scores to male candidates as opposed to their female counterparts. The designers at Amazon inadvertently encountered an unintended consequence whereby their AI system, despite their original intentions, exhibited a preference for selecting males over females. This preference was learned by the AI system through the utilization of an algorithm that was inherently biased towards a gender imbalance in its design parameters.

In response to the growing concern over algorithmic bias, governmental bodies have initiated the formulation of policies aimed at addressing this issue. In the month of April in the year 2019, a bill known as the Algorithmic Accountability Act was formulated by two senators from the United States. This legislative proposal entails the establishment of a requirement by the Federal Trade Commission for companies to conduct thorough evaluations of their artificial intelligence algorithms and training data (Corrigan, 2019). The primary objective of these evaluations is to identify and rectify any existing flaws that may result in biased or discriminatory outcomes. The proposed legislation would designate AI systems deemed "high-risk" based on their utilization of sensitive personal data, such as information pertaining to an individual's race, gender, sexual orientation, religion, genetic and biometric attributes, and criminal history, among other factors (Barbanel, 2019). The certainty regarding the passage of the aforementioned phenomenon remains considerably uncertain.

In a notable departure from conventional methodologies, the British government made a significant announcement in March 2019, wherein it declared the collaboration between its Centre for Data Ethics and Innovation, established in 2017, and its Race Disparity Unit (Department for Digital, Cultural Media and Sport (DCMS), 2019). This collaborative effort aims to undertake a comprehensive research program with the objective of investigating the potential biases inherent in artificial intelligence (AI) systems, specifically pertaining to the utilization of ethnicity as a determinant in decision-making processes within the justice system of the United Kingdom (DCMS, 2019).

## IPR Policy in the Age of Generative AI

As we traverse the profound ramifications of Generative Artificial Intelligence (AI) on the intellectual property domain, it becomes imperative to formulate a comprehensive framework for Intellectual Property Rights (IPR) policy in order to effectively tackle the intricate ethical and legal quandaries presented by this paradigm-shifting technology. From an ethical standpoint, it is imperative for the policy to acknowledge and duly consider the distinctive intricacies inherent in the realm of AI-driven creativity (Somaya & Varshney, 2020). This necessitates placing significant emphasis on fostering a harmonious and synergistic relationship between human developers and autonomous systems, thereby ensuring a collaborative and mutually beneficial environment. The establishment of unambiguous principles regarding the ascription of authorship and the ethical utilization of artificial intelligence in artistic endeavours assumes paramount significance within a progressive intellectual property rights framework (Mittal, 2023). The formulation of the policy ought to actively foster transparency within the realm of artificial intelligence (AI) development procedures, thereby guaranteeing the integration of accountability and ethical deliberations into the very essence of innovative practices (Chin, 2023).

In the realm of jurisprudence, the policy in question must deftly manoeuvre the complex landscape of delineating ownership entitlements pertaining to works generated by artificial intelligence. The existing legal frameworks in place may require modification in order to acknowledge non-human entities as valid creators, thereby promoting a harmonious equilibrium between the promotion of innovation and the protection of the rights of human creators. The development of the policy should include the incorporation of clear guidelines regarding potential violations, thus establishing robust mechanisms to prevent any infringements by AI-generated content on existing copyrights or trademarks. The paramount importance resides in the privacy implications inherent in intellectual property rights (IPR) frameworks, specifically in relation to the extensive datasets manipulated by Generative Artificial Intelligence (AI) systems (Collins, 2011). Policies should incorporate provisions that delineate the necessary procedures for obtaining informed consent and mitigating unauthorized encroachment upon individuals' personal data. In sectors such as healthcare and finance, where the widespread adoption of artificial intelligence (AI) for decision-making is observed, it is of utmost importance to establish policies that enforce stringent measures to protect and uphold the inherent rights associated with individual privacy (Collins, 2011).

Moreover, it is of utmost importance that the policy concerning intellectual property rights (IPR) be formulated in a manner that fosters a dynamic and progressive milieu, wherein conscientious and ethical innovation is encouraged. The current undertaking necessitates the implementation of mechanisms that foster ongoing dialogue among policymakers, industry stakeholders, and ethicists (Somaya & Varshney, 2020). The primary aim is to alter the policy framework in light of the perpetually evolving terrain of Generative AI, thereby encompassing the inherent complexities and potentialities it entails. By embracing a holistic viewpoint that encompasses the ethical, legal, and privacy dimensions, a proficient framework for intellectual property rights (IPR) pertaining to Generative Artificial Intelligence (AI) holds the promise of serving as an exemplar for ground breaking progress while concurrently maintaining societal responsibility.

## OPTIMAL LEGISLATIVE APPROACHES TO FOSTER ETHICAL AI DEVELOPMENT: A WORLDWIDE PERSPECTIVE

When scrutinizing legislative strategies intended to promote the ethical development of Artificial Intelligence (AI) on a global scale, a number of noteworthy models surface, each offering unique contributions to the broader dialogue. The General Data Protection Regulation (GDPR), 2016 of the European Union represents a ground-breaking endeavor that highlights the essential correlation between safeguarding data and promoting ethical artificial intelligence (AI) (Quinn, 2021). The General Data Protection Regulation lays down stringent principles pertaining to transparency, user consent, and the right to explanation, thereby presenting a comprehensive structure to address and alleviate potential hazards linked to decision-making driven by artificial intelligence (AI) (Quinn, 2021). It is further noteworthy to highlight the legislative propositions that have emerged within the United States. Specifically, the Algorithmic Accountability Act and the Algorithmic Justice and Online Platform Transparency Act of 2022 stand as prominent examples of the nation's dedication to rectifying bias and discriminatory tendencies inherent in artificial intelligence algorithms (Edelson, 2022). In order to ensure that AI systems conform to ethical norms, especially in industries that have a significant impact on society, the ideas outlined above stress the critical importance of openness and responsibility (Casciola, 2022).

Canadian law takes a privacy-first approach, with laws like PIPEDA (Personal Information Protection and Electronic Documents Act) of 2000 serving as prime examples (Jaar & Zeller, 2009). Organizations are obligated to get express permission before collecting, using, or disclosing personal information, as pointed out by the Personal Information Protection and Electronic Documents Act (PIPEDA), which emphasizes the critical need of gaining informed consent. Protecting people from possible invasions in an AI-driven world is the primary goal of legislation that prioritizes privacy rights, which is in line with ethical imperatives (Zimmer, 2018). On the contrary, it is noteworthy to mention that the Model AI Governance Framework established by Singapore serves as an exceptional paradigm. Although it does not possess the status of a legally binding requirement, it offers valuable and pragmatic counsel for entities that find themselves grappling with the intricate ethical dimensions associated with artificial intelligence. Singapore's approach to AI innovation is characterized by its commitment to fundamental principles such as fairness, transparency, and accountability. By incorporating these key considerations into its framework, Singapore fosters a culture of responsible AI development (Thong, 2021). This approach is characterized by its flexibility, allowing for adaptability to changing circumstances, while remaining firmly rooted in principled guidelines (Goh & Leon, 2020). The significance of ethical artificial intelligence (AI) in governmental decision-making is underscored by Australia's AI Ethics Framework, which has been meticulously developed by the Commonwealth Scientific and Industrial Research Organisation (CSIRO) (Dawson et al., 2019). The present non-binding framework delineates a set of principles pertaining to transparency, fairness, and accountability, thereby manifesting Australia's unwavering dedication to upholding ethical considerations in the realm of artificial intelligence (AI) applications. The inherent voluntariness of this framework underscores the paramount importance of industry self-regulation, thereby affording a certain degree of adaptability while concurrently fostering the adoption of responsible artificial intelligence (AI) practices (Dawson et al., 2019).

The aforementioned legislative instances, characterized by their varied origins and focal points, exhibit an inherent interconnectedness stemming from a collective dedication to the cultivation of ethical artificial intelligence (AI). The impact of the General Data Protection Regulation (GDPR) on international data protection norms is evident in the contemplation of privacy-focused legislations such as the Personal Information Protection and Electronic Documents Act (PIPEDA). The concurrent focus on transparency and accountability, as observed in both the European and North American paradigms, signifies a widespread agreement on these fundamental tenets at the global level. Through the synthesis of various legislative approaches, a comprehensive perspective on the development of ethical artificial intelligence (AI) is revealed. This perspective encompasses essential elements such as data protection, privacy rights, transparency, and accountability, which are considered fundamental components of a globally responsible AI ecosystem.

## CONCLUSION

In conclusion, it is imperative to acknowledge the manifold possibilities and challenges that arise within the intricate and dynamic legal landscape pertaining to generative artificial intelligence. The present study has demonstrated the significance of acquiring a thorough understanding of the legal dimensions that underpin the governance of Generative AI in order to ensure its responsible and ethical advancement and deployment.

The primary concern that necessitates considerable scrutiny pertains to the domain of intellectual property and copyright vis-à-vis content generated by artificial intelligence. The intricate legal considerations surrounding the attribution, acknowledgment, and protection of AI-generated creations assume great significance within a societal context wherein artificial intelligence possesses the capacity to autonomously generate musical compositions, artistic expressions, and literary works. To properly include the unique dynamics created by Generative AI, the present copyright restrictions need a thorough reevaluation and possible revision. The field of data privacy and security is quite prominent in the legal system. Important legal questions arise from the use of generative AI in conjunction with personally identifiable information (PII) due to the data's inherent vulnerability to abuse. Passage of laws that place a premium on protecting individuals' rights and ensuring the security of data is necessary to address problems like consent, data protection, and responsibility in relation to data breaches. Given the continuous growth of AI-generated content and its effects on public discourse and consumer behavior, it is crucial to recognize the increased significance of the legal aspect related to liability. Developers, operators, and consumers of artificial intelligence (AI) must be fully aware of their legal obligations in the event that AI deployment causes harm, spreads false information, or violates trust. Due to the far-reaching effects of Generative AI, its international legal repercussions must be carefully considered. Given the pressing need to achieve legal harmonization across many jurisdictions, it is more important than ever to tackle the problem of cross-border data flows and establish models of international governance.

So, it's safe to say that Generative Artificial Intelligence (AI) presents a complicated web of legal issues. In order to build a prudent framework for the worldwide regulation of Generative Artificial Intelligence (AI), it is crucial to fully grasp the complex legal issues at play, as this paper highlights. It is critical that legal systems adapt to deal with complex issues like data protection, intellectual property, liability, and global governance in a responsible and ethical way if we want to encourage innovation and creativity to flourish. Therefore, this essay is an argument in favor of establishing legal frameworks that effectively balance the vast capabilities of Generative AI with the primary goal of protecting human rights and improving society.

## REFERENCES

Atlantic Council. (2023). Annex 6: Learning from Cybersecurity, Preparing for Generative AI. In *Scaling Trust The On Web* (pp. 1–12). Atlantic Council. https://www.jstor.org/stable/resrep51651.26

Barbanel, J. (2019, April 29). *A look at the proposed Algorithmic Accountability Act of 2019.* IAPP. https://iapp.org/news/a/a-look-at-the-proposed-algorithmic-accountability-act-of-2019/

Beebe, B. (2017). Bleistein, the problem of aesthetic progress, and the making of American Copyright Law. *Columbia Law Review, 117*(2). https://columbialawreview.org/content/bleistein-the-problem-of-aesthetic-progress-and-the-making-of-american-copyright-law/

Brittain, B. (2023, February 6). *Getty Images lawsuit says Stability AI misused photos to train AI.* Reuters. https://www.reuters.com/legal/getty-images-lawsuit-says-stability-ai-misused-photos-train-ai-2023-02-06/

Cath, C. (2018). Governing artificial intelligence: ethical, legal and technical opportunities and challenges. *Philosophical Transactions: Mathematical, Physical and Engineering Sciences, 376*(2133), 1–8. https://www.jstor.org/stable/26601838

Chin, C. (2023). *Navigating the Risks of Artificial Intelligence on the Digital News Landscape.* Center for Strategic and International Studies (CSIS). https://www.jstor.org/stable/resrep53077

Choudhary, V. & Ali, Aamir S. M. (2023). ChatGPT and Copyright Concerns. *Economic and Political Weekly, 58*(16), 4–5.

Clelia Casciola, C. (2022). Artificial Intelligence and Health Care: Reviewing the Algorithmic Accountability Act in Light of the European Artificial Intelligence Act. *Vermont Law Review, 47*(1), 127–155. https://lawreview.vermontlaw.edu/wp-content/uploads/2023/03/06_Casciola_Book1_Final-copy.pdf

Collins, N. (2011). Trading Faures: Virtual Musicians and Machine Ethics. *Leonardo Music Journal, 21*, 35–39. https://www.jstor.org/stable/41416821. doi:10.1162/LMJ_a_00059

Corrigan, J. (2019, April 11). *Lawmakers Introduce Bill to Curb Algorithmic Bias.* Nextgov.Com. https://www.nextgov.com/artificial-intelligence/2019/04/lawmakers-introduce-bill-curb-algorithmic-bias/156237/

Daly, A., Devitt, S. K., & Mann, M. (2021). AI Ethics Needs Good Data. In P. Verdegem (Ed.), *AI for Everyone?: Critical Perspectives* (pp. 103–122). University of Westminster Press. https://www.jstor.org/stable/j.ctv26qjjhj.9 doi:10.16997/book55.g

Dastin, J. (2018, October 9). Amazon scraps secret AI recruiting tool that showed bias against women. *Reuters.* https://www.reuters.com/article/idUSKCN1MK0AG/

Dawson, D., Schleiger, E., Horton, J., McLaughlin, J., Robinson, C., Quezada, G., Scowcroft, J., & Hajkowicz, S. (2019). *Artificial Intelligence: Australia's ethics framework - a discussion paper.* Commonwealth Scientific and Industrial Research Organisation, Department of Industry, Innovation and Science (Australia). https://apo.org.au/node/229596

Department for Digital. Cultural Media and Sport (DCMS), (2019, March 20). Investigation launched into potential for #AI bias in algorithmic decision-making in society. *FE News*. https://www.fenews.co.uk/skills/investigation-launched-into-potential-for-ai-bias-in-algorithmic-decision-making-in-society/

Dixit, P. (2023, March 29). This chatbot will use the n-word and teach you how to build a bomb. *BuzzFeed News*. https://www.buzzfeednews.com/article/pranavdixit/freedomgpt-ai-chatbot-test

Edelson, L. (2022, April 29). *Platform Transparency Legislation: The Whos, Whats and Hows*. Lawfare. https://www.lawfaremedia.org/article/platform-transparency-legislation-whos-whats-and-hows

Engelke, P. (2020). *AI, Society, and Governance: An Introduction*. Atlantic Council. https://www.jstor.org/stable/resrep29327

European Commission. (2019, April 8). *Ethics guidelines for trustworthy AI*. EC. https://digital-strategy.ec.europa.eu/en/library/ethics-guidelines-trustworthy-ai

Ghose, A. & Ali, Aamir S. M. (2023). Amplifying Music with Artificial Intelligence. *Economic and Political Weekly*, *58*(17), 4–6.

Gil, R., Virgili-Gomà, J., López-Gil, J.-M., & García, R. (2023). Deepfakes: Evolution and trends. *Soft Computing*, *27*(16), 11295–11318. doi:10.1007/s00500-023-08605-y

Goh, Y., & Leon, N. R. (2020). The innovation of Singapore's AI ethics model framework. In L. Hui & B. Tse (Eds.), *AI Governance in 2019: A Year in Review (Observations of 50 Experts in the World* (pp. 77–78). Shanghai Institute for Science of Science.

Gupta, M., Parra, C. M., & Dennehy, D. (2022). Questioning racial and gender bias in AI-based recommendations: Do espoused national cultural values matter? *Information Systems Frontiers*, *24*(5), 1465–1481. doi:10.1007/s10796-021-10156-2 PMID:34177358

Hao, K. (2019, February 4). This is how AI bias really happens – and why it's so hard to fix. *MIT Technology Review*. https://www.technologyreview.com/2019/02/04/137602/this-is-how-ai-bias-really-happensand-why-its-so-hard-to-fix/

Haugh, B. A., Kaminski, N. J., Madhavan, P., McDaniel, E. A., Pavlak, C. R., Sparrow, D. A., Tate, D. M., & Williams, B. L. (2018). Strategy 3 Proposed Changes. In *RFI Response: National Artificial Intelligence Research and Development Strategic Plan* (pp. 3–6). Institute for Defense Analyses. https://www.jstor.org/stable/resrep22865.6

Henley, T. B. (1990). Natural Problems and Artificial Intelligence. *Behavior and Philosophy*, *18*(2), 43–56. https://www.jstor.org/stable/27759223

Jaar, D., & Zeller, P. E. (2009). Canadian Privacy Law: The Personal Information Protection and Electronic Documents Act (PIPEDA). *International In-house Counsel Journal*, *2*(7), 1135-1146. https://www.iicj.net/subscribersonly/09june/iicj4jun-dataprotection-patrickzeller-guidancesoftware-USA.pdf

Katyal, S. K. (2022). Democracy & Distrust in an Era of Artificial Intelligence. *Daedalus*, *151*(2), 322–334. https://www.jstor.org/stable/48662045. doi:10.1162/daed_a_01919

Loving, T. (2023, March 30). Current AI copyright cases – part 1. *Copyright Alliance.* https://copyrightalliance.org/current-ai-copyright-cases-part-1/

Margetts, H. (2022). Rethinking AI for Good Governance. *Daedalus, 151*(2), 360–371. https://www.jstor.org/stable/48662048. doi:10.1162/daed_a_01922

Mittal, A. (2023). ChatGPT and the Legal and Ethical Problems of Copyright in India. *Economic and Political Weekly, 58*(32), 62–63.

Quinn, P. (2021). Research under the GDPR – a level playing field for public and private sector research? *Life Sciences, Society and Policy, 17*(1), 1–33. doi:10.1186/s40504-021-00111-z PMID:33397487

Simonite, T. (2018, January 11). When it comes to gorillas, google photos remains blind. Wired. https://www.wired.com/story/when-it-comes-to-gorillas-google-photos-remains-blind/

Somaya, D., & Varshney, L. R. (2020). Ownership Dilemmas in an Age of Creative Machines. *Issues in Science and Technology, 36*(2), 79–85. https://www.jstor.org/stable/26949112

Tallman, I., Leik, R. K., Gray, L. N., & Stafford, M. C. (1993). A Theory of Problem-Solving Behavior. *Social Psychology Quarterly, 56*(3), 157–177. doi:10.2307/2786776

Thomas, D. C., Liao, Y., Aycan, Z., Cerdin, J.-L., Pekerti, A. A., Ravlin, E. C., Stahl, G. K., Lazarova, M. B., Fock, H., Arli, D., Moeller, M., Okimoto, T. G., & van de Vijver, F. (2015). Cultural intelligence: A theory-based, short-form measure. *Journal of International Business Studies, 46*(9), 1099–1118. https://www.jstor.org/stable/43653785. doi:10.1057/jibs.2014.67

Thong, J. L. K. (2021, June 29). Mapping Singapore's journey and approach to AI governance. *Digital Asia.* https://medium.com/digital-asia-ii/mapping-singapores-journey-and-approach-to-ai-governance-d01f76bbf5c6

Tobin, S. (2023, June 1). *Getty asks London court to stop UK sales of Stability AI system.* Reuters. https://www.reuters.com/technology/getty-asks-london-court-stop-uk-sales-stability-ai-system-2023-06-01/

Weinhardt, M. (2020). Ethical Issues in the Use of Big Data for Social Research. *Historical Social Research. Historische Sozialforschung, 45*(3), 342–368. https://www.jstor.org/stable/26918416

West, D. M. (2018, September 13). *The role of corporations in addressing AI's ethical dilemmas.* Brookings. https://www.brookings.edu/articles/how-to-address-ai-ethical-dilemmas/

Yeung, K. (2020). Recommendation of the council on artificial intelligence (OECD). *International Legal Materials, 59*(1), 27–34. doi:10.1017/ilm.2020.5

Zimmer, B. (2018). *Towards Privacy by Design: Review of the Personal Information Protection and Electronic Documents Act: Report of the Standing Committee on Access to Information, Privacy and Ethics.* House of Commons, Canada. https://www.ourcommons.ca/Content/Committee/421/ETHI/Reports/RP9690701/ethirp12/ethirp12-e.pdf

Zittrain, J. L. (2006). The Generative Internet. *Harvard Law Review, 119*(7), 1974–2040. https://www.jstor.org/stable/4093608

## KEY TERMS AND DEFINITIONS

**Algorithm Bias:** Algorithmic bias refers to repeated mistakes in a computer system that unfairly favour one user group over another.

**Artificial Intelligence:** The theory and development of computer systems that can execute human activities, including visual perception, voice recognition, decision-making, and language translation.

**ChatGPT:** ChatGPT is an AI-powered natural language processing tool that lets you conduct human-like chatbot chats and more. Language models can answer inquiries and help with email, essay, and code writing.

**Ethical AI:** The ethical principles of ethical AI include individual rights, privacy, non-discrimination, and non-manipulation. Ethical AI prioritises ethics in deciding AI usage.

**Generative AI:** Generative AI is an artificial intelligence that can generate text, pictures, audio, and synthetic data.

# Chapter 11
# Navigating the Legal Landscape of AI-Induced Property Damage:
## A Critical Examination of Existing Regulations and the Quest for Clarity

**Akash Bag**
https://orcid.org/0000-0001-8820-171X
*Adamas University, India*

**Astha Chaturvedi**
*Parul University, India*

**Sneha**
https://orcid.org/0000-0002-0158-2503
*KIIT University, India*

**Ruchi Tiwari**
*Parul University, India*

## ABSTRACT

*This chapter dissects the proposal for an AI Liability Directive by the European Parliament and the Council. The Directive aims to adapt civil liability rules for artificial intelligence (AI) systems. Two main types of non-contractual liability are scrutinized: fault-based and strict liability. The core of the chapter revolves around the proposed AI Liability Directive. It dissects key provisions and highlights conflicting perspectives from scholars and associations. It also looks at the advantages and disadvantages of these rules and concludes by summarizing its findings and discussing how they might impact future policies related to AI responsibility.*

DOI: 10.4018/979-8-3693-1565-1.ch011

## INTRODUCTION

Artificial intelligence (AI) is showing up in more and more aspects of our daily lives, such as self-driving cars and predictive analytics. Artificial Intelligence (AI) can potentially revolutionize our understanding of the world. It can lead to benefits like greater efficiency in factories through automation and help with hospital treatments employing robots (Cataleta, 2020). However, there are unintended repercussions to the broad usage of AI, such as data breaches, biased decision-making algorithms, and privacy violations. This raises questions about whether the rules can keep up with the rapid technological advancements since ambiguous regulations breed uncertainty and stifle artificial intelligence's tremendous benefits and creative potential (Franke, 2019). Liability is a key concern in this situation. Maintaining liability is a legal principle that is necessary for efficiently handling problems. Liability is becoming increasingly necessary as the hazards connected with AI systems rise to ensure those harmed get compensation. Assigning blame to AI systems is difficult because many moving parts and stakeholders are involved, especially now that the system can learn and make decisions without human oversight (Karnow, 1996).

Because of this lack of transparency and the general public's ignorance of AI systems, it is challenging to comprehend how a specific result was arrived at. As a result, it becomes difficult to decide who should be held accountable when AI systems make judgments on their own. Businesses and organizations may use this intricacy to absolve themselves of liability for harm caused by AI (Karnow, 1996). In theory, victims are entitled to reimbursement under current tort laws for harm brought on by AI. But because AI systems are different, it can be very difficult, if not impossible, to show a mistake and establish a causal relationship. The European Union (EU) has created the AI Liability Directive to solve this problem. This order creates a presumption of causality and adds steps to provide victims with easier access to evidence when working with high-risk AI systems. The aim is to ensure those who suffer harm due to AI get the same protection as those who suffer harm from other non-AI technology.

The chapter's primary goal is to examine the difficulties and consequences of enforcing the proposed AI Liability Directive in the European legal system and the European Economic Area (EEA). Its objective is to evaluate how well the Directive complies with the European legal system. The legal foundation for artificial intelligence (AI) is covered in part 2, along with legislative building pieces like the General Data Protection Regulation, the planned Artificial Intelligence Act, and the Product Liability Directive. Part 3 examines the liability subject, including strict liability laws and fault-based liability. It also addresses the difficulties and drawbacks of implementing conventional liability laws on artificial intelligence (AI) systems, considering the complexity, multi-actor engagement, growing autonomy, and ethical concerns. The chapter also discusses the possible advantages of AI liability laws, including how they could foster innovation, increase public confidence in AI systems, and guarantee moral AI operations.

The proposed AI Liability Directive is the subject of part 4, which thoroughly examines its main clauses and a range of viewpoints from academics and associations. Concerns regarding its efficacy, possible influence on innovation, and the requirement for more clarity are all addressed. The chapter also examines possible drawbacks to the suggested Directive. Part 6 concludes with some final thoughts, summarizing the main discoveries and discussing how they can affect AI liability laws in the future.

## THE BACKGROUND OF THE STUDY: BUILDING BLOCKS OF AI LIABILITY

### Product Liability Directive

A report to evaluate the impact of emerging digital technologies on safety and responsibility regulations was to be prepared by the European Commission, according to the Communication on AI for Europe, which was adopted in April 2018. This case focuses on the liability issue, previously informed by expert groups' participation, stakeholder feedback, and the Product Liability Directive. Now, a layer of protection that went beyond the conventional fault-based liability scheme was provided by the 1985 Product Liability Directive. A claimant in a fault-based liability system must demonstrate a clear and direct connection between the harm they have caused and the error of another party (Wendehorst, 2020). Certain nations, on the other hand, have strict liability laws that exempt the injured party from having to establish blame. It suffices to assign blame if damage can be linked to a product flaw (Sousa Antunes, 2020).

Even more was done under the Product Liability Directive. It established a system of strict liability for manufacturers, which meant they would be held legally responsible if a fault in their product caused harm. To protect customers, this was done, and it extended accountability to everyone involved in the production chain, including importers and the maker of the finished product (Reich, 1986). If more than one party was partially at fault for the same harm, the victims could sue each other for full damages. Today's challenge comes with the development of AI systems, which frequently function inside intricate networks of interconnected objects called the Internet of Things (IoT) (Rose et al., 2015). Although the Product Liability Directive's definition of "*product*" is rather broad, it might require more explanation to handle the complications these evolving technologies bring forth adequately.

It can be difficult to demonstrate the existence of an AI system problem, quantify the harm it has caused, and establish a causal relationship between the problem and the harm experienced. The absence of precise guidelines regarding who bears responsibility for certain AI-related problems further complicates matters.

In September 2022, the Commission proposed amending the Product Liability Directive, which outlines the current regulations that hold manufacturers legally accountable for defective products. These modifications, which apply to the European Economic Area (EEA), are intended to give companies precise legal guidance. They can invest in cutting-edge items confidently and guarantee that those affected will be compensated. It all comes down to balancing promoting innovation and safeguarding customers.

### General Data Protection Regulation

Because AI systems frequently require personal data to function, the General Data Protection Regulation (GDPR) is significant in the context of AI liability. Following the GDPR's guidelines is essential when creating and implementing AI systems to ensure they abide by the law. This covers the training and development processes for AI algorithms. 2018 saw the implementation of the GDPR, which was approved in 2016 and has a bearing on the European Economic Area (EEA). It applies to all businesses, regardless of location, that handle the personal data of EU people. This law establishes guidelines for protecting people's personal information while allowing it to be shared freely. Its goals are to give people more control over their data and, by establishing a single framework, to streamline regulatory standards within the EU (Deac, 2018).

Article 22 of the GDPR addresses automated decision-making procedures. People have the right to be free from choices made only through automated processing that could have serious legal repercussions or other similar effects. For example, Article 22 of the GDPR applies if a bank employs an automated system to determine a person's eligibility for a loan without involving them or allowing them to challenge the decision. Protecting people's rights in the era of automated artificial intelligence is paramount (Finlayson-Brown & Bossotto, 2021). The paper stresses the significance of human oversight, responsibility, and openness when creating and utilizing AI systems that impact people's lives. It is vital to comprehend that Article 22 of the GDPR does not apply to every automated decision-making process.

The GDPR's Chapter VIII now addresses liability, sanctions, and remedies. According to Article 82, the party liable for a breach of the GDPR, referred to as the controller or processor, may demand compensation from anybody who is harmed—physical or non-physical—by the violation. If the processor violates the controller's legal orders or neglects to comply with GDPR responsibilities, they may also be liable for any damages resulting from their conduct. Although the GDPR defines harm broadly and guarantees the right to full compensation, it leaves out important information regarding liability for damages. In an indication of further attempts to address legal difficulties regarding data privacy and compensation, the Supreme Court of Austria requested clarification on the interpretation of Article 82 of the GDPR to the Court of Justice of the European Union (CJEU) on May 12, 2021 (Deac, 2018).

In a recent instance (C-300/21), an Austrian citizen requested damages for psychological distress he suffered as a result of a business using his political views as data without his permission. According to a ruling by the European Court of Justice (CJEU), a person is not automatically entitled to compensation for just breaking the General Data Protection Regulation (GDPR) (Goutzamanis, 2023). It's important to remember, though, that there is no set threshold for the degree of emotional suffering necessary to be eligible for compensation. The GDPR must have been broken, there must be harm (either bodily or mental), and there must be a direct connection between the harm and the GDPR infringement, according to the court's three requirements for a right to compensation (Lombardo, 2022). It's significant to note that the GDPR does not provide a hard threshold for the degree of emotional suffering required to qualify for compensation. Each EU member state can determine the amount of compensation that should be given.

## White Paper on AI and the Safety and Liability Report

The European Union released a document known as the "*White Paper on AI*" in February 2020. The following were the goals of this paper (Godinho et al., 2021):

1. *Define AI:* The definition of artificial intelligence (AI) and its significance were covered first.
2. *Highlight Benefits:* The benefits of AI were highlighted, including how it would improve citizens' lives and make European industries more competitive. Additionally, it discussed how AI might aid in addressing problems like climate change.
3. *Policy Recommendations:* The paper made several recommendations for how to guarantee that AI is developed in the EU in a trustworthy and safe manner.

The EU sought to improve its data and technological capacities to accomplish these goals. Their goal was to lead the world in data. The EU also asked its member states for advice on handling AI. They published a follow-up study regarding the legal implications of robotics, IoT, and AI. Concerns about AI systems that are capable of acting independently of continuous human supervision were covered in this paper. It raised concerns about comprehending and gaining access to artificial intelligence's often intricate and costly internal workings (Godinho et al., 2021).

## The Proposed Artificial Intelligence Act

On April 21, 2021, the European Parliament and Council introduced a proposal dubbed the "*Artificial Intelligence Act*." This proposal is a component of a larger project that aims to advance the quality and credibility of European artificial intelligence (AI) systems (Mukherjee et al., 2023). It's important to note that the proposed AI Liability Directive and the updated Product Liability Directive are related documents to which this one is linked. The proposed AI Act's primary goal is to create uniform guidelines for AI in a variety of applications, with a focus on harm prevention. It presents the idea of "*strict liability*" for AI systems that carry a high risk (Duffourc & Gerke, 2023). This implies that developers and producers will be held accountable for any harm caused by a high-risk AI system, regardless of who is at fault. Article 33(8) of the proposed AI Act states suppliers of high-risk AI systems must have suitable insurance or financial arrangements to meet potential liability claims. The proposed AI Act also establishes a framework for independent evaluations of AI systems to ensure they comply with all regulations, including liability-related ones. The purpose of this technique is to make AI system vendors comply.

The proposed AI Act addresses concerns about AI system safety and transparency while highlighting the significance of human oversight. Its objective is to lessen threats to public safety and basic rights while easing the implementation of cutting-edge AI systems, which are growing increasingly complex, adaptive, autonomous, and learning-based. Concern over responsibility gaps and their possible detrimental effects on consumer acceptability and society trust is developing as these sophisticated AI systems proliferate (Duffourc & Gerke, 2023). On May 11, 2023, the European Parliament's top committees passed the AI Act at the committee level to solve these problems. This makes it possible for the AI Act to be adopted by the Parliament in June 2023. An important modification to the AI Act's compromised wording is the addition of "*foundation models*." Although many of these models are anticipated to be used, little is known about their operation, possible drawbacks, and potential benefits because of their emergent characteristics (Buiten et al., 2023).

## Digital Services Act and the E-commerce Directive

The goal of the 2000 E-commerce Directive was to provide guidelines for the unrestricted flow of digital services throughout the European Union (EU). It concentrated on removing obstacles to online commerce, including cross-border entertainment, advertising, and product sales. Regulations about commercial communications, transparency, and information obligations for online services were outlined in this Directive. It also brought legal immunity for intermediary service providers—but only for strictly automatic, passive, and technical services. This implied that service providers would not be held responsible for content they were ignorant of or had no control over. The EU enacted the Digital Service Act in 2022, updating and redefining the regulations outlined in the E-commerce Directive (Schwemer, 2022). Creating a trustworthy, dependable, and safe online environment is the main goal of the new Act. It covers topics such as disseminating illicit material and the dangers disinformation and other damaging materials pose to society (Schwemer, 2022). The legislation defines "*illegal content*" as fairly broad and includes everything from terrorist content to copyright infringement. Service providers are not eligible for liability exemptions under the Digital Service Act if they knowingly work with service recipients to engage in illicit activities. As soon as they become aware of any illegal activity or content, they must also take immediate action to stop or restrict access to them (Duivenvoorde, 2022).

### Resolution on Civil Liability and The Proposals for AI Liability Rules

The European Parliament took a significant vote on the application of artificial intelligence (AI) on October 20, 2020. They proposed a new legislative structure using Article 225 of the Treaty on the Functioning of the European Union (TFEU). This paradigm would hold people responsible when AI systems are hurt or damaged (Laux et al., 2022). The Parliament acknowledged that artificial intelligence (AI), whether utilized in virtual or physical operations, could occasionally go wrong and cause issues. The behaviors of those who developed, implemented, or used the AI system may be to blame for these issues. The Parliament made a significant point of emphasizing that AI systems shouldn't be endowed with human-like consciousness or legal rights. They are merely human-made tools. One major worry they had was that the complexity of AI systems could make it difficult to identify the origin of an issue. The Parliament proposed a remedy to deal with this: hold everyone involved in developing, maintaining, or managing the danger of AI systems accountable (Ebers et al., 2021).

The European Parliament believed that uniform and unambiguous AI system regulations were important for the European Union. They desired legislation to guarantee the responsible and safe application of AI technology while keeping up with its rapid advancements. A unified legal system would make it easier for everyone to understand their obligations and rights. On September 28, 2022, two years later, the European Commission made two significant changes to liability laws to make them more modern (Schwemer, 2022). Initially, it suggested revising the regulations that make product manufacturers answerable for faulty goods. Second, it recommended unifying national liability laws about artificial intelligence (AI) to facilitate compensation claims for anyone injured by AI-related problems. These suggestions seek to guarantee the same level of protection for individuals injured by AI systems or products as in other circumstances.

## RESEARCH GAP

The research presented offers a thorough examination of the proposed AI Liability Directive within the EU but leaves several critical areas underexplored. Notably, it lacks a comparative legal analysis with non-EU jurisdictions, which could enrich the understanding of global best practices in AI liability. Furthermore, the impact of the Directive on small and medium-sized enterprises (SMEs), essential for fostering innovation while ensuring compliance, remains insufficiently addressed. Additionally, there's a gap in assessing the Directive's adaptability to the rapid technological evolution of AI and its implications for consumer protection and trust. An interdisciplinary approach to developing AI safety standards, encompassing ethical, technical, and legal perspectives, is also missing. Lastly, the discussion does not extend to the need for global governance and international cooperation in regulating AI, a critical aspect given AI's transnational nature. Bridging these gaps could significantly enhance the legal framework for AI liability, ensuring it is robust, adaptable, and inclusive.

## LIABILITY

We explore the moral and legal issues brought on by AI systems' growing capabilities. As AI grows more robust, concerns about its legal standing and the possible repercussions of its activities surface. The chapter specifically addresses the idea of giving artificial intelligence (AI) systems legal personality, even though these systems are essentially distinct from people and cannot conduct business or sign insurance contracts (Nanos, 2023). At first, this idea can seem absurd. Evaluating AI decision-making in a way comparable to human decision-making is one suggested strategy to alleviate these worries.

***Fault-Based Liability:*** This regulation is considered the default norm in EU and EEA member states. It suggests that the individual who caused the harm must have done it carelessly or intentionally. In this sense, neglect is not using the care a reasonable person would have used in the same situation. Usually, it is up to judges or assessors to decide if carelessness happened. This might not be easy when evaluating AI systems because the assessors might not have the same technical expertise as those who created the technology. Producers may avoid accountability if they carry out their responsibilities and behave responsibly. However, setting unduly high standards of care can stifle creativity and lead to ambiguity in the law.

***Strict Liability:*** This idea is a stricter version of liability in which the party in charge is nevertheless liable for damages brought about by a product, even without carelessness or proof of wrongdoing. When working with sophisticated technology like artificial intelligence (AI), where it may be difficult to determine who is at fault or negligent, strict liability might be very important. Stricter liability laws may encourage the creation of safer technology. While there are worries that legal liability could discourage innovation, it can also motivate businesses to produce more dependable goods (Wagner, 2021).

### Challenges and Shortcomings of AI Liability Rules

In the ever-evolving landscape of technological advancements, particularly with the advent of Artificial Intelligence (AI), a pressing concern arises: how can we effectively assign liability when things go awry? This concern has surfaced due to the intricate nature of AI systems, often characterized as 'black boxes,' which can be exceedingly challenging to scrutinize when issues arise. Adding to this complexity is the involvement of numerous actors, making it expensive and, at times, nearly impossible to pinpoint who should bear responsibility in the event of harm caused by AI systems. The interconnected nature of AI systems further compounds these challenges, their reliance on external data sources, vulnerabilities to cybersecurity breaches, and their increasing autonomy through machine learning and deep-learning capabilities.

#### Complexity

The idea of complexity is essential to the field of artificial intelligence (AI), both in the development and functioning of AI systems. The underlying workings of AI systems are frequently obscured by their complexity, making them incomprehensible to users. The main problem is that these systems' decision-making and information-processing processes are opaque. Due to this lack of transparency, end users find it challenging to comprehend the exact parameters, information, guidelines, and available options that drive an AI system's decision-making process (Zech, 2021). In contrast to traditional software, AI systems don't follow clear, step-by-step instructions or have every conceivable result preprogrammed. Rather, they are built with broad objectives and determine the best course of action to accomplish those objectives by learning from the data they are fed (Koch et al., 2022). Because of its dynamic nature, even AI developers and programmers find it difficult to fully understand how the AI system comes to a certain conclusion or choice. This complexity poses a big challenge when things go wrong and a problem when AI performs excellently.

## Multiple Actors And Connectivity

In AI systems, managing many actors becomes difficult because of the systems' increasing inherent complexity, raising practical liability difficulties. Regarding AI systems, several parties, including data providers and hackers, may be involved in harming the system without the primary operator knowing about it or anticipating it. The operation of the AI system may be flawed due to this lack of cooperation and transparency, which could have negative effects and cause damage. The fact that more and more parties are involved in developing and managing AI systems only exacerbates the situation. Assigning blame for harm to a particular element, a well-defined cause, or a single actor becomes extremely difficult. Rather, the damage comes from the ecology that surrounds the AI system as a whole. Assigning shared liability is one way the legal system in the Product Liability Directive addresses this difficult issue. This strategy might not completely consider all the pertinent players when working with an AI system (Zech, 2021).

## Increased Autonomy

One significant challenge is AI systems' growing autonomy and machine learning capabilities. These artificial intelligence (AI) systems are now advanced enough to learn from data and make decisions independently, frequently without direct human assistance. This increased capacity for self-learning and adaptation lessens the need for programmers to foresee and account for every event that might arise while the system is in use. Although this development is promising, it implies that programmers have less power and influence over AI behavior. As a result, these systems' learning abilities might produce unanticipated or unexpected effects, making assigning blame for them challenging (Zech, 2021).

## Ethical Risks

It's critical to encourage diversity and inclusiveness across the board in the AI system to produce trustworthy AI. If this isn't done, the system may include inadvertent historical biases. If these prejudices are not addressed, they may result in prejudice and discrimination against particular people or groups (Müller, 2023). While addressing their basic requirements, AI systems should be created to respect and protect people's physical and mental health and cultural and personal identities. Furthermore, these systems ought to avoid employing unjust biases and work to improve the fairness of society as a whole.

The 2019 AI Index Report emphasized three ethical issues surrounding AI systems: interpretability, fairness, and openness. Interpretability is crucial, particularly for people whom an AI judgment or forecast can impact. People need to know why the system produced a specific result, which builds confidence in the technology (Zhang et al., 2021). AI system producers should not use datasets with discriminating biases to maintain justice. The mechanism via which an AI system makes decisions is called transparent (Hyden, 2022). It is strongly related to explainability in that, to facilitate traceability and improve transparency, the datasets and procedures that inform the decisions made by the AI system must be carefully documented. The ethical and accountable deployment of these technologies is essential to ensuring their safe and efficient use. As a result, all stakeholders engaged in creating and applying AI systems must take the necessary precautions to reduce the dangers involved (Waltl & Vogl, 2018). This guarantees that victims can hold the person causing the harm accountable if damages arise. But, the key to maximizing the potential of AI systems while preventing biases and unforeseen effects is to promote moral behavior and the responsible use of these systems.

## Advantages Of AI Liability Rules

### Innovation

The establishment of a systematic framework for the responsible development and application of AI technology is greatly aided by AI liability regulations. These regulations frequently demand that manufacturers and developers of AI give responsible innovation priority by following particular ethical and safety guidelines. Artificial Intelligence (AI) can create a revolutionary phase in multiple domains by enhancing or replacing human talents. It can significantly improve the innovation process' efficacy and efficiency. One of its most significant advantages is AI's ability to analyze large datasets and find patterns and insights humans might miss. Making decisions with greater knowledge and efficiency is facilitated by this ability. AI can also speed up innovation by reducing the time and expenses involved in the development process. Because of this agility, businesses can launch products onto the market faster and with less risk (Füller et al., 2022).

Technology has frequently taken on a supporting role in the creative process, doing routine and repetitive activities that people can find difficult or boring. Data handling across many databases and formats, data processing, and outcome replication are common responsibilities in this work. But with the development of AI, technology's place in creativity has changed, making it more than just a tool—rather, an active partner (Kakatkar et al., 2018). AI technologies give people greater access to creative potential and more in-depth data analysis. They are capable of pattern recognition, prediction, and efficient information management. AI systems can also continually evolve through learning from outside input. Artificial intelligence (AI) powered chatbots, which have completely changed how companies interact with their clients, provide a concrete illustration of this change. Natural language processing-driven chatbots help businesses personalize conversations, increase consumer engagement, expedite workflows, and make better decisions. They also increase production by using data analysis and automation. Chatbots are extremely helpful in customer service, where they may start discussions with clients, help with post-purchase assistance, and answer questions quickly and accurately. They also extract insightful information from customer data, which helps companies grow their clientele and improve their pricing policies (Wang et al., 2023).

### Trust

Building public confidence in AI technologies requires the establishment of clear liability regulations for harms committed by these systems. This is a crucial stage since the public's perception of new technologies is frequently distorted by unfavorable beliefs and anxieties stoked by false information from the media and scientific representations. Misconceptions about artificial intelligence (AI) are common in the public domain, with many people harboring fears that these systems would eventually replace human workers and endanger human lives. In light of these concerns, how to foster confidence in AI systems and their results emerges (Lee & See, 2004). The foundation of all human interactions is trust, enabling people to depend on one another and expose themselves to another's behavior. Things get trickier when it comes to trusting AI, though. Determining if AI is behaving autonomously or adhering to a predetermined set of behaviors is difficult. Because of this ambiguity, people may find it challenging to trust AI systems (Bartneck et al., 2021).

### Ethical AI

It is widely acknowledged that AI systems have the potential to produce harmful outcomes if they are implemented without careful consideration of their impact on individuals, specific communities, and society as a whole. These harmful impacts can manifest as bias, discrimination, violations of privacy, and increased surveillance. In ethical discussions surrounding AI, the goal is to ensure that AI is developed, deployed, and used in a way that aligns with societal values and norms. The European Parliament has emphasized the importance of addressing ethical concerns related to AI through a robust regulatory framework that reflects the principles and values of the European Union (EU). This framework should provide legal certainty for businesses and citizens alike. Mandatory measures should be implemented to prevent practices from compromising fundamental rights to achieve this objective. Additionally, the Parliament has stressed that ethical principles must not remain theoretical but should be translated into concrete legal provisions.

AI liability rules promote ethical AI by fostering responsibility, transparency, and innovation while preventing harm. These rules offer a structured framework for the continued development of AI systems that are both safe and beneficial for society. Much of the discourse on this subject revolves around distinguishing right from wrong in developing and applying AI systems. The General Data Protection Regulation (GDPR) addresses these ethical considerations in Article 5, emphasizing that the processing of personal data must be lawful, fair, and transparent. It should serve a specific purpose and be processed with appropriate security measures, while those responsible should be held accountable. Implementing these accountability principles encourages the responsible development and use of ethical AI. AI liability rules can further incentivize the rational allocation of resources and fair distribution of risks.

## THE PROPOSED AI LIABILITY DIRECTIVE

A major directive from the European Parliament and the Council regulating the application of non-contractual civil liability regulations relating to artificial intelligence (AI) was proposed on September 28, 2022. The European Commission has proposed strengthening and updating the EU's liability laws while enacting new laws to deal with harm caused by artificial intelligence (AI) systems. The justification for this Directive is its capacity to improve legal certainty and harmonize the legal environment in the European Union (EU) and the European Economic Area (EEA). It is essential to recognize this Directive's status as a directive to appreciate its significance fully (Duffourc & Gerke, 2023). In this sense, a directive refers to legislation outlining shared goals that all EU member states must pursue.

However, it provides freedom in achieving these goals, enabling member states to modify their strategy while guaranteeing conformity with the overall objectives. The main goal here is to create fair protection for people who are harmed by AI systems, similar to the protections provided for people harmed by traditional technology (Reuters, 2023a).

The Directive's notable aspects include adding a "*rebuttable presumption of causality*." This clause substantially lowers the standard of proof for anybody claiming to have been harmed by an AI system. Practically speaking, this suggests that these claimants won't have to fight as hard to prove a direct connection between the injury caused by the AI system and themselves. Furthermore, this action helps reduce the legal uncertainties that companies working on AI development must face. The possibility of ambiguity resulting from disparate national civil liability laws is reduced by offering a more precise framework for possible liability. In addition, the regulation gives national courts the power to order the release of data about AI systems that are considered high-risk and may be harmful (Duffourc & Gerke, 2023). This shows that efforts are being made to guarantee accountability and transparency in creating and applying AI technologies.

We shall go into greater detail about the AI Liability Directive's rules and its scope in the next sections of this chapter. We will also discuss opposing views, such as doubts about the Directive's effectiveness, possible effects on innovation, and the urgent need for further clarification. This investigation will advance a more thorough comprehension of the suggested direction and its ramifications. We will also discuss the important procedures of incorporation for EEA countries and transposition for EU member states in the adoption of EU directives. Finally, we will look at the main issues raised by the regulation, illuminating the subtleties and complexity of its application.

## Key Provisions

### Disclosure of Information

One major obstacle in the liability claims associated with AI is the difficulties victims encounter in acquiring the information required to support their claims. This difficulty is directly addressed by the proposed AI Liability Directive, especially in Article 3, which is seen as the cornerstone of efficient compensation. This chapter aims to give people a way to find the guilty parties and collect proof to back up their claims for losses brought on by high-risk AI systems. There is a significant difference between this and a comparable rule for information disclosure stated in Article 13 of the proposed AI Act (Ziosi et al., 2023). Information that can help victims achieve the burden of proof must be recorded and logged following the AI Act's regulations on evidence disclosure. These regulations apply only to high-risk AI systems, including those that risk infringing upon fundamental rights like non-discrimination and equality. By enabling them to use this data to support their compensation claims, the regulations outlined in the AI Liability Directive, on the other hand, aim to empower those who have suffered harm. Crucially, if the guilty party, referred to as "*the defendant*," unjustifiably withholds information, certain facts—like biased standards or unfair treatment—would be assumed (Ziosi et al., 2023). This clause makes requesting compensation more efficient, particularly when showing a direct connection between the harm caused and the AI system is difficult.

The Directive's Article 3(1) gives national courts the power to order the disclosure of pertinent data from a variety of sources, such as AI system providers, people who fall under the provider's obligations under the proposed AI Act (Article 24 or Article 28(1)), or users who comply with the AI Act. To start a request for evidence, the claimant must provide enough information and proof to support their assertion. If a claimant cannot collect evidence from the defendant, they may, according to Article 3(2), request evidence disclosure from providers or users who are not defendants, but only after exhausting all reasonable means to obtain evidence from the defendant. In addition, as stated in Article 3(3), the court may also issue orders to preserve evidence. It is imperative to acknowledge that specific constraints exist on the disclosure or retention of evidence, which underscore the notion of proportionality and the protection of legitimate interests (Hacker, 2022).

The crucial concept of proportionality in the disclosure and preservation of evidence is outlined in Article 3(4). It stipulates that national courts must only reveal relevant and required information to bolster a prospective lawsuit or damages claim. The courts must consider the legitimate interests of all parties, including third parties, when determining whether an order is proportionate. This includes measures to protect confidential information, such as public or national security topics and trade secrets, as specified in Article 2(1) of Directive (EU) 2016/943 (Deac, 2018). Furthermore, Member States must guarantee that national courts have the authority to take particular actions to preserve confidentiality throughout legal proceedings if a trade secret or putative trade secret the court has deemed confidential is ordered to be disclosed. In addition, those who are required to reveal or keep records must be able to contest these orders via the proper legal channels.

Despite being seen as a vital component of the Directive, Article 3 has come under investigation. "*Eurochambres*," the Association of European Chambers of Commerce and Industry, has voiced concerns regarding its invasiveness. The pre-litigation period is assumed by Article 3, which is not a trait shared by the majority of European frameworks for civil procedure, as Eurochambres points out. Significant action is required in light of this notion. Furthermore, Eurochambres points out that there should be more transparency, especially for users and suppliers turned down for disclosure (Hacker, 2022). Moreover, Eurochambres suggests that provision 3(4) be changed to a more distinct, complete provision with stricter safeguards for trade secrets. They also stress the importance of addressing security and access control measures and safeguarding sensitive data.

Professor Philipp Hacker highlights a divergence from well-established case law in non-discrimination law in his assessment of Article 3. He specifically emphasizes that the court cannot consider the defendant's unwillingness to submit information when determining whether or not there may be non-compliance. This is similar to the Court of Justice of the European Union's (CJEU) earlier decision in Case C-415/10, which maintained that a prima facie case of discrimination could be evaluated based partly on the potential offender's failure to furnish relevant information (*Galina Meister v Speech Design Carrier Systems GmbH*, 2012). All the same, the proposed AI Liability Directive seems to support leaving out denial from the non-compliance evaluation because future claimants may eventually obtain the information through a court order (Hacker, 2022).

Concerns are expressed by the European Association of Co-operative Banks (EACB) over disclosing pertinent evidence in Article 3(1). They push for clarifications, arguing that only material directly relevant to the high-risk AI system should be disclosed. They highlight the many obstacles that could make some datasets too sensitive to be released, including corporate secrets, intellectual property rights, and private banking and insurance data. For example, disclosing methods for identifying fraudsters—including those engaged in money laundering and supporting terrorism—could be extremely dangerous. Additionally, the EACB believes that the wording in Article 3(4) is overly ambiguous and suggests defining confidential information more precisely. They emphasize that enough time should be given for information disclosure in light of the high-risk nature of some AI systems. To summarize, the EACB recommends restricting the range of pertinent evidence, revising Article 3(4) to safeguard private information, and incorporating a clause guaranteeing adequate time for compliance.

Speaking on behalf of the European Engineering Industries Association, Orgalim raises concerns about possible modifications to the AI Act that could impact on the AI Liability Directive. They recommend delaying talks on the AI Liability Directive until the AI Act is signed into law. Orgalim believes that Article 3 is problematic because it includes common law ideas that might not be appropriate for use in the EU and EEA. They contend that although courts in EU and EEA nations are already empowered to require the release of particular documents to bolster damage claims, Article 3(1) broadens the reach of this authority to include more evidence and goes beyond what is often seen in European civil law systems. Orgalim argues that this will result in more drawn-out and costly legal actions, subjecting manufacturers to a deluge of information requests. They also raise questions about how manufacturers can be affected by information dumps that contain sensitive data. Orgalim concludes that Article 3 of the Directive ought to be removed.

## Presumption of Causality

The proposed AI Liability Directive, with its second notable provision, Article 4, introduces important principles for establishing causality and liability in cases involving AI systems. Article 4 revolves around the presumption of causality, stating that if it is reasonably likely that a fault influenced the output produced by an AI system or led to the AI system's failure, resulting in damage, then there is a presumption of causality. This means that if someone can show that a defendant did not fulfill their duty to reduce risk, such as providing high-quality training and testing data, it can be presumed that this failure caused the damages for which compensation is sought. However, there's a concern that enforcement measures by national courts could lead to delays and additional expenses for litigants, making it more challenging for injured parties to seek justice (Füller et al., 2022).

Article 4 is divided into seven paragraphs designed to create an effective basis for claiming compensation when a fault arises from the lack of compliance with a duty of care under national law. This measure applies to both high-risk and non-high-risk AI systems. An important development is highlighted in Case C-340/21, where the Advocate General proposed an opinion that holds controllers liable for unlawful access to personal data by third parties if they fail to implement suitable technical and organizational measures as required by the GDPR (*VB v Natsionalna agentsia za prihodite*, 2021).

To address the challenge of proving causality between non-compliance and AI system outputs or failures leading to damages, Article 4(1) lays down a targeted rebuttable presumption of causality. To apply this presumption, the claimant or the court must demonstrate the defendant's fault, the reasonable likelihood of the fault affecting the AI system's output or failure, and that this output or failure caused the damage. Articles 4(2) and 4(3) deal with different types of claims against providers and users of high-risk AI systems, outlining specific requirements for seeking damages.

Article 4(4) excludes the presumption of causality when the defendant can demonstrate that the claimant has reasonable access to sufficient evidence and expertise to prove the causal link, motivating the defendant to provide access to such evidence. Article 4(5) introduces a requirement for the presumption of causality to apply to non-high-risk AI systems only when establishing a causal link becomes excessively challenging for the claimant. Article 4(6) addresses scenarios where non-professional users interfere with AI systems and sets conditions for the presumption of causality to apply (Koch et al., 2022).

Lastly, Article 4(7) grants defendants the right to rebut the presumption of causality under Article 4(1). However, concerns have been raised about the lack of clarity in the conditions for the presumption of a causal link and the differentiation between illegality and fault in national civil liability regimes. There are suggestions to clarify and update the proposal, particularly regarding harm caused by prohibited AI systems and the relationship between the AI Act and the AI Liability Directive.

## Evaluation And Targeted Review

Article 5 of the proposed AI Liability Directive outlines provisions for assessing and reviewing the Directive's implementation. The European Commission will review and submit a report to the European Parliament, Council, and the European Economic and Social Committee. This report will scrutinize the impact of Articles 3 and 4 in achieving the Directive's objectives. It will evaluate whether the strict liability rules for claims against operators of certain AI systems are appropriate, assess the need for insurance coverage, and consider how these rules affect the adoption of AI systems. Article 5(3) also mandates the Commission establish a monitoring program that regularly collects essential data and evidence. This program will provide information on incidents related to AI systems, helping determine if additional measures, like a strict liability regime or mandatory insurance, are necessary.

The AI, Algorithmic, and Automation Incident and Controversies ("*AIAAIC*") repository is an independent and publicly accessible dataset documenting incidents and controversies linked to AI, algorithms, and automation. In Table 1, we observe a striking increase in reported AI incidents and controversies in the AIAAIC database, with 26 times more incidents reported in 2021 compared to 2012 (Duffourc & Gerke, 2023). This sharp rise underscores two key trends: AI's growing integration into the real world and a heightened awareness of the ethical risks tied to AI. Eurochambres appreciates adopting a minimally interventionist approach and views Article 5 of the proposed AI Liability Directive as a mechanism for reviewing and reassessing the need for more stringent and harmonized regulation.

## Transposition

The process of "*transposition*," which entails each member state modifying and incorporating a directive's provisions into its national legal system, is how proposed EU directives are implemented in EU member states. Experts from the European Free Trade Association (EFTA) evaluate whether an act adopted by the EU has a bearing on the European Economic Area (EEA), which comprises all EU member states and EFTA nations like Iceland, Norway, and Liechtenstein (Reis et al., 2020). The EEA aims to establish a single economic area where people and legal entities have equal access to all regional rights and opportunities, including the freedom of movement for people, capital, products, and services. If EEA regulations conflict with other laws, the EFTA nations agree to enact new legislation that precedes EEA regulations. A Joint Committee Decision (JCD) is written when input on an EU Act is received from EEA EFTA nations. After clearance from pertinent subcommittees and EFTA experts, it is sent to the Commission for inter-service consultation (Hacker, 2022). The Commission takes the EU's stance; if the draft JCD has major revisions, it is forwarded to the Council for approval. The JCD becomes binding on the EEA upon approval by the Joint Committee of the EEA, and member states are required to incorporate the directives into their national laws through a procedure known as "*incorporation*."

The proposed AI Liability regulation's Article 7 specifies that EU member states must put the regulation into effect within two years of it being passed. They must notify the Commission of the Directive's principal provisions and cite it in their national legal frameworks. Member states are allowed to refer to the Directive in whatever way they see fit, and they are required to offer explanation documents outlining the specifics of how each provision is being implemented. This helps the Commission determine relevant legislative requirements, regardless of the transposition technique selected (Mukherjee et al., 2023). We have examined the suggested articles of the AI Liability Directive and varying viewpoints from organizations and academics throughout this chapter. Eurochambres suggested explicitly addressing the protection of sensitive information and underlined the need for clarity when requests for disclosure are denied. The EACB recommended additional time provisions and voiced worries about possible difficulties in sharing sensitive AI knowledge. There were concerns regarding the timing of these talks when Orgalim suggested postponing them until the AI Act is finalized (Goutzamanis, 2023).

Professor Philip Hacker addressed harm caused by banned AI systems and drew attention to the lack of a thorough method for shifting the burden of proof concerning Article 4 of the proposed AI Liability Directive. The EACB recommended codes of conduct to apply high-risk AI requirements to non-high-risk AI systems. Orgalim, however, voiced concerns regarding Article 4, arguing that it puts AI system suppliers at more risk. Divergent views were aroused by Article 5. Orgalim recommended that businesses evaluate their insurance requirements, while the EACB opposed insurance laws because they would affect competition and prices. Eurochambres saw Article 5 as a minimally intrusive mechanism for reviewing and reevaluating the strictness of regulations. Article 6 of the proposed AI Liability Directive changes Directive (EU) 2020/1828 on collective actions for consumer protection; it permits claimants to bring collective claims in a less explicitly defined situation. These various viewpoints and recommendations highlight how complicated AI liability is and how important it is to have constant communication and cooperation to create useful frameworks.

## Potential Challenges of the Proposed AI Liability Directive

The article discusses the challenges posed by different legal traditions in the national laws of European Economic Area (EEA) member states in implementing the proposed AI Liability Directive. The Directive aims to establish a legal framework for holding AI systems liable for damages they may cause. However, the diversity of legal traditions across EEA member states complicates its implementation. Many questions remain unanswered in artificial intelligence (AI), and governance solutions are often untested. This rapidly evolving landscape provides an opportunity for regulatory innovation and highlights the necessity for a legal framework in an uncertain field. Different EEA countries have varying rules concerning which damages trigger liability. Some countries treat the material and immaterial damage equally, while others only allow compensation for immaterial damage when expressly provided for by law. The Nordic countries follow the latter approach, but exceptions exist for specific immaterial harm, such as injury to reputation or cases of intentional or gross negligence. The line between fault-based and strict liability often blurs in jurisdictions permitting recovery for immaterial damages. However, compensation is only possible in Portugal if the loss is severe.

The proposed AI Liability Directive addresses this issue by defaulting to the country's law where the damage occurs. However, prior EU cases have introduced some ambiguity, emphasizing the need for consistency and predictability in cross-border jurisdiction determinations. The article also raises concerns about possibly disclosing sensitive information, such as trade secrets and confidential data, in AI liability cases. Article 3 of the AI Liability Directive allows victims to request the disclosure of information about high-risk AI systems, but only to the extent necessary to prove non-compliance with the proposed AI Act's requirements. Balancing the interests of both parties, the Directive emphasizes protecting sensitive information while ensuring access to necessary evidence. However, this leaves room for interpretation by national courts, which must consider various factors when deciding on disclosure measures, including the legitimate interests of the parties involved and any potential harm.

## DISCUSSION: WHAT'S MISSING FROM THE NEW EU AI ACT

Recently, Mistral AI, a French firm in artificial intelligence, released a huge, notable language model called Mixtral 8x7B. This model has proven to be more effective than GPT 3.5 in several benchmark tests, even though it is smaller. The success of Mixtral 8x7B is ascribed to its novel architecture, which blends eight expert models. This publication highlights the most urgent issues facing AI policy today, even if it represents a substantial technical achievement. It also highlights the holes the AI Act has not filled, a recent legislative attempt. These difficulties include four critical facets: the requirement for basic AI safety standards to be mandated, the dilemma about open-source models, the environmental consequences of AI technology, and the pressing need for additional public investment in AI to supplement the AI Act (Hacker, 2023). These are real, pressing issues that require immediate attention; they are not just theoretical issues. Recent releases such as Google's Gemini, Mixtral 8x7B, and Claude 2.1 demonstrate the rapid advancement of AI technology and call for a prompt and deliberate governmental response (Hacker, 2023).

While the AI Act is a positive beginning, it does not fully address fundamental challenges, leaving the European Union (EU) vulnerable in important areas of AI development and governance. However, after protracted talks, the EU created minimal requirements for all foundation models—referred to as general-purpose AI models in the AI Act. Furthermore, stricter rules were applied to systemically risky high-impact foundation models. Nonetheless, it is interesting that these minimal requirements are considered incredibly inadequate. They essentially include things like restricted copyright rights and openness. H stricter requirements apply when high-impact models are trained with more than 10^25 FLOPs (floating-point operations, roughly comparable to calculation steps) (Reuters, 2023b). But currently, only GPT-4 and possibly Gemini, plus a couple of other models, can cross this barrier . It seems clear that smaller, more powerful models are the direction of AI development. Even "*smaller*" models—like Bard and ChatGPT—present significant cybersecurity and AI safety hazards that cannot be left to self-regulation, particularly for models with 10^24 FLOPs or less. By comparison, it's like playing in the Champions League but following non-binding guidelines instead of the stringent Disinformation Code of Practice requirements, which Twitter left behind when it didn't suit its new owner's tastes.

Four key aspects of these challenges include mandatory fundamental AI safety standards, the difficulty around open-source models, the environmental effects of AI technology, and the urgent need for greater public investment in AI to support the AI Act. These are not merely theoretical problems; they are urgent problems that must be addressed now. The latest releases of AI technologies, including Google's Gemini, Mixtral 8x7B, and Claude 2.1, show how quickly the field is developing and demand a thoughtful and timely response from the government. Although the AI Act is a good start, it falls short of addressing the underlying issues, exposing the European Union (EU) to critical AI development and governance domains. However, after lengthy negotiations, the EU did establish minimum requirements for all foundation models, which the AI Act refers to as general-purpose AI models. Moreover, higher regulations were imposed on high-impact foundation types that posed a systemic danger. It is noteworthy, however, that these bare minimums are seen as utterly insufficient. They consist of things like openness and limited copyright rights (Hacker, 2023). More stringent constraints apply when training high-impact models with more than 10^25 FLOPs (floating-point operations, roughly equivalent to calculation steps). However, only GPT-4 and probably Gemini, along with a few other models, can pass over this barrier. The trajectory of AI development is towards more compact and potent models. It is inappropriate to rely solely on self-regulation for even "*smaller*" models like Bard and ChatGPT, as they pose serious risks to cybersecurity and AI safety. This is especially true for models with 10^24 FLOPs or fewer. In contrast, it's like participating in the Champions League but according to non-binding criteria rather than the strict guidelines of the Disinformation Code of Practice—a document that Twitter abandoned when its new owner didn't like it.

The current technology regulatory framework is insufficient to solve the challenges posed by the most powerful technology available today. When limitations forbid "*foundation models*" (FMs), those utilizing these models later on are responsible for adhering to the rules. This approach is problematic because it focuses on repairing deployment problems, which is much less efficient than addressing the foundation model itself, the source of the issue. This theory aligns with traditional legal and economic theory, which maintains that the "*least-cost avoider*" prevails. Therefore, it is essential to provide well-considered guidelines appropriate for foundation models. Self-control in this situation is not only dangerous but also ineffectual. One of its main weaknesses is that the AI Act does not offer a comprehensive framework for AI safety that incorporates all base models. Content moderation, cybersecurity, and the mandated content evaluation by outside experts—also referred to as "*red teaming*"—should all be covered by this framework. Because these huge language models can generate massive amounts of content, they are highly useful tools in many high-impact domains, including education and medicine.

However, as the Dutch Cyber Security Centre notes, this can lead to a flood of misinformation, hate speech, cyber threats, and even the facilitation of chemical and biological terrorism in the lack of robust security measures. Insufficient cybersecurity measures can disseminate these hazards across the artificial intelligence ecosystem and generate prospects for malevolent entities independent of state backing. To solve these problems, industry best practices, such as red teaming and safety layers, must be implemented to prevent exploitation by malicious actors. Unfortunately, the AI Act does not mandate that foundation models take these safety measures. One potential strategy is implementing a strong, decentralized content moderation system that adheres to the principles of the Digital Services Act (DSA). If this method is not incorporated into the AI Act shortly, it may be considered for a revised DSA. The field of generative artificial intelligence should be promptly included in the trusted flagger and notice-and-action mechanism provisions of the Digital Services Act, as defined in Article 16 and beyond. This innovation is critical to creating a more efficient and decentralized system for identifying and removing harmful, dangerous, or poisoned content produced by AI systems as major global election cycles (in the US, EU, and elsewhere) draw near.

This approach will enhance current industry standards by including community-driven oversight, maybe involving registered non-governmental organizations (NGOs). Given the potential risks associated with the rapid advancement of AI technology, proactive and ambitious regulatory measures are necessary. It's critical to recognize that, contrary to what some may claim, sensible regulation of foundation models (FMs) does not stifle innovation. As per a recent study, even for moderately advanced FMs (with processing capacity below 10^24 FLOPs), like Bard or ChatGPT, the expected compliance expenses amount to roughly 1% of the entire development expenditure. This little investment can and should be made by all AI industry participants, including smaller European providers such as Mistral and Aleph

Alpha, to adhere to basic industry best practices for AI safety. Finding a balance between the benefits of open-source innovation and public safety is challenging. Making open-source models like Mixtral 8x7B, similar to the Falcon family or Meta's Llama 2, promotes transparency and accessibility. Without a doubt, open-source methods benefit the area of artificial intelligence by encouraging diversity, accessibility, and competition while preventing monopolization tendencies. However, as these approaches gain traction, concerns about the risks associated with open-source grow. Unrestricted use of such potent models could lead to malicious applications, inciting terrorism and producing malware. Furthermore, it is possible for safety precautions to be easily and inadvertently removed from open-source models.

The EU has to reconsider its stance on open-source AI models in light of these concerns. Currently, these models are not regulated until they pose a risk to the system. It is proposed that complete open sourcing should be forbidden above a given performance threshold, potentially even lower than the current one (10^23 FLOPs or comparable to the GPT 3.5 benchmark performance, for instance). This ban will remain in place until we have more information about adding safety features to these gadgets. Instead, it is best to implement a hosted, controlled access system to track and regulate usage.

Most importantly, this approach encourages innovation to be applied correctly rather than suppressing it. A mechanism similar to Article 40 of the Digital Services Act (DSA) that allows access to qualified researchers should be implemented in addition to existing restrictions. Facilitating independent verification of stress tests and benchmarks guarantees that open models are closed responsibly via hosted access and that closed models are opened to permitted researchers. Adhering to the principle of "*trust, but verify*," this approach ensures that oversight relies on resource-constrained regulatory authorities and model suppliers and encompasses the broader academic community.

Ultimately, this well-rounded approach aims to maintain public safety and accountability while encouraging creativity in the rapidly evolving field of artificial intelligence. The impact of enormous AI models on the environment must be addressed immediately since it poses an existential safety issue. A step in the right approach towards controlling AI sustainability is the legal endeavor known as the AI Act. However, its provisions are a little limited. The environment can benefit greatly from AI applications, but the massive amounts of water and computing power required to train and run these large-scale AI models contribute to climate change. This makes achieving sustainability more challenging on a global scale, especially when immediate action is required. By 2027, AI models are expected to consume the same energy as an entire country, like the Netherlands or Argentina. Future iterations of the AI Act must include additional provisions to ensure that the industry grows in an environmentally responsible manner. This means stricter measures, such as mandating sustainability impact studies and extending carbon trading schemes to cover data centers and other high-energy-consuming IT operations, are needed.

Now, let's talk about the economic side of things. In the European AI framework, public investment is a key component that is overlooked in the effort to foster AI innovation. Prominent cases from throughout the globe are used to illustrate this idea. Norway has committed heavily to the advancement of AI, the UK is investing heavily in AI supercomputing, and the US is spending heavily under President Biden's 2024 budget to maintain its position as the industry leader in AI. China, Saudi Arabia, and even private companies like G42 are making significant investments in the advancement of AI. The EU must combine resources and make them visible to invest significantly more in AI to compete globally. The objective is to surpass current AI capabilities, as exemplified by models like GPT-4. By making this investment, we intend to establish strategic autonomy in a field critical to the twenty-first century and open the door for long-term AI advancements.

It is crucial to realize that no EU member state can solve this issue independently. The AI Act agreement should have accompanied a major announcement of EU and collective Member State funding for AI research and implementation. This covers computational infrastructure, talent retention, and chip manufacturing. Avoiding strategic dependencies can be likened to the challenges faced by the oil and gas supply business in preserving Europe's position in the AI industry. Regarding producing state-of-the-art AI models, Europe now lags considerably behind, with very few exceptions. Given the current state of the world, this divide poses a geopolitical problem. A major flaw of the AI Act is that it does not provide a thorough framework for AI safety that includes all base models. This framework should address content moderation, cybersecurity, and the required content assessment by outside specialists (sometimes known as "red teaming"). These large language models are highly effective tools in various high-impact fields, including education and medicine, because they can produce enormous volumes of content. But as the Dutch Cyber Security Centre points out, in the absence of strong security measures, this can result in a deluge of false information, hate speech, cyber threats, and even encouragement of chemical and biological terrorism.

Inadequate cybersecurity safeguards have the potential to spread these threats throughout the AI ecosystem and create opportunities for hostile actors, whether or not a state supports them. Adopting industry best practices—such as red teaming and installing safety layers to thwart misuse by malevolent entities—is essential to addressing these issues. Regretfully, not all foundation models are required by the AI Act to implement these precautions. One possible approach is introducing a robust, decentralised content moderation system in the Digital Services Act (DSA) spirit. Should this technique not be included in the AI Act anytime soon, it might be considered for an updated DSA. To create a more effective and decentralized system for identifying and removing toxic, harmful, or dangerous content generated by AI systems, the provisions outlined in Articles 16 and beyond of the Digital Services Act—which include trusted flaggers and a notice-and-action mechanism—should be expeditiously extended to the domain of Generative AI. This is especially important as we approach crucial global election cycles (in the US, EU, and elsewhere). This method would improve industry practices by incorporating community-driven supervision, maybe involving registered non-governmental organizations (NGOs).

Proactive and ambitious regulatory measures are required due to the rapid growth of AI technology and the possible risks that come with it. It's important to understand that reasonable regulation of foundation models (FMs) does not impede innovation, despite what some may suggest. According to a recent study, the estimated compliance costs equal to just approximately 1% of the total development expenditures, even for moderately advanced FMs, such as Bard or ChatGPT, which are not as powerful as GPT-4 or Gemini (with processing power below 10^24 FLOPs). All AI industry participants, including smaller European providers like Mistral and Aleph Alpha, can and should make this modest investment to comply with fundamental industry best practices for AI safety. Striking a balance between the advantages of open-source innovation and public safety is a difficult task. Transparency and accessibility are encouraged by the choice to make models like Mixtral 8x7B open-source, akin to the Falcon family or Meta's Llama 2. Unquestionably, open-source approaches positively impact AI by promoting accessibility, diversity, and competitiveness while thwarting monopolization tendencies. But as these models get stronger, the dangers of open source become more worrying. Unrestricted access to such powerful models may result in nasty usage, such as malware creation and the encouragement of terrorism. Moreover, safety measures might be readily and unintentionally deleted from open-source models.

Given these worries, the EU must reevaluate its position on open-source AI models. These models are not subject to regulation at the moment unless they present systemic risks. It is suggested that full open sourcing should be prohibited over a specific performance barrier, possibly even below the present one (10^23 FLOPs or equivalent to GPT 3.5 benchmark performance, for example). This ban would stay in effect until we learn more about incorporating safety safeguards into these devices. Rather, a hosted access controlled access system should be implemented, enabling monitoring and managing usage. Crucially, this strategy promotes the proper application of innovation rather than its stifling. In addition to these regulations, a system like Article 40 of the Digital Services Act (DSA) that permits access to qualified researchers should be implemented. The goal is to facilitate independent verification of stress tests and benchmarks to ensure that open models are closed responsibly via hosted access, and that closed models are opened to approved researchers. Finally, this balanced approach aims to promote innovation while protecting public safety and accountability in the rapidly evolving AI landscape. It does this by ensuring that oversight is not solely dependent on model providers and resource-constrained regulatory bodies but involves the larger academic community, adhering to the principle of "trust, but verify."

It is now urgently necessary to address the environmental impact of massive AI models, much like it is an existential safety risk. The legislative effort known as the AI Act is a positive move in the direction of regulating AI sustainability. Its provisions are, nevertheless, a little restricted. Although AI applications have much to offer the environment, the enormous amounts of computational power and water needed to train and implement these large-scale AI models hurt climate change. This makes it more difficult to attain sustainability on a global scale, particularly at a time when prompt action is necessary. According to projections, AI models will use as much energy by 2027 as an entire nation, such as Argentina or the Netherlands. The AI Act must be expanded upon in subsequent versions to guarantee that the AI sector develops in an environmentally conscious way. This calls for a more stringent strategy that may include requiring sustainability impact studies and expanding emissions trading programs to include data centers and other high-consumption IT operations.

Turning to the economic side, public investment is a crucial element that the European AI framework ignores in its attempt to promote AI innovation. This point is illustrated with notable cases from throughout the world. Norway has made a major commitment to the development of AI, the UK has made large investments in AI supercomputing, and the US, under President Biden's 2024 budget, has made large financial commitments to uphold its leadership in AI. China, Saudi Arabia, and even private businesses such as G42 are investing heavily in the development of artificial intelligence. To properly compete globally, the EU needs to invest more in AI, combining resources and making them visible. The goal is not just to match but to exceed existing AI capabilities, as demonstrated by models such as GPT-4. With this investment, we hope to secure strategic independence in technology essential to the twenty-first century and pave the route for sustainable AI innovations.

Understanding that each EU member state cannot handle this problem independently is imperative. A significant announcement of EU and collective Member State financing for AI research and deployment should have gone hand in hand with the AI Act agreement. This includes things like chip manufacturing, talent retention, and computer infrastructure. Preventing strategic dependencies is crucial to safeguard Europe's standing in the AI space, as it may be compared to the difficulties encountered in the oil and gas supply sector. With few exceptions, Europe lags far behind in generating cutting-edge AI models. In the present global setting, this gap presents a geostrategic dilemma.

## CONCLUSION

The chapter's first section covers the urgent topic of responsibility in the context of artificial intelligence (AI). It highlights the need for an all-encompassing legal framework to address the possible harm that artificial intelligence systems may inflict. The introduction sets the scene by summarizing the state of AI technology today and its significant social effects. AI has had both beneficial and negative effects; while it has greatly advanced technology, it has posed serious ethical issues. The article's consideration of the difficulties and unanswered legal questions surrounding AI is one of its main points of emphasis. A thorough analysis of the current liability regimes and legal frameworks reveals weaknesses and inadequacies in their capacity to handle the complexity of artificial intelligence adequately. It becomes clear that established legal frameworks are finding it difficult to stay up with the quick advancements in artificial intelligence.

A thorough examination is conducted on the concept of liability, specifically concerning AI. This investigation covers the potential benefits of present AI liability standards and their challenges and drawbacks. The question of who should be responsible when AI systems cause harm is intricate and constantly changing. After that, the proposed AI Liability Directive becomes the chapter's main topic. This Directive attempts to establish a much-needed legal framework to give clarity and uniformity in AI-related liability matters throughout the European Union (EU) and the European Economic Area (EEA). The main objective is to balance promoting AI technology innovation and ensuring that people and companies are sufficiently safeguarded if they sustain harm due to AI systems. The chapter emphasizes how crucial it is to create artificial intelligence (AI) liability regulations to guarantee that all parties engaged in creating and utilizing AI systems take the required safety measures to reduce risks and give victims a way to pursue damages compensation. Although the creation of AI impact assessment reports is viewed as a step in the right direction towards reducing potential harm, protecting fundamental rights and the rights of AI producers and developers must be balanced carefully. Determining acceptable risk levels and evaluating possible harm can be complex, making this balancing act difficult.

Article 4a of the proposed AI Act 196 introduces a provision for AI Impact Assessment Reports. This implies that a plan for monitoring and reducing the risks of deploying an AI system must exist before implementation. The person in charge of implementing the AI system should hold off on using it if such a plan cannot be found. This legislation aims to guarantee the responsible deployment of AI systems, considering all possible outcomes. Furthermore, in the event of loss or damage, the Commission has proposed a method that would oblige creators of AI algorithms to reveal the datasets' design parameters and metadata. However, because of how general it is, this suggestion causes considerable anxiety. There is a lack of clarity regarding what information needs to be recorded and what obligations producers have to reveal. These issues require additional explanation to create a transparent and equitable structure for accountability in the event of loss or damage associated with AI.

The chapter explores the difficulties raised by the AI Liability Directive and the planned AI Act, especially concerning the possibility of legal ambiguity. Different member states within the European Union (EU) may interpret and apply these directives differently, which could lead to ambiguity. The final Directive's degree of flexibility, which permits modifications to account for variances among various legal jurisdictions, determines how difficult this task will be to overcome—remembering that the country's legal procedural norms and traditions can greatly impact how these guidelines are understood and implemented.

The prospect of "*forum shopping*," in which a party to a case looks for a jurisdiction or court with the most advantageous rules or laws to their advantage, is one major problem brought up in the article. The availability of several locations for legal action may encourage the party at fault to look for the best legal setting. Although some view it as advantageous, leaving national courts to interpret directives might result in inequities and legal confusion because different courts may come to different judgments. The article points out that there have been other instances of legal confusion regarding the proposed AI Liability Directive. Article 4, which permits victims to show that non-compliance with pertinent responsibilities is reasonably expected to have caused damage, is a crucial part of this regulation. This clause imposes a heavy responsibility on users and providers of AI systems to ensure their AI systems follow all applicable laws and regulations. If this isn't done, there might be a shift in the burden of evidence, requiring the person in charge of the AI system to prove they didn't hurt anyone.

The proposed AI duty Directive raises several important questions and issues, mainly around the division of duty between AI providers and users. These are highlighted in the article. The necessity of access to evidence for a fair trial is emphasized in the Directive, namely in Article 3. The precise distribution of culpability between providers and users is unclear, though. It may be challenging to decide who should be held accountable if AI-related harm occurs due to this lack of clarity, which may create legal confusion. One of the article's major concerns is that parties outside the purview of the proposed AI Act cannot be asked for evidence. A harmonized approach and legal certainty in AI liability issues are the Directive's goals; nonetheless, this limitation raises worries that parties may avoid liability merely by not falling under the purview of the AI Act. The article offers some possible remedies to these concerns, including asking the European Court to issue preliminary rulings or advisory opinions interpreting the Directive and its substantive requirements that must be incorporated into national legislation.

The difficulty Iceland has putted the planned AI Liability Directive into practice is also mentioned in the article. Iceland enjoys procedural autonomy under the EEA Agreement, meaning it is allowed to create its procedural laws. However, several of the Directive's provisions—like lowering the burden of proof for victims and enabling national courts to require the disclosure of evidence on high-risk AI systems—might be at odds with Iceland's current legal system. While acknowledging the potential risks and benefits of AI technology, Iceland is unsure if the Directive in its current form will have the desired impact on the nation. The paper concludes by pointing out that the vague elements in the proposed AI Liability Directive prevent it from accomplishing its intended objectives. It doesn't offer precise instructions on how consumers and AI suppliers should split liability. Furthermore, the Directive's scope raises questions about how well it would operate with Iceland's legal system, possibly going against established procedural and liability rules. The article highlights the necessity for additional clarification to guarantee the proper implementation of the Directive's provisions under the EEA and Icelandic law. Ultimately, resolving these concerns and offering precise guidance will determine how well the Directive works to advance legal clarity and a unified approach to AI liability.

## REFERENCES

Bartneck, C., Lütge, C., Wagner, A., & Welsh, S. (2021). *An Introduction to Ethics in Robotics and AI.* Springer. doi:10.1007/978-3-030-51110-4

Benhamou, Y., & Ferland, J. (2020). *Artificial Intelligence & Damages: Assessing Liability and Calculating the Damages* (SSRN Scholarly Paper 3535387). https://papers.ssrn.com/abstract=3535387

Buiten, M., de Streel, A., & Peitz, M. (2023). The law and economics of AI liability. *Computer Law & Security Report*, *48*, 105794. doi:10.1016/j.clsr.2023.105794

Cataleta, M. S. (2020). *Humane Artificial Intelligence: The Fragility of Human Rights Facing AI*. East-West Center. https://www.jstor.org/stable/resrep25514

Deac, A. (2018). Regulation (Eu) 2016/679 Of The European Parliament And Of The Council On The Protection Of Individuals With Regard To The Processing Of Personal Data And The Free Movement Of These Data. *Perspectives of Law and Public Administration, 7*(2), 151–156. https://ideas.repec.org//a/sja/journl/v7y2018i2p151-156.html

Duffourc, M. N., & Gerke, S. (2023). The proposed EU Directives for AI liability leave worrying gaps likely to impact medical AI. *NPJ Digital Medicine*, *6*(1), 1. doi:10.1038/s41746-023-00823-w PMID:37100860

Duivenvoorde, B. (2022). The Liability of Online Marketplaces under the Unfair Commercial Practices Directive, the E-commerce Directive and the Digital Services Act. *Journal of European Consumer and Market Law, 11*(2). https://kluwerlawonline.com/api/Product/CitationPDFURL?file=Journals\EuCML\EuCML2022009.pdf

Ebers, M., Hoch, V. R. S., Rosenkranz, F., Ruschemeier, H., & Steinrötter, B. (2021). The European Commission's Proposal for an Artificial Intelligence Act—A Critical Assessment by Members of the Robotics and AI Law Society (RAILS). *J, 4*(4), 4. doi:10.3390/j4040043

Finlayson-Brown, J., & Bossotto, L. (2021, August 9). *Italian data protection supervisory authority fines two food delivery companies for non-compliant algorithmic processing*. Allen Overy. https://www.allenovery.com/en-gb/global/blogs/data-hub/italian-data-protection-supervisory-authority-fines-two-food-delivery-companies-for-non-compliant-algorithmic-processing

Franke, U. (2019). *Harnessing Artificial Intelligence*. European Council on Foreign Relations. https://www.jstor.org/stable/resrep21491

Füller, J., Hutter, K., Wahl, J., Bilgram, V., & Tekic, Z. (2022). How AI revolutionizes innovation management – Perceptions and implementation preferences of AI-based innovators. *Technological Forecasting and Social Change*, *178*, 121598. doi:10.1016/j.techfore.2022.121598

Godinho, I., Flores, C., & Marques, N. (2021). CONSULTATION ON THE WHITE PAPER ON ARTIFICIAL INTELLIGENCE - A EUROPEAN APPROACH. *ULP Law Review*, *14*, 157–167. doi:10.46294/ulplr-rdulp.v14i1.7475

Goutzamanis, Y. (2023). Closing the Floodgates on Privacy Class Actions: Lloyd v Google LLC. *The Modern Law Review*, *86*(1), 249–262. doi:10.1111/1468-2230.12744

Hacker, P. (2022). *The European AI Liability Directives – Critique of a Half-Hearted Approach and Lessons for the Future* (*SSRN* Scholarly Paper 4279796). doi:10.2139/ssrn.4279796

HackerP. (2023). What's Missing from the EU AI Act: Addressing the Four Key Challenges of Large Language Models. *Verfassungsblog*. doi:10.17176/20231214-111133-0

Hyden, H. (2022). Regulation of AI: Problems and Options. *The Swedish Law and Informatics Research Institute*, (pp. 295–314). Law Pub. doi:10.53292/208f5901.9118259e

Kakatkar, C., Bilgram, V., & Füller, J. (2018). *Innovation Analytics: Leveraging Artificial Intelligence in the Innovation Process* (SSRN Scholarly Paper 3293533). doi:10.2139/ssrn.3293533

Karnow, C. E. A. (1996). Liability for Distributed Artificial Intelligences. *Berkeley Technology Law Journal, 11*(1), 147–204. https://www.jstor.org/stable/24115584

Koch, B. A., Borghetti, J.-S., Machnikowski, P., Pichonnaz, P., Ballell, T. R. de las H., Twigg-Flesner, C., & Wendehorst, C. (2022). Response of the European Law Institute to the Public Consultation on Civil Liability – Adapting Liability Rules to the Digital Age and Artificial Intelligence. *Journal of European Tort Law, 13*(1), 25–63. doi:10.1515/jetl-2022-0002

Laux, J., Wachter, S., & Mittelstadt, B. (2022). *Trustworthy Artificial Intelligence and the European Union AI Act: On the Conflation of Trustworthiness and the Acceptability of Risk* (SSRN Scholarly Paper 4230294). doi:10.2139/ssrn.4230294

Lee, J. D., & See, K. A. (2004). Trust in Automation: Designing for Appropriate Reliance. *Human Factors, 46*(1), 50–80. doi:10.1518/hfes.46.1.50.30392 PMID:15151155

Lombardo, G. (2022). The AI industry and regulation: Time for implementation? In R. Iphofen & D. O'Mathúna (Eds.), *Ethical Evidence and Policymaking* (1st ed., pp. 185–200). Bristol University Press. doi:10.2307/j.ctv2tbwqd5.15

Maliha, G., Gerke, S., Cohen, G., & Parikh, R. B.MALIHA. (2021). Artificial Intelligence and Liability in Medicine: Balancing Safety and Innovation. *The Milbank Quarterly, 99*(3), 629–647. doi:10.1111/1468-0009.12504 PMID:33822422

Mukherjee, S., Coulter, M., Chee, F. Y., Mukherjee, S., & Chee, F. Y. (2023, December 14). Explainer: What's next for the EU AI Act? *Reuters*. https://www.reuters.com/technology/whats-next-eu-ai-act-2023-12-14/

Müller, V. C. (2023). Ethics of Artificial Intelligence and Robotics. In E. N. Zalta & U. Nodelman (Eds.), *The Stanford Encyclopedia of Philosophy* (Fall 2023). Metaphysics Research Lab, Stanford University. https://plato.stanford.edu/archives/fall2023/entries/ethics-ai/

Nanos, A. (2023). *Criminal Liability of Artificial Intelligence* (SSRN Scholarly Paper 4623126). doi:10.2139/ssrn.4623126

Reich, N. (1986). Product safety and product liability—An analysis of the EEC Council Directive of 25 July 1985 on the approximation of the laws, regulations, and administrative provisions of the Member States concerning liability for defective products. *Journal of Consumer Policy, 9*(2), 133–154. doi:10.1007/BF00380508

Reis, J., Santo, P., & Melão, N. (2020). Artificial Intelligence Research and Its Contributions to the European Union's Political Governance: Comparative Study between Member States. *Social Sciences (Basel, Switzerland), 9*(11), 11. doi:10.3390/socsci9110207

Reuters. (2023a). *Artificial Intelligence Liability Directive: Legislation tracker.* Practical Law. https://uk.practicallaw.thomsonreuters.com/w-037-5533?transitionType=Default&contextData=(sc.Default)&firstPage=true

Reuters. (2023b, December 9). *EU clinches deal on landmark AI Act—Reaction*. Reuters. https://www.reuters.com/technology/eu-clinches-deal-landmark-ai-act-2023-12-09/

Rodríguez de las Heras Ballell, T. (2023). The revision of the product liability directive: A key piece in the artificial intelligence liability puzzle. *ERA Forum, 24*(2), 247–259. 10.1007/s12027-023-00751-y

Rose, K., Eldridge, S. D., & Chapin, L. (2015). *THE INTERNET OF THINGS: AN OVERVIEW Understanding the Issues and Challenges of a More Connected World*. Semantic Scholar. https://www.semanticscholar.org/paper/THE-INTERNET-OF-THINGS-%3A-AN-OVERVIEW-Understanding-Rose-Eldridge/6d12bda69e8fcbbf1e9a10471b54e57b15cb07f6

Schwemer, S. F. (2022). *Digital Services Act: A Reform of the e-Commerce Directive and Much More* (*SSRN* Scholarly Paper 4213014). doi:10.2139/ssrn.4213014

Sousa Antunes, H. (2020). *Civil Liability Applicable to Artificial Intelligence: A Preliminary Critique of the European Parliament Resolution of 2020* (*SSRN* Scholarly Paper 3743242). doi:10.2139/ssrn.3743242

VB v Natsionalna agentsia za prihodite, Case C-340/21 (ECJ 2021). https://eur-lex.europa.eu/legal-content/EN/TXT/?uri=CELEX%3A62021CN0340

Wagner, G. (2021). *Liability for Artificial Intelligence: A Proposal of the European Parliament* (SSRN Scholarly Paper 3886294). doi:10.2139/ssrn.3886294

Waltl, B., & Vogl, R. (2018). Increasing Transparency in Algorithmic- Decision-Making with Explainable AI. *Datenschutz Und Datensicherheit - DuD, 42*(10), 613–617. doi:10.1007/s11623-018-1011-4

Wang, X., Lin, X., & Shao, B. (2023). Artificial intelligence changes the way we work: A close look at innovating with chatbots. *Journal of the Association for Information Science and Technology, 74*(3), 339–353. doi:10.1002/asi.24621

Wendehorst, C. (2020). Strict Liability for AI and other Emerging Technologies. *Journal of European Tort Law, 11*(2), 150–180. doi:10.1515/jetl-2020-0140

Zech, H. (2021). Liability for AI: Public policy considerations. *ERA Forum, 22*(1), 147–158. 10.1007/s12027-020-00648-0

Zhang, D., Mishra, S., Brynjolfsson, E., Etchemendy, J., Ganguli, D., Grosz, B., Lyons, T., Manyika, J., Niebles, J. C., Sellitto, M., Shoham, Y., Clark, J., & Perrault, R. (2021). *The AI Index 2021 Annual Report* (arXiv:2103.06312). arXiv. https://doi.org//arXiv.2103.06312 doi:10.48550

Ziosi, M., Mökander, J., Novelli, C., Casolari, F., Taddeo, M., & Floridi, L. (2023). *The EU AI Liability Directive: Shifting the Burden From Proof to Evidence* (SSRN Scholarly Paper 4470725). doi:10.2139/ssrn.4470725

## ADDITIONAL READING

Agrawal, A., Gans, J. S., & Goldfarb, A. (2019). Artificial Intelligence: The Ambiguous Labor Market Impact of Automating Prediction. *The Journal of Economic Perspectives*, *33*(2), 31–50. https://www.jstor.org/stable/26621238. doi:10.1257/jep.33.2.31

Brattberg, E., Csernatoni, R., & Rugova, V. (2020). *Assessing the EU's Approach To AI*. Carnegie Endowment for International Peace. https://www.jstor.org/stable/resrep25784.7

Cuéllar, M.-F. (2019). A Common Law for the Age of Artificial Intelligence: Incremental Adjudication, Institutions, and Relational Non-Arbitrariness. *Columbia Law Review*, *119*(7), 1773–1792. https://www.jstor.org/stable/26810848

Johnson, D. G. (2015). Technology with No Human Responsibility? *Journal of Business Ethics*, *127*(4), 707–715. https://www.jstor.org/stable/24702822. doi:10.1007/s10551-014-2180-1

Karnow, C. E. A. (1996). Liability for Distributed Artificial Intelligences. *Berkeley Technology Law Journal*, *11*(1), 147–204. https://www.jstor.org/stable/24115584

Shneiderman, B. (2016). The dangers of faulty, biased, or malicious algorithms requires independent oversight. *Proceedings of the National Academy of Sciences of the United States of America*, *113*(48), 13538–13540. https://www.jstor.org/stable/26472631. doi:10.1073/pnas.1618211113 PMID:27911762

## KEY TERMS AND DEFINITIONS

**AI Liability Directive:** Proposed EU rules for assigning responsibility and compensating those harmed by AI systems.

**Artificial Intelligence Act:** Proposed EU legislation introducing "strict liability" for high-risk AI systems, making developers and producers liable for harm.

**Digital Services Act (DSA) and E-commerce Directive:** Regulations addressing online content, including illegal material and disinformation.

**General Data Protection Regulation (GDPR):** EU law governing personal data protection, crucial for AI systems using personal data.

**Product Liability Directive:** EU regulation holding manufacturers accountable for defective products, including AI-related ones.

**Resolution on Civil Liability and AI Liability Rules:** European Parliament's proposal to establish clear AI-related liability rules for those involved in AI systems.

**White Paper on AI:** EU document defining AI, outlining benefits, and providing policy recommendations for its safe development.

# Chapter 12
# Navigating the Quandaries of Artificial Intelligence-Driven Mental Health Decision Support in Healthcare

**Sagarika Mukhopadhaya**
*Adamas University, India*

**Akash Bag**
https://orcid.org/0000-0001-8820-171X
*Adamas University, India*

**Pooja Panwar**
*National Law School of India University, India*

**Varsha Malagi**
*Manipal University, India*

## ABSTRACT

*The integration of artificial intelligence (AI) into mental health services is examined in this chapter, highlighting the potential advantages of intelligent decision support systems in reducing the workload of medical personnel and enhancing patient care. However, there are serious worries due to the delicate nature of healthcare and the moral dilemmas brought on by possible malpractice or neglect. Five reoccurring ethical issues are identified and analyzed in this chapter, which includes interviews with healthcare professionals and AI researchers. These challenges are handling inaccurate suggestions, negotiating moral dilemmas, preserving patient autonomy, addressing the liability conundrum, and building trust. The chapter thoroughly analyzes these issues through empirical data and a literature study, illuminating the convoluted ethical terrain at the nexus of AI and mental health.*

DOI: 10.4018/979-8-3693-1565-1.ch012

## INTRODUCTION

Artificial Intelligence (AI) is a general term that refers to using a system or machine with human behavior to mimic the thoughts and actions of a human (Hamet & Tremblay, 2017; Humerick, Matthew, 2018). AI can perform advanced tasks previously considered only humans and handle large amounts of data that are too complex for a human to handle (Pannu, 2015). AI can be used in many areas, such as security, healthcare, transport, industrial automation, and agriculture (Pannu, 2015). The potential of AI is considered endless, and many industries are eager to tap into its possibilities (Pannu, 2015). One of these is healthcare, where AI has the potential to become a new asset. Healthcare generally uses outdated technologies and systems (Chowdhury, 2012), and while there is great interest in using AI, it currently has the biggest role in research. As each industry has specific needs, AI systems must be developed with these needs in mind, as the general complexity of healthcare can bring some challenges to the use of AI (Morley et al., 2020). Healthcare must identify these challenges, as the consequences can endanger people's health and well-being (Morley et al., 2020). One use for AI is as an intelligent decision support system (eng: Intelligent Decision Support Systems, IDSS). With the help of intelligent decision support, users can enter data and then get a result that can be used to make a decision (Tariq & Rafi, 2012).

A care context in great need of support is the treatment of mental illness (Baker & Kirk-Wade, 2023; Kakuma et al., 2011). Mental illness is increasing above all among the ages 10–34 at the same time as there is a global shortage of healthcare professionals in the field of, for example, psychiatrists, psychologists, and psychotherapists (Kakuma et al., 2011; Richter et al., 2019). Diagnosing mental illness is also complex as patients can suffer from both somatic illnesses, which refer to physical and bodily illnesses, and psychological problems in parallel (Davenport & Kalakota, 2019). Intelligent decision support has great potential in healthcare for mental illness, where there is an increased need for early interventions for diagnosis, as there are currently no resources and personnel with relevant skills. With the help of intelligent decision support, the burden on the staff can also be reduced (Davenport & Kalakota, 2019). If AI is to be used as decision support in mental illness, its ethical challenges must be identified and investigated.

When using complex AI systems, it is not uncommon for its users to not understand how the system works (Barredo Arrieta et al., 2020). AI also processes large amounts of personal and personal data where ethical complications can arise (Pannu, 2015). As interest in AI has increased in recent years, there has also been an increased need to understand ethics, accountability, and additional difficulties with justice and responsibility (Kokciyan et al., 2021). Due to its complexity and high opacity, ethical challenges arise in the use of AI that need to be explored (Brey, 2012). Due to the complexity of the challenges and the importance of understanding them well, this essay aims to answer the question: *What ethical challenges arise when using intelligent decision support in healthcare for mental illness?* The study aims to investigate and understand the complex ethical challenges with AI as decision support in mental health care. The use of intelligent decision support may expose both physicians and patients to undesired consequences, and an understanding of the ethical challenges that may arise may facilitate the safe use of the system. Ethical challenges identified from the literature are further investigated through an empirical study to understand which ethical issues become relevant when using AI in mental health care.

## RELATED LITERATURE

This section begins with an overview of AI and intelligent decision support in healthcare. Next, the need for mental health care is explored for illness. Finally, it highlights potential ethical challenges in healthcare for mental illness and the use of AI and intelligent decision support.

### Artificial Intelligence

Intelligent decision support is fundamentally based on AI technology. There are different definitions of what AI means. One of the first definitions was added in the 1950s by Alan Turing, who created the Turing test to define a system as an AI and see if the system can think like a human (Oppy & Dowe, 2021). Another definition coined in 2007 is that AI involves the science behind creating intelligent machines (Schuett, 2019), and a third definition of AI is the study of intelligent agents (Schuett, 2019). However, the most common and recurring definition of AI is intelligent systems and machines that have a human-like behavior to mimic a human's thoughts, behaviors, and actions (Hamet & Tremblay, 2017; Humerick, Matthew, 2018), and it is this definition that is used in this study.

#### Intelligent Decision Support

Decision support, or decision-support systems (DSS), focuses on information systems that support its users when making decisions (Fernandes et al., 2020; Tariq & Rafi, 2012). The purpose behind its use is to assist decision-makers and ensure that important details are not overlooked (Ngai et al., 2014; Tariq & Rafi, 2012). Decision support is only intended to assist and does not replace the human decision-maker (Fernandes et al., 2020; Tariq & Rafi, 2012). This study focuses on intelligent decision support, so-called Intelligent decision support systems (IDSS). Unlike regular decision support, intelligent decision support includes an intelligent factor applied in the system with the help of AI (Fernandes et al., 2020; Tariq & Rafi, 2012). Tariq and Rafi (2012) believe that a correctly designed intelligent decision support should be able to take raw data, documents, and learned behavior and then decide based on the data it has received. The advantage of intelligent decision support is its ability to handle complex tasks efficiently, allowing it to use decision support in contexts previously considered complex, for example, certain parts of healthcare or finance (Fernandes et al., 2020; Sachan et al., 2019).

Intelligent decision support helps increase the user's work performance and helps them perform better in their work together (Tariq & Rafi, 2012). An increased use of traditional decision support, without an intelligent factor, can give users a better ability to use intelligent decision support (Tariq & Rafi, 2012). Tariq and Rafi (2012) believe that there are three different components of intelligent decision support:

1. *A subsystem for databases* that contain relevant data for the decision support to be able to do its job
2. *A model management subsystem* that includes financial, statistical, and other types of models that convert models and help the user by building models
3. *Finally, a User interface Subsystem* that oversees all aspects of the communication between the user and components of the decision support.

Intelligent decision aids are Machine Learning systems that automate calculations or analyses to support decisions in various professions (Sachan et al., 2019; Tariq & Rafi, 2012; Fernandes et al., 2020).

## Machine Learning

The AI technology known as Machine Learning (ML) is often used for intelligent decision support. ML refers to self-learning AI systems trained to solve tasks as efficiently as possible. While traditionally programmed systems are fed data and algorithms to generate a result, ML systems are trained by inputting data and results to generate their algorithms (Choi et al., 2020). By having the ML system perform the same task multiple times, the system learns the fastest and most efficient way to perform the task and generate correct results. The complexity of ML systems can vary; therefore, they are divided into Shallow Learning (SL) and Deep Learning (DL). These two categories refer to the depth of the algorithms' data calculation, where DL refers to those algorithms that have extra deep and complex calculations, and SL refers to those that do not land within the framework of DL (Xu et al., 2021). SL refers to ML systems that use fewer parameters and require less training data than systems with DL (Xu et al., 2021). These systems are competent enough to perform complex tasks but do not generate sufficiently accurate results for all use areas, such as finance (Sachan et al., 2019). DL is a further development of ML where the system develops algorithms to solve abstract problems at a high level.

Unlike SL systems, DL systems work more like the human nervous system, using nodes that work individually and communicate with each other (Choi et al., 2020). This functionality allows DL systems to solve complex tasks that cannot be solved with known algorithms without understanding the problem. DL systems are generally more accurate than traditional SL systems, but their complexity makes them opaque and difficult for users to understand (Sachan et al., 2019). SL systems provide good results in simple situations that require simple algorithms, but in finance and healthcare, where the problems are often complex and require precise results, SL systems are not always sufficient (Sachan et al., 2019). DL systems are often more suitable in these cases, as they can solve complex problems with high precision (Sachan et al., 2019).

## Mental Illness

Mental illness is constantly growing among young adults, and many of those diagnosed also continue to require treatment for ten years or more (Jurewicz, 2015). In line with this, care costs have also increased by 30% in the last ten years. Healthcare for mental illness has specific improvement needs (Jurewicz, 2015). These needs are to reduce the burden on the care staff and reduce the waiting time for care queues (Jurewicz, 2015). There is also a need for early diagnosis, better identification of somatic and addictive diseases, and reliable diagnosis. It is of particular importance to reduce the burden on healthcare specialists in the treatment of mental illness, as the number of healthcare professionals does not increase in line with the need for treatment (Jurewicz, 2015). There is a global shortage of health professionals for the treatment of mental illness, which places a great burden on those who practice the profession. This shortage is significantly greater in low-income countries, creating a gap between countries and regions (Kakuma et al., 2011). Reduced waiting time and early diagnosis are needed that build on each other. Waiting times for treatment have increased in both children and young adults (Jurewicz, 2015). A report from the National Board of Health and Welfare shows that only 29% of young people receive a first assessment within seven days and that in large regions, this is below 15% (Jurewicz, 2015). These long waiting times lead to a late start of treatment and, thus, a later diagnosis.

The time until the patient receives treatment is as important as before the first visit (Jurewicz, 2015). Late diagnosis risks worsening the patient's mental illness and thus complicates treatment (Pedrelli et al., 2015). Early first visits can facilitate the assessment of the severity of the patient's condition and the possibility of offering the right treatment as early as possible (Pedrelli et al., 2015). Therefore, it is important to increase the availability of primary assessments of patients' need for care (Pedrelli et al., 2015). Through prevention and early intervention, both the growth and medication of mental illness can be reduced (Pedrelli et al., 2015). Identification of somatic and addiction diseases, as well as reliable diagnosis, are also needs that build on each other. It is important to be able to identify somatic and addictive diseases more easily, as it is common for mental illnesses to appear in connection with somatic diseases, as well as addictive diseases such as alcohol or drug abuse (Pedrelli et al., 2015). Many mental illnesses and syndromes can also show similar symptoms in earlier stages (Jurewicz, 2015). This can make it difficult to give the correct initial diagnosis (Jurewicz, 2015).

The risks of misdiagnosis are, therefore, high, as patients in psychiatry have a higher mortality and, thus, a lower life expectancy than the general population due to increased somatic morbidity and suicidal tendencies (Pedrelli et al., 2015). As the diagnosis is complex, supplementary diagnostic interviews are recommended to increase the reliability of the diagnosis (Jurewicz, 2015). Because of these complex challenges in diagnosing mental illness, it is challenging to develop an appropriate system that can facilitate the treatment of mental illness. However, with the growing need for care of mental illness, the need for support in this area also increases drastically, which motivates attempts to use intelligent decision support in healthcare for mental illness (Kakuma et al., 2011).

## Artificial Intelligence in Healthcare

AI is new in healthcare, and its use is currently not widespread, but it does have a major role in research on innovation in healthcare (E. E. Lee et al., 2021). Introducing various innovation opportunities with AI in healthcare can reduce stress and strain on healthcare professionals (Jin et al., 2023). India's innovation authority presents a SWOT analysis where, among other things, they summarize innovation opportunities with AI in the Indian healthcare system (Vinnova, 2018). In the analysis, they present, among other things, that AI can contribute to new and interesting jobs, an improved work environment due to less workload, and increased efficiency(Basu et al., 2020). AI in healthcare can also mean an efficient use of resources and address the problem of high costs and the lack of inexperienced employees in the field (Jurewicz, 2015). Using an AI system in healthcare for mental illness can also encourage patient participation in their investigation, reduce errors and costs, and increase efficiency in healthcare (D. Lee & Yoon, 2021). Regarding AI systems used within care contexts, the market should develop specialized systems for each care context, as different care contexts have complex needs (D. Lee & Yoon, 2021). The inclusion of algorithms that touch on ethnic and cultural information about patients can also allow the system to become smarter (D. Lee & Yoon, 2021). By letting the system study different cases, bias can be reduced (D. Lee & Yoon, 2021).

## Intelligent Decision Support in Healthcare

Intelligent decision support has great potential in healthcare through the ability to assist and strengthen healthcare staff's ability to make well-informed decisions (Lei et al., 2023). Healthcare decision support is currently mostly simple and comes, for example, in the form of color-coded record systems, but innovations have led to systems that, for example, provide the opportunity to adapt the medical dose to a patient's conditions (Chowdhury, 2012). There is a further need for new types of decision support, and intelligent decision support can be seen as the next step in development. Kokciyan et al. (2021) believe that decision support can save time effectively. As caregivers often need routine information, decision support can contribute to detailed treatment plans that complement those proposed by human actors (Kökciyan et al., 2021). The increased competence offered by intelligent decision support provides the opportunity for use in complex contexts in healthcare, for example, in close treatment of patients (Fernandes et al., 2020). As many tasks in healthcare are complex and misjudgment can lead to unwanted consequences, a competent and accurate system is required (Morley et al., 2020). DL systems are the best-suited system for complex tasks where accuracy is needed (Sachan et al., 2019). The problem is that the complexity of a DL system makes its execution of tasks opaque to users (Sachan et al., 2019). This, in turn, creates ethical challenges when using the systems.

## Ethics

The study assumes that ethics is the teaching of morality about what is right and wrong or good and bad (Dewey, 2016; Kazim & Koshiyama, 2021). Ethics is used to systematically explain a position on right and wrong behavior or behavior (Dewey, 2016). Ethics can guide people's decision-making to avoid unwanted consequences (Aita & Richer, 2005).

## Basic Ethical Principles in Healthcare

Ethical approaches are constantly changing in healthcare as laws have changed in recent decades (Summers, 2009). In addition, healthcare laws also differ from country to country, reducing the relevance of specific laws as a basis for study. Summers (2009) believes that an understanding of theories about ethics is important for correct decision-making regarding patients and organizations. These theories provide a logical basis for ethical decisions in healthcare practice (Summers, 2009). The four basic ethical principles are non-maleficence, beneficence, autonomy, and justice (Summers, 2009). The first two principles, non-maleficence, and beneficence, emerged as early as the time of Hippocrates, while autonomy and justice emerged in the early 20th century, and they are accepted as important and relevant principles of clinical ethics even today (Varkey, 2020). Non-maleficence is the principle of not harming. The complexity of this principle concerns the meaning of the term "*harm*" (Summers, 2009). There are many philosophical understandings of what "*harm*" can mean, but the consensus is that it refers to choices that lead to less good alternatives or whose execution goes against an actor's best interests (Summers, 2009). For healthcare, the term "harm" worsens the patient's condition (Summers, 2009).

Charity is the principle of doing only good. Charity usually comes from non-maleficence: "*Not harm, only do good*" (Summers, 2009). Charity is more than just avoiding harm; it refers to an obligation of altruism (Summers, 2009). Healthcare professionals are expected to maintain altruistic social interactions that are not expected in other contexts, and this is seen as an implicit responsibility to appear professional (Summers, 2009). Non-maleficence and beneficence are the basic principles for how healthcare should be carried out, but the assessment of what is "*harm*" and what is "*benefit*" lies not only with healthcare professionals but also with the patient (Summers, 2009). Autonomy means that a person is self-governing. In the case of healthcare, this means that the patient has the right to be involved in the treatment process (Summers, 2009). Two requirements must be met for a patient to assume a decisive role in his treatment (Summers, 2009). The patient must be competent enough to make their own decisions and simultaneously be free from coercion (Summers, 2009). Specific competence is required to fulfill the requirement of autonomy, but the definition of competence is partly complex in healthcare (Summers, 2009). A patient without physical limitations may have the competence to perform simple tasks but lack the competence to understand the consequences of their treatment, while another patient may have this competence but lack the competence to perform simple physical tasks (Summers, 2009).

Competence may also be insufficient if the patient does not have the opportunity to make decisions without coercion (Summers, 2009). Justice is the last fundamental principle and implies the importance of fair treatment. Generally, we can consider something unfair if we have a good reason to think it is morally wrong (Summers, 2009). However, it becomes complex when deciding whether something is morally wrong (Summers, 2009). There are two main types of justice in healthcare: procedural justice and distributive justice. Procedural justice concerns whether fair processes were available and whether these processes were carried out (Summers, 2009). Distributive justice concerns resource distribution between cases (Summers, 2009). Discrimination is considered if these justices are not met, which can lead to ethically undesirable consequences (Summers, 2009). The need for the treatment of mental illness is mostly linked to the principles of justice, non-maleficence, and autonomy. Distributive justice is a major concern in mental illness as the staff is overburdened, which leads to many patients not receiving treatment and, thus, directly into the challenge of non-maleficence. This problem will only worsen as the number of psychiatrists and psychologists decreases concerning patients (Richter et al., 2019; Kakuma et al., 2011). Since the diagnosis of mental illness is complex due to co-morbidity with possible somatic or addictive diseases, a concrete understanding of the diagnosis can be difficult for the patient to achieve, which affects patient autonomy.

## Ethical Challenges with Intelligent Decision Support

Intelligent decision support can solve many problems, but its use has many risks, especially in healthcare. A problem that underlies several ethical challenges is the backbox effect. The blackbox effect occurs when the system is so opaque that the user does not have the opportunity to understand how the system makes its calculations or arrives at its assessments (Durán & Jongsma, 2021). This phenomenon is most common in complex systems, such as those using DL algorithms, as these are too complex to offer any level of interpretability (Sachan et al., 2020). This leads to ethical challenges regarding bias and trust in the AI system (Durán & Jongsma, 2021). With opaque systems, for example, it is almost impossible to identify unethical calculations and assess whether the system has a bias, but in the case of measures against this, such as the development of so-called Explainable AI, the accuracy of the system is usually sacrificed to achieve greater transparency (Durán & Jongsma, 2021). While researchers such as Arrieta et al. (2020) work to promote the development of explainable AI, Durán and Jongsma (2021), argue for developing reliable algorithms and instead want to promote trust in black-boxed systems. These alternative solutions lead to ethical challenges.

With less complex systems, such as those using SL algorithms, a level of interpretability allows users to interpret how the system has arrived at its result (Sachan et al., 2019). However, these SL systems do not achieve the same complexity and competence DL systems achieve (Sachan et al., 2019). The advantage of SL systems is that users can have more trust in the system and higher self-confidence in their decisions (Sachan et al., 2019). The disadvantage is that the system's low precision does not make it suitable for certain areas of use, such as finance (Sachan et al., 2019). A DL system is better suited in these cases, as precision is one of the most important aspects (Sachan et al., 2019).

## Ethical Challenges When Using Intelligent Decision Support in Healthcare

Intelligent decision support in a clinical setting generates specific ethical challenges. These challenges comprise a combination of the critical areas in healthcare and the shortcomings in intelligent decision support. Based on the literature, five potential ethical challenges have been identified: Dealing with incorrect recommendations, dealing with moral dilemmas, achieving patient autonomy, the responsibility dilemma, and building trust. Dealing with incorrect recommendations is about the risk of misdiagnosis. This means there is a risk of doctors giving an incorrect diagnosis or recommendation without the doctor questioning the system (Morley et al., 2020; Duran & Jongsma, 2021). In cases where the intelligent decision support is so complex that the user cannot interpret its calculations, the user may have difficulty knowing if the system has made an error (Morley et al., 2020). Above all, it is difficult for the user to know which parameters the system may have missed when recommending (Morley et al., 2020).

The next challenge is the risk of the doctor being exposed to moral dilemmas. In cases where an intelligent decision support diagnosing a serious illness in a patient whose symptoms do not appear, the doctor must decide whether he should risk over- or under-treating a patient, where both consequences risk harming the patient (Durán & Jongsma, 2021). In these cases, it is important to be able to interpret how the system arrived at its recommendation (Duran & Jongsma, 2021). When using intelligent decision support, patient autonomy can be harmed as the black box effect can make the treatment process difficult for the patient to understand (Duran & Jongsma, 2021). The patient may lose the opportunity to receive a treatment that matches their values, as the AI system's values may, for example, weigh the patient's lifespan over freedom from pain (Durán & Jongsma, 2021). The empathy between caregivers and patients cannot be recreated with algorithms. Therefore, it is important to strive to maintain interpersonal interaction (Morley et al., 2020). As patients expect AI systems to facilitate their care but at the same time does not take over the caregiver's task and final decision-making, the human contact between patient and caregiver must not be lost (Braun et al., 2020).

The liability dilemma concerns who bears the responsibility in the event of negligence or injury to a patient after using intelligent decision support. As many actors are involved from the development to the use of the system, it is unclear who bears moral and legal responsibility when errors occur (Braun et al., 2020). One perspective is that unless the systems replace healthcare professionals but only enhance their decision-making abilities, the healthcare professionals bear ultimate responsibility (Braun et al., 2020). The counterargument against this perspective is the system's competence to make complex assessments that exceed the human ability of healthcare professionals (Braun et al., 2020). The higher the system's complexity, the less reasonable it is to expect the healthcare staff to feel safe questioning its recommendations (Braun et al., 2020; Duran & Jongsma, 2021). Trust can be challenging to achieve with intelligent decision support (Braun et al., 2020). Braun et al. (2020) state that AI systems must be developed with ease of use and adequate risk-benefit analysis to create user trust. The blackbox effect

*Table 1. Compilation of potential ethical challenges in the use of intelligent decision support for the treatment of mental illness*

| Potential ethical challenges | Description | References |
|---|---|---|
| Dealing with Incorrect Recommendations | Due to the *Blackbox effect*, it is difficult for users to identify incorrect recommendations. | Braun et al., 2020<br>Duran & Jongsma, 2021<br>Morley et al., 2020 |
| Dealing with Moral Dilemmas | In cases where the system has identified a serious illness/syndrome that the doctor has not identified, the doctor is faced with a dilemma as to whether to over- or under-treat the patient. | Duran & Jongsma, 2021 |
| Achieve patient autonomy | It may be more difficult to include the patient in their care if the system's recommendations do not consider the patient's values or if the system is difficult to understand | Duran & Jongsma, 2021<br>Morley et al., 2020 |
| Ansvarsdilemmat | Many actors are involved in the system's development and use, and it is unclear who should bear the ultimate responsibility. | Braun et al., 2020<br>Duran & Jongsma, 2021<br>Morley et al., 2020 |
| Build Trust | *The blackbox effect* makes it difficult for users to build trust in the AI system, as they cannot determine its security. | Duran & Jongsma, 2021<br>Braun et al., 2020 |

makes it difficult for users to create trust in the AI system themselves, as they cannot determine how secure the system is (Duran & Jongsma, 2021). Measures must, therefore, be taken to assure users of the system's security (Braun et al., 2020). We summarize potential ethical challenges in the table below based on previous research.

## METHODOLOGY

This study investigates the ethical challenges of utilizing AI as decision support in mental health care. Employing a qualitative approach, the research explores the complex intersection of healthcare and AI, recognizing the scarcity of AI in Indian healthcare and aiming to comprehend the challenges and needs of healthcare professionals. The study delves into human behavior, attitudes, and values through interviews to identify ethical challenges. A literature survey supplements the study, critically examining materials from Google Scholar, PubMed, and PsycInfo and reports from the National Board of Health and Welfare and the Indian Innovation Authority. The focus spans statistics on mental illness in India, global AI usage, and key concepts: AI, ethics, healthcare, and decision support. The chosen articles were meticulously analyzed to extract essential information, forming the basis for a nuanced understanding of the ethical implications surrounding integrating AI into mental health care decision support.

This study collected empirical data through semi-structured interviews to investigate the intersection of AI in healthcare and ethical challenges. The method allowed for predetermined questions with room for free respondent expression and facilitated follow-up questions. The selection of participants involved a stratified target-driven sample, including healthcare professionals (psychiatrists, psychologists, psychotherapists, school counselors) and AI researchers/specialists. A snowball sample was utilized due to difficulty finding interviewees, resulting in eight participants, four from each subcategory. A pilot study was conducted to evaluate question quality and order, involving a test subject with basic AI knowledge for accessibility. Wording adjustments were made based on the pilot study feedback. The individual and group interviews were guided by interview scripts tailored to each sample group, covering topics such as

*Table 2. Central concepts and keywords for the literature study*

| Key concepts | Keyword |
|---|---|
| AI (Artificial Intelligence) | Ethics of AI, ML (Machine Learning), DL (Deep Learning), SL (Shallow Learning) |
| Ethics | Ethics in Healthcare, Ethics of AI |
| Health care | Mental health |
| Decision support | DSS (Decision Support System), IDSS (Intelligent Decision Support System) |

AI as decision support, ethical challenges, and user interactions. Ethical principles were communicated to participants before each interview, and notes and audio recordings were used for transcription. The study aimed to provide comprehensive insights into the perspectives of AI researchers and healthcare professionals on the role of AI in decision support and associated ethical challenges.

## Analysis of the Methodology

Initially, the interviews were transcribed, and then a thematic analysis was carried out, which was done top-down. Thematic analysis is a method for finding and identifying recurring concepts and themes from data collection (Bryman, 2018). The analysis has been applied to identify and find recurring patterns and insights from the interviews. This can lead to a deeper understanding of nuances and gray areas within the ethical challenges, which is relevant to this study. The analysis used the four basic ethical principles (non-maleficence, beneficence, autonomy, and justice) as an ethical lens. Using the basic principles instead of the potential ethical challenges was considered relevant, as this provided the opportunity to investigate and identify new ethical challenges not identified based on the literature. The principles are also cornerstones of ethics in healthcare and were therefore considered suitable for use as an ethical lens in this study. The basic ethical principles provided the opportunity to understand the nuances of ethics in the collected material. Connecting the insights to the basic ethical principles could create a deeper understanding. By conducting a top-down analysis, it was possible to analyze the codes based on already predetermined themes (the four ethical basic principles). There was also an opportunity to identify subcategories in each theme to create further understanding and clarity.

*Table 3. Table of interview participants and their professional role and length of completed interviews*

| Respondent | Occupational role | Type of interview | Interview length |
|---|---|---|---|
| A | AI Researcher | Intercultural Group | 1 hour |
| B | AI Researcher | Intercultural Group | 1 hour |
| C | AI Researcher | Intercultural Group | 1 hour |
| D | AI Researcher | Individual interview | 1 hour |
| 1 | Psychologist | Individual interview | 32 min |
| 2 | Psychologist | Individual interview | 30 min |
| 3 | Specialist in Internal Medicine | Individual interview | 35 min |
| 4 | Emergency Physician | Individual interview | 45 min |

*Table 4. Examples of codes in clusters that received the designations "responsibility" and "who is responsible? Color-coded after the interviews."*

| Responsibility | Who is responsible? |
|---|---|
| A challenge is to know who is responsible for mistakes | The decision lies with the doctor |
| Accountability is an ethical challenge. | The doctor must have the final assessment. |
| Understanding who should be accountable is challenging | The doctor cannot blame AI for the decision since it is a support |
| Reality is complex and nuanced. | People who produce requirement specifications should be responsible. |
| The thoughts about responsibility are a bit black-and-white | Always the doctor in charge |
| The question of responsibility is an interesting one | The authority who approved the system should be responsible |
| A lot of responsibility | If autonomous, I do not know who to blame. |

## Coding and Thematization

After transcribing the interviews, each transcription was read through to pick out and highlight important and relevant insights. Insights were then coded on post-it notes in a *Miroboard t*o summarize insights from the interviews in one word or a short sentence. The codes from each interview were written on post-it notes in their specific color to distinguish the participants from each other more easily and to see more insights from each participant. Similar codes were clustered together to see which codes dealt with similar insights more simply. For example, several responsibility-related codes were clustered, where many interviewees contributed insights. (See Table 4). Codes were combined if they dealt with similar insights, and themes were created for these codes. Under these themes, new codes fit in, which were not necessarily the same as the other codes but indicated similar insights to the theme created.

# RESULTS

The results section presents the results based on the collected empirical data. The result is presented to each selection group separately (AI researchers and healthcare professionals).

## AI Researchers

From the AI researchers, insights were identified on the following potential ethical challenges: access to sensitive data, incorrect recommendations, transparency, trust, and the liability dilemma.

For AI, access to sensitive data was considered a major challenge. As the data in healthcare is sensitive, it was considered challenging to justify the need for the data to gain access to it. Rules, policies, and GDPR must be considered when using the data to avoid privacy breaches. This step is important because the data will be used to learn the ML system. A DL system can be developed if sufficient data can be used. With smaller amounts of available data, an SL system can be developed. "*Getting access to the data, it's sensitive data, so the first and most important challenge was to get all the players involved.*" - Respondent D. "*Deep learning achieves in some cases, not all of the time, achieves some gain in performance compared to conventional machine learning or shallow models.*" - Respondent A

"*... there are a lot of regulations, and it depends where you live. There are very different scenarios in India, the US, and China. In the EU, the regulations, the GDPR and other regulations.*" - Respondent C

When using intelligent decision support and other AI systems, there is always the risk of incorrect recommendations. In cases where errors occur, all AI researchers believe that these should be reported and analyzed. In this analysis, it can be clarified where the error occurred, with the doctor or with the system, and what it is due to, incorrect input or other. It may also be that the system is not faulty but has not received enough data to make a reasonable decision. This may depend on whether it is a DL or SL system, as DL systems generally need more data. The AI system does not have to make a "*wrong*" decision, but it only allows giving more reasons and reasons behind a decision. "*I think there are different ways to approach it. What we like is feedback, so if it's very similar to what you use with Google Translate, it allows it to improve and improve through time.*" - Respondent C. "*It's just the usual flag to look into something. The team that built that system, data scientists, data curators, and everyone joined hands and tried to understand what was happening. It's then analyzed. Either explained or fixed, these are the two possible options. Either the doctor is not right, or the algorithm is incorrect.*"- Respondent D.

All AI researchers considered trust challenging to achieve as many factors affect doctors' trust in a system. First, one of the researchers felt that keeping the system's limitations in mind was important. The system should be considered an advanced tool supporting decision-making, not an all-knowing system. The system makes its calculations based on information entered, and the system rarely receives all relevant information, resulting in the system not being 100% accurate. Although it should not be expected to have 100% accuracy, the system should be accurate enough to be considered reliable. One interviewee suggested that if the system is reliable, doctors will use it. The system should be trained in relevant data and comprehensive enough for its use area. One interviewee also believed that the system should be designed according to the doctor's needs, as all doctors have different needs. Understanding the system's limitations and being able to question the system is important to create trust but to be able to question the system requires a level of transparency. One interviewee believed that it is only a matter of time before healthcare professionals gain confidence in AI despite the black box effect. "*It's related to the context that the target user expects the model to use to decide. So maybe this will allow the decision support to be transparent to the target user so that they can trust that decision made by the tool.*" - Respondent B.

"*This is a way of increasing the trust in the decisions made by (the system) because if the doctors were the target users that will use this decision support system, they understand the risk factors that this AI is using to give me a decision. Then I can have more trust.*" - Respondent B "*The problem is that not all the AI models can be opened.*" - Respondent C. "*Doctors today use ultrasounds all the time. They trust them. But I hardly doubt any doctor understands what a Fourier transformation is. Suppose these decision support systems are redundant. Yeah, and they get the trust, doctors will use it.*" - Respondent D Transparency was considered important in healthcare but cannot always be achieved. DL systems that can use several million parameters go far beyond human interpretation, and doctors cannot be expected to understand how these systems make their calculations. Instead, an interviewee believes that it becomes important for doctors to understand what kind of data the system is trained on to determine whether the data itself can be trusted. The level of insight needed varies between situations, as not all data is always relevant to physician decision-making. The AI researchers also highlighted that doctors are often busy and don't always have time to review all the data, but they content themselves with the most important details. One interviewee suggested that some physicians are satisfied with a prediction score that gives them insight into the likelihood of readmission, while some physicians desire more detailed information.

"*For instance, if you train a model on data from Canada and then apply it to the Indian population, I won't trust that algorithm. These two populations are quite different, and you haven't evaluated it on the Indian population.*"- Respondent D.

"*It's just filtering out from all the hundreds of patients, a few tens of patients on which you might need to focus a bit more or to understand them. Many doctors are happy with this, too. So it depends. I have worked with so many doctors, and it depends on the user himself what they want out of it.*" - Respondent D Regarding the responsibility dilemma, empirical evidence showed that it is unclear who is responsible. The programmer is not the main person to contact in case of problems with the system. Their task is to develop the system and to ensure that research is carried out on the system. Their job is also to check the AI's training data to make sure it makes decisions based on the right and similar data it was trained with. "*Usually, the programmers are not the direct point of contact. I would say we usually do a little research before implementing.*" - Respondent C "*How do we know that the data we are training in healthcare is correct? That's the biggest challenge in healthcare in decision support systems.*" - Respondent D Several researchers saw the responsibility dilemma as a major challenge. As many actors are involved in developing an AI system, it is unclear where the moral responsibility lies. Since doctors are the ones who ultimately make the decision, doctors were considered to bear the greatest responsibility, but the emphasis was placed on the fact that the system must be approved for use by a third-party organization that also bears responsibility in the event of errors and conflicts. One interviewee mentioned Medical Device Regulation (MDR) as an example of such a third-party organization.

"*This AI wants to help fill the gaps that doctors are feeling, but it's still challenging who is responsible. The doctor cannot say, 'Well, this is the AI's fault, but I didn't make any choice'. It's not like that at the moment. It's just a decision support.*" - Respondent C "*The doctor is still responsible because the doctor is seeing the patient.*" - Respondent C "*The authority who approved the system. We trust a lot on the marking system.*" - Respondent D Three of the interviewees highlighted that the role of decision support is to give other aspects to a doctor's decision and investigate whether the doctor changes his decision based on the assistance of intelligent decision support. "*It's interesting to log and document that to see if it was good that the system influenced the doctor or not because it could be in both directions.*" - Respondent A "*When you look at the regulations and those things still, the doctor is responsible for making the decisions. And so this is why we call it a clinical decision support system, not just a decision maker. So they're just trying to provide more facts and not make the final clinical diagnosis.*" - Respondent C "*Well, the idea is you will use this information as a doctor to intervene, maybe, make some interventions.*" - Respondent D.

To summarize, insights were identified on the following potential ethical challenges from the AI researchers: access to sensitive data, incorrect recommendations, transparency, trust, and the liability dilemma. Access to sensitive data can be challenging, affecting what kind of system can be developed. Incorrect recommendations were considered to be influenced by the system's usage and its learning data. Any recommendations should be flagged and analyzed to create an understanding or correct the system. Trust was considered challenging to achieve as many factors influenced it. Transparency and understanding the system's limitations were thought to influence trust, but it was also thought to be only a matter of time. Transparency is important but not always achievable. A basic understanding of the data source on which the system was trained is important to trust the system. In addition, different care contexts have different needs for transparency.

## Nursing staff

From the healthcare professionals, insights were identified regarding the following potential ethical challenges: responsibility dilemma, trust, transparency, patient autonomy, moral dilemmas, and incorrect recommendations. A recurring answer from the interviewees was that decision support is only support and that it is not a system that makes decisions; it is the doctor who makes decisions. Many circumstances influence a decision, and these circumstances are invoked when "*errors*" arise. Two of the interviewees believed that intelligent decision support can, if used correctly, increase the quality of a doctor's decision as the system can consider other aspects that the doctor did not think of. This is because a doctor can't be completely up-to-date, even in his field, which is something that intelligent decision support can facilitate for the doctor to understand the patient better. Decision support can also help the doctor understand more about the patient to make the right decision. "*When you have a very tired doctor who has to take care of a patient compared to if you come in the middle of the day, it is clear that there is a much greater risk that you will miss something; it is completely human, so we can then have some kind of support that helps us just see: 'have you thought about this?'. I think both increases the quality and makes it safer for the patient, therapists, and doctors.*" - Respondent 3 "*Decision support is still something that we are obliged to maintain our responsibility and think about.*"- Respondent 4

The healthcare staff with experience in AI had a relatively clear view of the liability dilemma. This was obvious as the doctor always takes responsibility in clinical contexts. It was also presented that doctors can bear responsibility for nurses who work under them. It was considered important always to have a human involved in the decision-making and ensure that the decision support is used only as a support, and not let it make final decisions. All the interviewees expressed that the doctor should always make the final decision in treatment and that they bear the greatest responsibility regardless of how they used the decision support. Therefore, the doctor should have the right and the opportunity to question the system and justify his decisions that go against the recommendations of the decision support, if necessary. In these cases, it is the doctor's role to make a risk and reasonableness assessment. It was considered that high requirements were required for AI systems in healthcare to be considered reliable, as it was believed that patients trust doctors more than digital systems. The interviewees also highlighted that a cultural change or new legislation will be required before AI can be considered reliable. "*I think it is important that it should still be the responsibility of the doctor, the person. We are used to it, and I think we should follow that tradition.*"- Respondent 4

"*Today, it is the caregiver's responsibility. It can be a doctor or a nurse. As a doctor, I can also be responsible for something a nurse performs, which I have prescribed." - Respondent 4 "How do you know that the person you meet is making the right decision? It's about trust somewhere. It's about trusting someone.*"- Respondent 3 "*I think that before that, there may need to be both legislation and cultural changes before we fully trust an AI.*" - Respondent 3 A challenge with using an AI as decision support in healthcare is getting the doctor to feel trust in the system. Trust in the doctor is always needed to a certain extent, and a patient cannot always know if what the doctor decides is right but must have confidence that the doctor knows his job. Two interviewees considered it important to have transparency and understand the process behind the AI system's recommendations. At the same time, they debated how much of the process was considered important to understand. Doctors today use tools and aids that they do not fully understand, which is acceptable as they know how to use them and can be trusted. As healthcare differs in different sectors, it is necessary to show different amounts of information about how the AI system arrived at its decision. In cases where the patient is in a critical condition, it may be enough only to know what the system thinks and not how it has decided.

"*In the best of worlds, I would like it to say 'admit the patient. These three variables carry the most weight.*" - Respondent 4. To achieve patient autonomy, transparency is required between healthcare provider and patient. Being transparent and including the patient in their care was considered particularly important in the case of mental illness, as it motivates the patient to participate in their treatment and help themselves. However, the psychologists did not consider it particularly challenging to include patients in their care at present. On the other hand, a psychologist mentioned that more work needed to be put into writing detailed journal entries for the patient to understand and to be included in cases where the patient took part in the journal entries. One of the interviewees considered that it is not always possible to be completely transparent with the patient, as the patient is not always conscious or does not have the opportunity to understand the meaning of a treatment fully.

"*Journal notes will be longer and perhaps a little more time-consuming for any note than it would otherwise be*" - Respondent 2. "*I am very open with my assessment and what I think they might need*" - Respondent 1. "*Always try to be involved in, like, whatever it is, patient visits, because I think they're still the ones who need to make some kind of change, like*"- Respondent 2. "*I probably do it partly by being transparent about different treatment options*"- Respondent 2. One of the interviewees expressed that in the case of moral dilemmas where the system's recommendation does not match the physician's thoughts, the physician should consider these options with patient safety. The interviewee also believed that the user should make up his mind before using the decision support. The deviation should then be reported in a process similar to that used in the case of human negligence. The doctor should, therefore, have an idea of what kind of recommendation they should receive before using decision support. After that, careful follow-up is needed to understand why the system recommended a certain alternative. "*I probably took two things into account. Firstly, I thought: What do I think about myself? Secondly, I would probably have chosen the option with the highest patient safety*" - Respondent 4

In the event of incorrect recommendations, good feedback mechanisms were also considered necessary to be able to write deviations in the system. The interviewee's reason is that it should work similarly to when a doctor neglects a patient who risks losing his ID. In case of an "*error,*" it is also necessary to show what kind of data the decision support has gained access to and double-check if it is correct. This also applies if the doctor asks one human or another doctor who gives a "*wrong*" answer to a question. The doctor should also ensure that the entered data is correct and, if possible, ask another doctor for advice. An interviewee ended the interview by discussing the unethical use of AI and that not using AI could be considered unethical. This can be seen as negligence. "*Perhaps it will be unethical not to use these supports in the future because it is like choosing to drive a car without wearing a seat belt; usually, it goes well, but at some point, it will crash, and then you have not taken the safety that exists.*" - Respondent 3

To summarize, insights were identified regarding healthcare professionals' potential ethical challenges: the responsibility dilemma, trust, transparency, patient autonomy, moral dilemmas, and incorrect recommendations. Regarding the responsibility dilemma, the doctor was considered to bear the greatest responsibility and should then have the opportunity to question the system. Confidence in the decision support was considered important, and transparency was considered an influencing factor. The necessary level of transparency was considered to vary between different care contexts, and it was considered important to present the most critical information for the situation's needs. At present, patient autonomy is not considered particularly challenging to achieve, but extra work needs to be put into writing explanatory journal entries in cases where the patient takes part in journal entries. In the case of moral dilemmas, it was important to identify the option with the lowest patient safety and follow up on the treatment to learn from the situation. In the event of incorrect recommendations, feedback mechanisms were considered necessary to be able to write deviations, and the data the system gained access should be double-checked.

## ANALYSIS

This section presents the analyzed material based on the empirical data collected. The analysis is presented in four different ethical themes based on Summer's (2009) four basic ethical principles (non-maleficence, charity, autonomy, and justice).

### Non-Malice

The interviews highlighted that the risk of causing harm is central to using intelligent decision support in research and healthcare. Procedures for incorrect recommendations were described clearly and similarly among several participants, and while some established routines are potentially transferable to the use of AI, this is not necessarily an obvious one. Lack of transparency can lead to unwanted neglect due to incorrect recommendations, which can cause harm. Two sides of transparency emerged from the material, divided into the categories Obligatory and Situational. Mandatory transparency is required in all care situations to understand the system. This understanding concerns the data source used to train the system and the limitations that the system has. To achieve non-evil, these points emerge as central to all usage situations, and lack of any of these points can lead to unwanted neglect. Understanding the data source was considered important as the patients treated and the care centers they were treated to can drastically differ between the training data and the context in which it is used. A system trained on training data from other countries can draw incorrect conclusions in Indian care when the patients and the available care in the training data cannot be replicated. Furthermore, doctors must understand the system's limitations as it allows the doctor to question the system.

The system can never achieve 100% accuracy, and this is acceptable according to all AI researchers, as it is still generally significantly higher than the general doctor's accuracy. The system is a tool that, to the best of its ability, conveys a recommendation based on the data that is available. There will always be information that does not fit within the system's parameters, which the doctor must then handle directly to reduce the risk of causing harm. It was considered that doctors need to ensure that the entered data is correct and, if possible, ask another doctor. However, whether this is always possible or not should be explored. Situational transparency refers to the level of understanding required for each specific situation to ensure the patient's safety and achieve trust with the physician. Not all ML models can be opened and explained, but this does not necessarily make them unsuitable for healthcare contexts. Doctors in certain fields need an indicator of where to look after errors, while doctors in other specialties need to analyze the causal relationships identified by the system. In cases where doctors only need a prediction rating, AI researchers believe that a DL system may be suitable if a sufficient amount of data is available, as the accuracy of the prediction is valued more than its transparency. In cases where the doctor needs to analyze the causal relationships in the recommendation, an SL system may be more suitable, as the transparency outweighs the lower accuracy. The doctor then has the opportunity to complete the system, and the risk of causing damage is reduced.

Doctors' need for trust in the system can be seen as a need to ensure the patient's safety and avoid risks of harming the patient. The study's results indicate trust is achieved by balancing transparency, reliability, and change. Transparency appears to have an important role in creating trust, as it allows doctors to ensure that no harm is done to the patient. As the need for transparency varies between different care contexts, transparency does not always provide a sufficient understanding of the system to create trust, for example, when using predictive ratings. Therefore, the system used must be reliable. If the system is reliable, the need for a high level of transparency is likely to be reduced, as a general understanding of the recommendation and knowledge of its reliability may be sufficient to ensure that no harm is done to the patient. Another important part of building trust highlighted in the results concerns change within culture and law. It is believed that with clear laws and requirements, ambiguity will be reduced in the use of AI in the healthcare context, and with cultural change, our view of AI may change. When computers began to be used in healthcare, there was a lot of reluctance, but after cultural change, transparency, and reliability, computers have become a standard resource in healthcare. An obstacle to using AI systems is confidentiality and access to sensitive data. As ML systems require training data, the amount of available data limits the depth the ML system can achieve. AI requires large amounts of data, and if this data is leaked to the public, it can cause harm.

## Charity

The basic ethical principle of charity means that something should only do as much good as possible without bringing anything bad. The interviewees highlighted that it was not a challenge to do good but rather that the fear of harming can be linked to the basic ethical principle of non-evil. Above all, it is highlighted in several interviews with AI researchers and healthcare professionals that AI is a positive aid that can help to do good in society. In one of the interviews, it was also highlighted that an AI can increase the quality of decisions if the decision is not only based on the decision support. This can provide higher patient safety if a doctor, for example, double-checks with intelligent decision support to see if the result is the same as the doctor originally intended. That charity was not discussed in the interviews because it is obvious only to do good with the existing conditions and that it is not seen as a challenge but as a matter of course to help patients. To do well, healthcare staff always need aids, everything from stethoscopes to EKG machines, X-rays, and computers. These aids are taken for granted so the healthcare staff can simultaneously do their job and help patients. It was also highlighted that it may be unethical not to use intelligent decision support in the future and may cause harm when the user has not taken the measures available to help a patient to the full. If intelligent decision support is available, this can be used to achieve charity.

## Autonomy

The patient needs to be included in their care to achieve patient autonomy. Including the patient in their care can motivate them to help themselves because they will undergo a change or treatment. Currently, it is not considered challenging to include patients in their care when treating mental illness, but it was mentioned that extra work needed to be put into writing explanatory journal entries in cases where the patient participates in the journal entries. As the journal is an aid used in mental illness, reasoning can be made about the potential increased work that may be required when explaining the recommendations of intelligent decision support. If intelligent decision support needs to be explained to the patient to achieve patient autonomy, this may complicate the care specialist's work process. This could have taken up more time the care specialist could have spent helping patients. Time is an important need in the treatment of mental illness, as the time before the first visit and diagnosis has a great impact on the patient's health. This can be seen as a step in the wrong direction. Even with opaque decision support, open communication and the possibility of choosing between treatments can be achieved for patients. According to the interviewees, it is not always possible to explain everything to a patient today, as the patient's knowledge in the field is significantly lower than the doctor's, but this is not considered an obstacle to including the patient in their care.

## DISCUSSION

Through the empirical work, insights into potential ethical challenges have been explored and analyzed. In this section, we discuss the insights of the empirical study based on the literature.

### Incorrect Recommendations

The empirical study showed that the healthcare staff did not see it as a risk that doctors make incorrect diagnoses because they do not dare to question the intelligent decision support. The healthcare staff participants clearly understood how the situations containing incorrect recommendations should be handled. It was considered that the healthcare staff should have their own opinion before using the decision support and that if the opinion and the recommendation do not match, the option with the highest patient safety should be followed. This is, therefore, not in line with Morley (2020), who highlights that incorrect recommendations can increase the risk of misdiagnosis as doctors use a system they do not understand and do not dare to question. While the empirical evidence emphasized the importance that doctors should have their own opinions before using the system, the complex nature of diagnosis in the treatment of mental illness can make it difficult to draw concrete conclusions before using diagnostic tools (Davenport & Kalakota, 2019). As early and reliable diagnosis is a major need in the treatment of mental illness, intelligent decision support should be designed to help the healthcare provider achieve this need (Jurewicz, 2015). Since one's opinion before using the system is not always possible, this cannot be used to reduce the risk that the care specialist makes incorrect diagnoses, which is important to highlight when implementing systems within this care context.

### Moral Dilemmas

The experience showed that moral dilemmas should be handled by the care specialist weighing his thoughts against the decision support recommendation to identify the option with the highest patient safety. The black box effect can make the recommendations of intelligent decision support difficult to question, and even in cases where the care specialist chooses between his own opinion and the recommendation of the decision support, it can be difficult to find the option with the highest patient safety (Duran & Jongsma, 2021). In these cases, the decision support recommendation must be interpreted to make a reasonable comparison between the alternatives (Duran & Jongsma, 2021). The decision support's transparency and interpretability depend on the algorithm's depth, and to achieve higher interpretability, SL must be applied. SL has a higher level of interpretability, but because the diagnosis of mental ill-health is complex and involves an assessment of various symptoms and factors, for example, somatic diseases, addictive diseases, and lifestyle choices, DL is more appropriate. However, this means that the healthcare specialist is less able to interpret and possibly question the system's recommendation, which can lead to a risk of incorrect recommendations (Jurewicz, 2015). In addition, DL algorithms require large amounts of training data, and since psychology and psychiatry are highly classified, it can be challenging to access sensitive data, which is important for learning the algorithm.

## Patient Autonomy

The blackbox effect can also impair the patient's understanding of the treatment process (Duran & Jongsma, 2021), an important component of clinical ethics (Summers, 2009). The experience shows that the psychologist participants did not see any difficulty in including the patients in their care, but they had no experience with advanced systems such as intelligent decision support. The empirical study indicated that extra emphasis needs to be placed on explaining the aids used to give the patient an understanding of the content in cases where the patient uses the aid. If the aid in question is an intelligent decision support, this can be seen as a challenge as the care specialist does not fully understand the aid. If the care specialist does not have a full understanding of the decision support, he should not be expected to be able to explain it to the patient, but when it was considered important to explain the aid used, this could have potentially harmed patient autonomy. Interpersonal interaction is not lost when using intelligent decision support, as the care specialist has the last word, and the decision support does not make decisions but only contributes recommendations. This interpersonal interaction may achieve patient autonomy without understanding the decision support. Doctors use a large selection of complex tools in healthcare, such as EKG machines, whose functions are beyond patients' understanding. Despite this, patient autonomy can be achieved through interpersonal transparency.

## The Liability Dilemma

The empirical study indicated nuances in the responsibility dilemma, but that the user bears the greatest responsibility, the physician in this case, which is in line with Braun et al. (2020), who highlights that users should bear the greatest responsibility as intelligent decision support does not make their own decisions. The nuances of the liability dilemma include the actors involved, the need for laws, and the black box effect. The researchers, the programmer, the care center, and the care specialist are all participating actors within its intelligent decision support life cycle, and although the care specialist is responsible for the use of the decision support, the contributions of the other actors should be considered. The third-party organization responsible for approving the decision support for clinical care also impacts the use of the decision support, and these were considered to be the next actors to bear responsibility after the user.

Empirical evidence indicated that laws were needed to end the liability dilemma, which could make healthcare professionals more comfortable using AI-based systems, but how these laws should be designed is beyond the scope of this study. The blackbox effect affects the liability dilemma as users experience difficulties questioning competent AI systems they do not understand (Braun et al., 2020; Duran & Jongsma, 2021). If the user is not expected to understand how the system works, laws regarding who bears responsibility in case of failure should consider this. Since intelligent decision support is only a small part of the number of different AI systems, it can be challenging to form universal laws for all systems, as all systems have different strengths, weaknesses, and challenges. Intelligent decision aids with SL algorithms are more interpretable than those with DL algorithms, and when using SL systems, the user can be expected to interpret the system. Then, the question arises whether the user should bear more responsibility for the more control he has over the system. The liability dilemma is complex, making it challenging to shape fair laws to deal with the problem.

## Reliance

To increase users' trust in intelligent decision support, three aspects were discussed in the empirical study: transparency, reliability, and change. To create trust in the decision support, the user must be assured of its security (Braun et al., 2020). The empirical study showed that transparency was important for building trust in AI, as understanding the system being used can assure the user of the system's security. The participating AI researchers reflected on the need for situational transparency and that not all AI models can be interpretable. Furthermore, not all information is relevant to the care specialist, but only the most important information should be presented. As care specialists in different contexts have different needs, the most important information is different for different care specialists. The empirical study, therefore, indicated that the decision support should be designed according to the needs of the care specialist and should consider what the care specialist needs to decide. The balance between transparency and performance was considered to depend on the needs of the context, and this balance can be difficult to achieve.

At the same time that the decision support needs to be adapted to the user, the user may also need to adapt to the decision support. In the empirical study, mandatory transparency was also raised, which involved the care specialist understanding the limitations of the decision support and the data source on which the decision support was trained. As many aspects of patients' lives can affect their mental health, it can be challenging to get relevant recommendations from intelligent decision support based on training data from a population of a different culture and standard of living. Different populations have different needs, and if intelligent decision support is trained on a certain population, it is not necessarily applicable to another population. The better the training data matches the patient target group of the decision support, the better the decision support can perform. It was also felt that change was important to create trust in AI. The empirical study indicated that legal and cultural changes are required to create trust in AI. It was also shown that it is only a matter of time before people trust AI, even the opaque DL systems.

Tariq and Rafi (2012) suggest that a habit of different types of common decision support can give users a better ability to use intelligent decision support. This habit can vary between different care contexts, as different specialties have different habits of working with decision support. The psychologist participants did not use many aids or decision aids in their profession, which may impact their willingness or reluctance to use intelligent decision aids. Whether their free hands make them more willing or their unfamiliarity with advanced aids makes them more reluctant needs to be explored further in future studies. Based on literature and empirical evidence, we have identified ethical challenges when using intelligent decision support: dealing with incorrect recommendations, dealing with moral dilemmas, achieving patient autonomy, the responsibility dilemma, and creating trust. These challenges are summarized in the table below (*see* Table 5).

A large part of the focus in the study concerns ethical challenges that discourage the use of AI in healthcare, where the basic ethical principle of non-maleficence is a leading factor. But just as Summers (2009) points out, the complexity within the principle of non-maleficence lies in the meaning of the term "*harm*." The general opinion is that it applies to designs that go against an actor's best interest (Sum-

*Table 5. Table with potential ethical challenges and description*

| Ethical challenge | Description |
|---|---|
| Dealing with incorrect recommendations | There is a risk of misdiagnosis if the healthcare specialist does not question incorrect recommendations. The complex nature of diagnosis in mental health treatment can be complicate this challenge. |
| Dealing with Moral Dilemmas | Finding the option with the highest patient safety is challenging when a recommendation is questioned. The low transparency of the intelligent decision support system can make the healthcare specialist less able to interpret and question the system's recommendations. |
| Achieve patient autonomy | Difficulties in including patients in their care when the complexity of the decision support is beyond the patients' and the care specialists' ability to understand. This places increased emphasis on interpersonal transparency. |
| The liability dilemma | Ambiguity regarding who should bear the greatest responsibility as many actors contribute to the performance of the decision support. The healthcare specialist bears the greatest responsibility but has the lowest understanding of the decision support. The complexity of the challenge makes it challenging to shape fair laws to deal with the problem. |
| Build Trust | Healthcare specialists find it difficult to trust AI systems because they cannot determine their security. Aspects that affect trust can be *transparency, trustworthiness,* and *change.* Trust places higher demands on users. Change is needed in the form of laws and cultural change. |

mers, 2009), but even there, debates can be raised regarding what an actor's "*best interest*" is. Then, for example, Morley et al. (2020), Duran and Jongsma (2021), and Braun et al. (2020) focus on the harmful use of AI as that which goes against an actor's "*best interest*," the deprivation of a higher standard of care can also be considered to "*harm*" patients. The empirical study indicated that the opportunities for improvement with AI in healthcare are so great that not taking advantage of them can be considered unethical. By not using AI in healthcare, we are missing out on an important aid, not only in the treatment of mental illness but in healthcare in general. AI can contribute to aids that can save lives, and if this aid is not used, not all measures have been taken to save a patient. Over time, reluctance to use AI in healthcare can be seen as negligence, which can be seen as an ethical challenge.

## Conclusion and Further Research

This section presents the purpose of the study and the ethical challenges that have been identified, as well as recommendations for further research. This study aimed to answer the research question: What ethical challenges arise when using intelligent decision support in mental health care? A qualitative approach has been carried out to determine the ethical challenges of intelligent decision support in care linked to mental illness. Four AI researchers and four healthcare providers were interviewed through an interview study. Through literature and empirical evidence, five ethical challenges have been identified in using intelligent decision support in mental health care: dealing with incorrect recommendations, dealing with moral dilemmas, achieving patient autonomy, the responsibility dilemma, and creating trust.

Dealing with incorrect recommendations can be challenging as diagnosis is complex in the treatment of mental illness. Dealing with moral dilemmas was considered an ethical challenge as the user may have difficulty deciding the option with the highest patient safety when using intelligent decision support. According to the literature, achieving patient autonomy and including the patient in their care was considered an ethical challenge. The responsibility dilemma was an active question in the interview study as there is no clear answer as to who is responsible for diagnosis/treatment. Creating trust is important when using intelligent decision support, and transparency is an important factor in seeing how the system has arrived at a decision, thus increasing the system's credibility.

From a societal perspective, intelligent decision support can raise the standard of care for mental illness by reducing the time before first visit and diagnosis. Since prevention and early interventions are important to reduce the societal growth of mental illness, this is of great interest to society. The potential that intelligent decision support can contribute can even be considered unethical not to exploit. Intelligent decision support can involve risks if used incorrectly. Hence, it is important to understand these risks so that they can be counteracted. There is a risk of harming a patient through the system giving incorrect recommendations. Being aware of the risk can be avoided, which is an important part of the basic principle of non-evil. Intelligent decision support has great potential, and it is important not to avoid and overlook this potential to benefit the patient's well-being, an important part of the basic principle of charity. The focus should be on addressing the ethical challenges that hinder the introduction of intelligent decision support, as it can raise societal standards of care and save lives.

Further research should explore physicians' current procedures for handling systems' erroneous recommendations and their transferability to the use of AI. Further research is also required regarding the necessary depth of ML algorithms for decision support in healthcare and the treatment of mental illness, as well as a deeper exploration of the impact of opaque decision support on patient autonomy. Any challenges in communicating the data source on which the decision support has been trained to the user should also be explored more deeply in further research. Any challenges in explaining intelligent decision support to patients and whether this is necessary to achieve patient autonomy should also be explored in further research. Further research is also needed regarding how users' technology habits and experience with decision support and AI affect their willingness or unwillingness to use intelligent decision support. Further research should also explore the patient's perspective on the use of intelligent decision support in the treatment of mental illness. Finally, further research would benefit from exploring opinions from different cultures and gathering empirical evidence outside of India. This way, possible new challenges that do not necessarily exist within an Indian healthcare context can be identified.

## REFERENCES

Aita, M., & Richer, M.-C. (2005). Essentials of research ethics for healthcare professionals. *Nursing & Health Sciences*, *7*(2), 119–125. doi:10.1111/j.1442-2018.2005.00216.x PMID:15877688

Baker, C., & Kirk-Wade, E. (2023). *Mental health statistics: Prevalence, services and funding in England*. Commons Library. https://commonslibrary.parliament.uk/research-briefings/sn06988/

Barredo Arrieta, A., Díaz-Rodríguez, N., Del Ser, J., Bennetot, A., Tabik, S., Barbado, A., Garcia, S., Gil-Lopez, S., Molina, D., Benjamins, R., Chatila, R., & Herrera, F. (2020). Explainable Artificial Intelligence (XAI): Concepts, taxonomies, opportunities and challenges toward responsible AI. *Information Fusion*, *58*, 82–115. doi:10.1016/j.inffus.2019.12.012

Basu, K., Sinha, R., Ong, A., & Basu, T. (2020). Artificial Intelligence: How is It Changing Medical Sciences and Its Future? *Indian Journal of Dermatology*, *65*(5), 365–370. doi:10.4103/ijd.IJD_421_20 PMID:33165420

Braun, M., Hummel, P., Beck, S., & Dabrock, P. (2020). Primer on an ethics of AI-based decision support systems in the clinic. *Journal of Medical Ethics*, *47*(12), e3. doi:10.1136/medethics-2019-105860 PMID:32245804

Brey, P. A. E. (2012). Anticipatory Ethics for Emerging Technologies. *NanoEthics*, *6*(1), 1–13. doi:10.1007/s11569-012-0141-7

Bryman, A. (2018). *Social Research Methods* (3rd ed.). OUP Oxford.

Choi, R. Y., Coyner, A. S., Kalpathy-Cramer, J., Chiang, M. F., & Campbell, J. P. (2020). Introduction to Machine Learning, Neural Networks, and Deep Learning. *Translational Vision Science & Technology*, *9*(2), 14. doi:10.1167/tvst.9.2.14 PMID:32704420

Chowdhury, J. (2012). Hacking Health: Bottom-up Innovation for Healthcare. *Technology Innovation Management Review*, 31–35.

Davenport, T., & Kalakota, R. (2019). The potential for artificial intelligence in healthcare. *Future Healthcare Journal*, *6*(2), 94–98. doi:10.7861/futurehosp.6-2-94 PMID:31363513

Dewey, J. (2016). *Ethics*. Read Books Ltd.

Durán, J. M., & Jongsma, K. R. (2021). Who is afraid of black box algorithms? On the epistemological and ethical basis of trust in medical AI. *Journal of Medical Ethics*. doi:10.1136/medethics-2020-106820

Fernandes, M., Vieira, S. M., Leite, F., Palos, C., Finkelstein, S., & Sousa, J. M. C. (2020). Clinical Decision Support Systems for Triage in the Emergency Department using Intelligent Systems: A Review. *Artificial Intelligence in Medicine*, *102*, 101762. doi:10.1016/j.artmed.2019.101762 PMID:31980099

Hamet, P., & Tremblay, J. (2017). Artificial intelligence in medicine. *Metabolism: Clinical and Experimental*, *69S*, S36–S40. doi:10.1016/j.metabol.2017.01.011 PMID:28126242

Humerick, M. (2018). Taking AI Personally: How the E.U. Must Learn to Balance the Interests of Personal Data Privacy & Artificial Intelligence. *Santa Clara High-Technology Law Journal*, *34*(4), 393. https://digitalcommons.law.scu.edu/chtlj/vol34/iss4/3

Jin, K. W., Li, Q., Xie, Y., & Xiao, G. (2023). Artificial intelligence in mental healthcare: A scoping review. *The British Journal of Radiology*, *96*(1150), 20230213. doi:10.1259/bjr.20230213 PMID:37698582

Jurewicz, I. (2015). Mental health in young adults and adolescents – supporting general physicians to provide holistic care. *Clinical Medicine*, *15*(2), 151–154. doi:10.7861/clinmedicine.15-2-151 PMID:25824067

Kakuma, R., Minas, H., van Ginneken, N., Dal Poz, M. R., Desiraju, K., Morris, J. E., Saxena, S., & Scheffler, R. M. (2011). Human resources for mental health care: Current situation and strategies for action. *Lancet*, *378*(9803), 1654–1663. doi:10.1016/S0140-6736(11)61093-3 PMID:22008420

Kazim, E., & Koshiyama, A. S. (2021). A high-level overview of AI ethics. *Patterns (New York, N.Y.)*, *2*(9), 100314. doi:10.1016/j.patter.2021.100314 PMID:34553166

Kokciyan, N., Sassoon, I., Sklar, E., Modgil, S., & Parsons, S. (2021). Applying Metalevel Argumentation Frameworks to Support Medical Decision Making. *IEEE Intelligent Systems*, *36*(2), 64–71. doi:10.1109/MIS.2021.3051420

Lee, D., & Yoon, S. N. (2021). Application of Artificial Intelligence-Based Technologies in the Healthcare Industry: Opportunities and Challenges. *International Journal of Environmental Research and Public Health, 18*(1), 271. doi:10.3390/ijerph18010271 PMID:33401373

Lee, E. E., Torous, J., De Choudhury, M., Depp, C. A., Graham, S. A., Kim, H.-C., Paulus, M. P., Krystal, J. H., & Jeste, D. V. (2021). Artificial Intelligence for Mental Healthcare: Clinical Applications, Barriers, Facilitators, and Artificial Wisdom. *Biological Psychiatry: Cognitive Neuroscience and Neuroimaging, 6*(9), 856–864. doi:10.1016/j.bpsc.2021.02.001 PMID:33571718

Lei, L., Li, J., & Li, W. (2023). Assessing the role of artificial intelligence in the mental healthcare of teachers and students. *Soft Computing*, •••, 1–11. doi:10.1007/s00500-023-08072-5 PMID:37362257

Morley, J., Machado, C. C. V., Burr, C., Cowls, J., Joshi, I., Taddeo, M., & Floridi, L. (2020). The ethics of AI in health care: A mapping review. *Social Science & Medicine (1982), 260*, 113172. doi:10.1016/j.socscimed.2020.113172

Ngai, E. W. T., Peng, S., Alexander, P., & Moon, K. (2014). Decision support and intelligent systems in the textile and apparel supply chain: An academic review of research articles. *Expert Systems with Applications, 41*(1), 81–91. doi:10.1016/j.eswa.2013.07.013

Oppy, G., & Dowe, D. (2021). The Turing Test. In E. N. Zalta (Ed.), *The Stanford Encyclopedia of Philosophy* (Winter 2021). Metaphysics Research Lab, Stanford University. https://plato.stanford.edu/archives/win2021/entriesuring-test/

Pannu, A. (2015). *Artificial Intelligence and its Application in Different Areas*. Semantic Scholar.https://www.semanticscholar.org/paper/Artificial-Intelligence-and-its-Application-in-Pannu-Student/9a4d9a755134e612854db1897c03adb3983413df

Pedrelli, P., Nyer, M., Yeung, A., Zulauf, C., & Wilens, T. (2015). College Students: Mental Health Problems and Treatment Considerations. *Academic Psychiatry, 39*(5), 503–511. doi:10.1007/s40596-014-0205-9 PMID:25142250

Richter, D., Wall, A., Bruen, A., & Whittington, R. (2019). Is the global prevalence rate of adult mental illness increasing? Systematic review and meta-analysis. *Acta Psychiatrica Scandinavica, 140*(5), 393–407. doi:10.1111/acps.13083 PMID:31393996

Sachan, S., Yang, J.-B., Xu, D.-L., Benavides, D., & Li, Y. (2019). An Explainable AI Decision-Support-System to Automate Loan Underwriting. *Expert Systems with Applications, 144*, 113100. doi:10.1016/j.eswa.2019.113100

Schuett, J. (2019). A Legal Definition of AI. SSRN *Electronic Journal*. doi:10.2139/ssrn.3453632

Summers, J. (2009). *Principles of Healthcare Ethics*. Eweb:321396. https://repository.library.georgetown.edu/handle/10822/953367

Tariq, A., & Rafi, K. (2012). Intelligent Decision Support Systems- A Framework. *Information and Knowledge Management*. Semantic Scholar. https://www.semanticscholar.org/paper/Intelligent-Decision-Support-Systems-A-Framework-Tariq-Rafi/98250a732c5e5e11f6c7c9d0bf69efdd1ee71f4b

Varkey, B. (2020). Principles of Clinical Ethics and Their Application to Practice. *Medical Principles and Practice*, *30*(1), 17–28. doi:10.1159/000509119 PMID:32498071

Xu, Y., Zhou, Y., Sekula, P., & Ding, L. (2021). Machine learning in construction: From shallow to deep learning. *Developments in the Built Environment*, *6*, 100045. doi:10.1016/j.dibe.2021.100045

## ADDITIONAL READING

Espejo, G., Reiner, W., & Wenzinger, M. (2023). Exploring the Role of Artificial Intelligence in Mental Healthcare: Progress, Pitfalls, and Promises. *Cureus*, *15*(9), e44748. doi:10.7759/cureus.44748 PMID:37809254

Fakhoury, M. (2019). Artificial Intelligence in Psychiatry. *Advances in Experimental Medicine and Biology*, *1192*, 119–125. doi:10.1007/978-981-32-9721-0_6 PMID:31705492

Jin, K. W., Li, Q., Xie, Y., & Xiao, G. (2023). Artificial intelligence in mental healthcare: A scoping review. *The British Journal of Radiology*, *96*(1150), 20230213. doi:10.1259/bjr.20230213 PMID:37698582

Lee, E. E., Torous, J., De Choudhury, M., Depp, C. A., Graham, S. A., Kim, H.-C., Paulus, M. P., Krystal, J. H., & Jeste, D. V. (2021). Artificial Intelligence for Mental Healthcare: Clinical Applications, Barriers, Facilitators, and Artificial Wisdom. *Biological Psychiatry: Cognitive Neuroscience and Neuroimaging*, *6*(9), 856–864. doi:10.1016/j.bpsc.2021.02.001 PMID:33571718

Lei, L., Li, J., & Li, W. (2023). Assessing the role of artificial intelligence in the mental healthcare of teachers and students. *Soft Computing*, 1–11. doi:10.1007/s00500-023-08072-5 PMID:37362257

Sun, J., Dong, Q.-X., Wang, S.-W., Zheng, Y.-B., Liu, X.-X., Lu, T.-S., Yuan, K., Shi, J., Hu, B., Lu, L., & Han, Y. (2023). Artificial intelligence in psychiatry research, diagnosis, and therapy. *Asian Journal of Psychiatry*, *87*, 103705. doi:10.1016/j.ajp.2023.103705 PMID:37506575

## KEY TERMS AND DEFINITIONS

**Artificial Intelligence (AI):** AI involves intelligent systems and machines that mimic human behavior. In the context of mental health, AI technology includes machine learning (ML), natural language processing (NLP), deep learning (DL), and emotion-AI, offering opportunities for early diagnosis, personalized treatment plans, and improved therapeutic interventions.

**Digital Innovations:** Digital innovations in mental health, facilitated by AI, include the development of applications and self-help tools. These tools aim to monitor and improve mental health, but the challenge lies in the lack of research-based data and continuous reviews, requiring users to assess the quality and trustworthiness of these applications.

**E-health:** E-health refers to using digital tools and technologies, particularly in healthcare, to exchange information and preserve mental, physical, and social well-being. In India, there is a priority on digital innovations, including services like online consultations with psychologists and doctors through methods such as chats and video calls.

**Mental Illness:** Mental illness is a complex and multifaceted concept encompassing a range of conditions, from temporary symptoms like worry and low mood to more serious disorders like depression and anxiety, all negatively impacting an individual's quality of life.

**Preventive Mental Health Work:** Preventive work in mental health aims to promote good mental health and reduce the need for psychiatric treatment. It encompasses efforts to stop physical ill-health, enhance work capacity, and ultimately increase societal financial income by addressing mental health challenges.

**Public Health Report:** The annual report on the development of public health, specifically in India, highlights a significant increase in both serious and milder mental health problems from 2006 to 2021, emphasizing the growing societal challenge posed by mental illness.

# Chapter 13
# Sustainable Islamic Financial Inclusion:
## The Ethical Challenges of Generative AI in Product and Service Development

**Early Ridho Kismawadi**
https://orcid.org/0000-0002-9420-5212
*IAIN Langsa, Indonesia*

## ABSTRACT

*The study investigates the convergence of digital transformation, artificial intelligence (AI), and Islamic finance. In particular, it examines the ethical consequences that may arise from the integration of Generative AI in the sustainable development of Islamic financial services and products. This research fills a void in the current body of knowledge by examining the ethical consequences of generative AI in the context of Islamic finance. Using an interdisciplinary framework that integrates Islamic finance and technological ethics, the study seeks to make scientific and practical contributions. At the intersection of AI technology and Islamic finance, it is anticipated that new theories will emerge, as well as ethical principles that will serve as a guide for technology developers, policymakers, and Islamic financial institutions. The study has the potential to lead in creating a sustainable and inclusive Islamic finance ecosystem by ethically integrating Generative AI.*

## INTRODUCTION

With the rapid advancement of the digital transformation era, Islamic finance undertakes a progressively significant role in fulfilling the prerequisites for sustainable financial inclusion(Al Shehab & Hamdan, 2021; Albalawee & Al Fahoum, 2023; Qudah et al., 2023; Tlemsani & Matthews, 2023). The implementation of artificial intelligence (AI) technology has significantly impacted the Islamic financial services industry, generating fresh opportunities for growth and enhanced accessibility and reach. However, there are a number of ethical challenges associated with this potential development that require

DOI: 10.4018/979-8-3693-1565-1.ch013

cautious consideration, particularly in regards to the use of Generative AI in the development of Islamic financial products and services.

Islamic finance plays a pivotal role in promoting sustainable financial inclusion in the context of the digital era.(Ali et al., 2020; Baber, 2020; Ibrahim, 2015; Tahiri Jouti, 2018) On the basis of sharia law, justice, and sustainability, the Islamic financial system is emerging as a potential solution to meet the evolving needs of society. As AI technology continues to advance, the possibility of enhancing the effectiveness and efficiency of Islamic financial services grows in significance.

The implementation of Generative AI within the realm of Islamic finance offers a substantial potential for optimising the product and service development process(Abbas & Hafeez, 2021; Al Shehab & Hamdan, 2021). Artificial intelligence (AI) possesses the capability to perform exhaustive data analysis, delineate market trends, and identify investment opportunities that conform to the tenets of Islamic finance. However, in the midst of these advancements, an ethical aspect has surfaced as a subject of apprehension.

Concerning the application and interaction of Generative AI with the tenets of Islamic finance, an ethical dilemma emerges. A substantial challenge exists in ensuring that decisions produced by AI algorithms comply with the ethical principles and sharia law that underpin Islamic finance. Furthermore, it is imperative to conduct a comprehensive evaluation of the possibility of bias or inequity in AI decision-making in order to guarantee authentic and enduring financial inclusion.

This research endeavours to fill a longstanding gap in the body of knowledge concerning Islamic finance, specifically with respect to the integration of artificial intelligence. Notwithstanding the rapid expansion of Islamic finance, there remains a scarcity of understanding concerning the ethical ramifications of generative AI within this specific field, thereby resulting in unanswered inquiries. As the significance of Islamic finance as a foundation for inclusive sustainable finance increases, so does the urgency to address this research gap.

Unique among studies is the concentration on the ethical ramifications of generative AI as they pertain to Islamic finance. To promote progress in the discipline, this research employs an interdisciplinary structure that combines two significant and interrelated areas: Islamic finance and technological ethics. By employing this specific methodology, the inquiry is expanded to encompass not only the implementation of AI technology, but also its ramifications on the tenets and ideals of Islamic finance. Both academically and practically, significant contributions are anticipated from the research. From an academic standpoint, this research possesses the capacity to produce novel theories and conceptions concerning the convergence of AI technology and Islamic finance. In the interim, the practical contribution of this work is the development of ethical principles that can be utilised by Islamic financial institutions, policymakers, and technology developers as a guide when integrating Generative AI into the creation of sustainable Islamic financial products and services. Built upon a robust ethical framework, this study is anticipated to serve as a trailblazer in the development of an all-encompassing and enduring Islamic financial ecosystem via the implementation of AI technology.

## GENERATIVE AI INTEGRATION

There are numerous positive effects associated with the incorporation of generative artificial intelligence into the development of Islamic financial services(Allam & Dhunny, 2019; Kharbat et al., 2021; Liu et al., 2022). Generative AI utilises market trends and data analysis to furnish profound insights during the development phase of novel products and services. This empowers Islamic financial institutions to devise inventive solutions that adhere to the tenets of Islamic finance. Furthermore, the degree of service customization is substantially enhanced as the system acquires the capability to produce more precise recommendations, incorporating personal inclinations and the Islamic profile of the client.

Utilising Generative AI further improves the quality of decisions made within the framework of Islamic finance(Checco et al., 2021; Lăzăroiu et al., 2022; Shin, 2021; Yigitcanlar & Cugurullo, 2020). Islamic principle-compliant risk assessment, fund allocation, and financial instrument selection are all tasks that can be facilitated by the system. Furthermore, the capacity of AI Generative to comprehend local cultures and contexts facilitates the adaptation of Islamic financial services to the specific requirements of local communities.

The accuracy of investment portfolio management forecasts and the speed with which risks are assessed increase efficiency. Islamic financial institutions optimise their operational processes by automating administrative duties and conducting real-time sharia compliance monitoring. Furthermore, the implementation of Generative AI is facilitating the realisation of innovations in Islamic financial services through the analysis of trends and provision of pertinent solutions.

This integration contributes to enhanced security and risk management practises. Generative AI has the capability to identify potential risks or atypical activities, concurrently enhancing the security of financial transactions and data in adherence to the tenets of Islamic finance. Therefore, the incorporation of Generative AI into Islamic financial services significantly influences the development of more innovative, sustainable, and compliant financial services in accordance with the tenets of Islamic finance. This development facilitates the pursuit of financial inclusion and progress.

## DATA ANALYTICS AND MACHINE LEARNING

In the context of Islamic finance, generative AI can be utilised to analyse vast amounts of data pertaining to consumer behaviour, financial transactions, and market trends(Ala'raj et al., 2021; Gu & Zhu, 2021; Waliszewski & Warchlewska, 2020). By utilising machine learning algorithms, the system is capable of discerning patterns and generating enhanced insights pertaining to investment opportunities, customer preferences, and financial risks that adhere to the tenets of Islamic finance.

By employing generative artificial intelligence (Generative AI) to analyse vast amounts of data within the domain of Islamic finance, prospects arise to acquire a more profound comprehension of market trends, consumer behaviour, and financial transactions that adhere to the tenets of Islamic finance. The implementation of machine learning algorithms enables the system to rapidly discern intricate patterns within data, thereby facilitating a more comprehensive comprehension of customer inclinations, financial vulnerabilities, and investment prospects in accordance with the tenets of Islamic finance(Anand & Mishra, 2022; Shieh, Nguyen, & Horng, 2023; Shieh, Nguyen, Chen, et al., 2023).

By taking into account Islamic principles, the use of Generative AI in Islamic financial services has a substantial positive effect on risk assessment, fund allocation, and the selection of financial instruments. As an illustration, generative AI systems are capable of analysing market trends and historical data to detect potential hazards associated with specific financial instruments that might contravene Islamic principles. The incorporation of sharia principles into risk assessment algorithms enables financial institutions to manage and evaluate investment risks with greater precision.

The implementation of Generative AI has increased the efficiency of investment portfolio management (Kalayci et al., 2020; Pallathadka, Mustafa, et al., 2023; Pallathadka, Ramirez-Asis, et al., 2023).

By utilising precise and timely data analysis, the system is capable of forecasting market fluctuations, enabling fund managers to optimise their portfolios in adherence to the tenets of Islamic finance. For instance, in the event that alterations in market conditions have the potential to impact specific financial instruments that adhere to Islamic principles, the system can promptly furnish suggestions for implementing modifications.

Furthermore, the incorporation of Generative AI into the Islamic finance industry contributes to enhanced security and risk management(Mallikarjuna & Rao, 2019; Rodríguez-Espíndola et al., 2022; Wong et al., 2022). Potential threats or anomalous behaviour that could compromise the integrity of the financial system can be identified by generative AI. As an illustration, should the system identify dubious transaction patterns or deviations from the tenets of Islamic finance. Such cases warrant further intervention by the authorities.

Furthermore, Generative AI augments the level of security surrounding financial transactions and data, ensuring adherence to the tenets of Islamic finance. By implementing algorithms powered by artificial intelligence that recognise potential security threats, financial institutions can safeguard consumer data and financial transactions against cybercriminals. To enhance operational efficiency, incorporating Generating AI into Islamic financial services guarantees that every decision and procedure adheres to the tenets of Islamic finance. Such services are more innovative, efficient, and compliant with the tenets of Islamic finance, thereby contributing to the development of the Islamic finance industry and sustainable financial inclusion.

## BUSINESS PROCESS OPTIMIZATION

Generative AI has a transformative effect on the manner in which financial institutions conduct business, make decisions, and deliver services to consumers. Utilising this technology presents novel prospects for enhancing operational efficiency, optimising risk management, and fostering innovation within the realm of finance(Kraus et al., 2022; Popkova & Sergi, 2020; Vorzhakova & Boiarynova, 2020).

The implementation of Generative AI must include the automation of mundane duties. Financial institutions rely on extensive data processing to carry out various administrative tasks, including the administration of application forms and the verification of documents. This responsibility can be assumed by generative AI, thereby enhancing operational efficacy. For instance, the system can process and validate documents automatically, thereby reducing the amount of time and effort required of human personnel.

The implementation of Generative AI enables financial institutions to more precisely assess consumer sentiment and feedback. By employing social media monitoring, mail, and customer service interactions, the technology is capable of deciphering and comprehending the subtleties of human language. This capability empowers financial institutions to promptly address evolving customer needs and concerns. Enhancing customer service efficacy is not the only objective; fostering stronger customer relationships is also vital.

The implementation of Generative AI permits more intelligent management of investment portfolios. This technological advancement enables financial institutions to deliver more individualised and efficient services to their clients by identifying investment opportunities and recommending optimal asset allocation in accordance with their financial objectives and risk tolerance.

By analysing market and consumer data, Generative AI significantly contributes to the development of new products and services(Dachs et al., 2023; Padigar et al., 2022; Zhang et al., 2021). These technologies can assist financial institutions in identifying deficiencies in financial services and developing inventive products that fulfil consumer demands by delivering profound insights. For instance, identifying unserved market segments and designing products to satisfy their specific requirements using generative AI.

In the presence of Generative AI, the customization of financial services becomes more feasible. By conducting an analysis of consumer behaviour data, this technology has the capability to deliver service solutions that are more personalised. The use of Generative AI to provide investment advice that is personally tailored to the customer's risk profile and financial objectives is a concrete example of how this enhances the overall customer experience.

## CUSTOMER SERVICE DAN CHATBOT

The implementation of generative artificial intelligence (AI) within the domain of customer service has emerged as a paradigm shift in the technological and business sectors. Generative artificial intelligence (AI), specifically by leveraging chatbots, offers efficacious resolutions for enhancing customer service responsiveness and quality (Amjad et al., 2022; Bandara et al., 2020; Jorzik et al., 2023; Schrettenbrunnner, 2020). As used here, a chatbot refers to a computer programme that is specifically engineered to engage in dialogues with humans via text or voice. The chatbot's capability to provide prompt and precise responses to customer inquiries renders it a critical tool in enhancing the rapport between customers and businesses, particularly financial institutions.

Efficiency gains are a fundamental aspect of integrating generative AI into customer service (Canhoto & Clear, 2020; Lo Piano, 2020; Prentice & Nguyen, 2020). The implementation of chatbots capable of automating responses to consumer inquiries can effectively reduce response time. This offers a twofold benefit: firstly, it enhances customer satisfaction through prompt responses; and secondly, it enables the allocation of human resources towards endeavours that demand greater emotional intelligence and interpersonal intricacy.

Furthermore, the implementation of generative AI in chatbots contributes to the enhancement of response precision. The capacity of machines to comprehend and evaluate human language enables chatbots to deliver responses that are more accurate and pertinent. This practise aids in mitigating the likelihood of human error and enhances consumer confidence in the rendered services. When a customer inquiry about a particular product or policy, for instance, the chatbot can furnish precise and current information by consulting a database that is perpetually updated.

One additional benefit of generative AI implementation is the capacity to customise the consumer experience. Chatbots have the capability to comprehend consumer preferences and interaction history in order to deliver responses that are more suitable and pertinent to individual requirements. As an illustration, in the event that a patron consistently inquiries regarding a specific service, the chatbot may propose supplementary products or services that could potentially appeal to that patron, thereby fostering a more targeted and gratifying interaction.

Generative AI has the potential to enhance the analysis of customer data. By collecting and analysing data from customer interactions, chatbots can acquire a more comprehensive understanding of customer preferences and requirements. Financial institutions may be able to identify market trends, develop more effective marketing strategies, and tailor their products and services to consumer demands with the aid of this data.

Notwithstanding the immense advantages that can be derived from the integration of generative AI in customer service, it is critical to remain cognizant of certain ethical dilemmas and factors that could emerge. For instance, the imperative to safeguard customer data, ensure transparent communication regarding chatbot interactions with customers, and contemplate the potential displacement of human labour by automation.

## DEVELOPMENT OF NEW PRODUCTS AND SERVICES

The implementation of generative artificial intelligence (AI) within financial institutions has the potential to significantly impact the creation of novel services and products(Kusiak, 2018; Makridakis, 2017). By utilising algorithms and machine learning models, the system analyses market trends, evaluates customer demand, and generates the necessary insights to develop innovative products that align with market demands. By analysing stock market data and predicting changes in investment trends, for instance, financial institutions can employ AI to make more informed investment decisions.

Furthermore, through the analysis of transaction data, customer service feedback, and digital interactions, AI is capable of assessing customer requests in order to ascertain their preferences and requirements. In order to develop novel products and services, financial institutions may employ AI with an in-depth comprehension of market trends and consumer requirements (Alonso et al., 2020; Borges et al., 2021). AI can assist in the improvement of product personalization by analysing past customer transaction data and recommending products that correspond to the financial profiles and objectives of specific individuals. Therefore, by employing generative AI to develop products and services, financial institutions not only enhance their ability to promptly adapt to market fluctuations, but also augment their prospects of success by utilising a more intelligent and knowledgeable strategy.

## ETHICAL CHALLENGES OF GENERATIVE AI

Concerning the application of generative artificial intelligence (AI) to Islamic finance, there are a number of ethical issues that require attention(Du & Xie, 2021; Floridi & Strait, 2020; Thamik & Wu, 2022). In the Islamic finance industry, generative AI systems may offer substantial improvements in terms of innovation and efficiency, but they also pose a number of ethical concerns that require careful consideration. Detailed below is an explanation of the ethical issues raised by the implementation of generative AI in Islamic finance:

### Sharia Conformity

One of the most important concerns is ensuring that the application of generative AI adheres to the tenets of Islamic or Sharia law. It is imperative that the advancement and execution of these technologies guarantee that AI policies, algorithms, and activities are in accordance with the tenets of Islamic finance. These tenets comprise the prohibition of riba (interest), excessive speculation, and the principle of equitable wealth distribution.

The application of generative artificial intelligence (AI) in the realm of Islamic finance presents a specific challenge in terms of guaranteeing compliance with the tenets of Sharia, which pertains to

Islamic law (Azmat & Ghaffar, 2021; Calder, 2019). In accordance with the tenets of Islamic finance including the proscription of riba (interest), the prohibition of excessive speculation, and the principle of equitable wealth distribution certain crucial issues warrant further examination and elaboration.

A fundamental tenet of Islamic finance is the abolition of usury, which refers to the payment of interest. When considering the application of generative AI, it is critical to guarantee that the algorithms and policies employed do not facilitate or entail interest-bearing transactions. It is imperative that the design of the algorithm prevents the generation or encouragement of transactions that directly or indirectly contravene the prohibition of usury. In the development of financial products, such as lending and financing, for instance, artificial intelligence must be capable of verifying that interest rate policies and payment schemes do not violate the principle prohibiting usury. This may entail the formulation of risk models and policies that take into account financial structures that adhere to the tenets of Islamic finance.

Islamic finance emphasises the prohibition of potentially unethical transactions and speculation(Chowdhury et al., 2020; de la O González et al., 2019). It is critical, in the context of generative AI, to ensure that algorithms do not facilitate or promote speculative transactions that may cause damage to customers or contravene the tenets of Islamic finance. For instance, when trading equities or other financial instruments, AI algorithms should be developed with the intention of thwarting speculative endeavours that lack a substantial foundation or contravene the principle of equitable wealth distribution.

Islamic finance places significant emphasis on the principle of equitable wealth distribution. When considering the application of generative AI, it is critical to ensure that the advantages it offers are not limited to a select few individuals or groups, but rather are distributed equitably throughout society. For instance, AI must be programmed to ensure that resource allocation and investment decisions in fund or investment management adhere to the principles of sustainability and fairness, so that wealth can be distributed in accordance with Islamic financial values.

Ensuring compliance with Islamic finance principles when employing generative AI necessitates the establishment of a stringent ethical and auditing framework. This entails the cooperation of software developers, regulators, Islamic finance specialists, and software experts to ensure that this technology not only complies with the tenets of Islamic finance but also delivers value additions that align with the objectives and values of Islamic finance. Therefore, the integration of generative artificial intelligence (AI) into the domain of Islamic finance may serve as a mechanism to promote progress while upholding ethical standards and integrity.

## Accountability and Transparency

Frequently, generative AI systems are intricate and challenging for humans to comprehend. Consequently, it is critical to ensure that algorithms and decisions generated by AI are transparent. This clarity is necessary not only to facilitate the explanation of AI decisions, but also to ensure Sharia compliance is accounted for.

In order to ensure adherence to Shariah principles, the utilisation of generative artificial intelligence (AI) in Islamic finance presents obstacles pertaining to transparency and accountability (Durán & Jongsma, 2021; Riedl, 2019; Rodríguez-Espíndola et al., 2020). There is a requirement for transparency in generative AI systems, which are frequently intricate and challenging for humans to comprehend, by providing an explanation of the decision-making process. This is crucial in order to establish comprehension and confidence among stakeholders, including customers, regarding the adherence of financial decision-making processes to Sharia principles.

Requiring assurance that the algorithms employed in generative AI adhere to Sharia principles such as the prohibition of usury and the principle of equitable wealth distribution—transparency necessitates a shared comprehension of the process. Furthermore, accountability underscores the duty of service providers or organisations to provide explanations for each decision or action produced by the AI system. This entails utilising a well-defined history of success to offer insight into the decision-making procedure and facilitate open assessment.

An approach that can be employed to enhance the transparency of AI decisions is the implementation of Interpretable Machine Learning (IML) methodologies (Chen et al., 2022; Luna et al., 2019). To illustrate, generative AI models may employ LIME (Local Interpretable Model-agnostic Explanations) techniques to assist in elucidating particular decisions. By generating simpler, more interpretable models that endeavour to replicate the decisions of complex AI models, LIME assists humans in comprehending the decision-making process's contributing factors.

The Explainable AI (XAI) framework can be implemented by developers to guarantee that the decisions produced by generative AI models are explicable. To illustrate, one could employ methodologies like SHAP (SHapley Additive Explanations) to ascertain the extent to which each feature contributes to the ultimate determination. Thus, the explanatory power of the model can be enhanced, which facilitates a greater comprehension of the decision-making process.

Transparency regarding the decision-making processes of generative AI can be enhanced through the stringent implementation of audit trails and recording. It is possible to meticulously document each stage of the decision-making process, encompassing data input, algorithmic steps, and ultimate decisions. It generates a trail that is auditable and amenable to analysis in order to gain insights into decision-making processes and determinants.

Construct a dashboard for transparency that end users and interested parties can access. These displays may offer visual representations of the decision-making processes of generative AI models, including an analysis of the primary influencing factors. This improves the end user's comprehension of the reasoning and process behind a decision.

Integrating Islamic certification and verification systems directly into generative AI algorithms is another crucial technical step in the context of Islamic finance. The implementation of Shariah authorities to test and verify AI models can contribute to increased levels of transparency and adherence to legal requirements. The transparency and accountability issues associated with the application of generative artificial intelligence in finance may be mitigated through the implementation of these methods. It is critical to ascertain that the decisions produced by AI systems adhere to the ethical principles and Islamic law that regulate Islamic finance, to being comprehensible to human beings.

In order to maintain clarity regarding Sharia principles and overcome the intricacy of algorithms, the participation of Islamic finance experts, Sharia experts, and other stakeholders is vital. This involvement can contribute to the integration of Islamic finance viewpoints throughout the implementation and development of generative AI systems. Moreover, in order to foster optimal methodologies and uphold the tenets of Islamic finance while advancing technological progress, it is imperative that regulators and industry work in tandem to establish industry standards and guidelines. Generative AI implementations that deliver optimal benefits while upholding the integrity of Sharia principles, the bedrock of Islamic finance, are feasible when supported by a well-defined regulatory framework and robust industry standards.

## Privacy Protection and Data Security

The application of Generative AI within the framework of Islamic finance gives rise to several ethical concerns that warrant consideration, particularly concerning data security and privacy protection(D. Li et al., 2023; Villegas-Ch & García-Ortiz, 2023; Xiao et al., 2023). In the realm of safeguarding consumer privacy, it is paramount that financial platforms conduct data collection in a sincere and transparent manner. In the scenario where a bank employs Generative AI to conduct risk analysis, for instance, customers ought to be furnished with comprehensive information pertaining to the nature of data gathered, its intended use, and the rationale behind its collection (Gao, 2022; Sun et al., 2022). The concentric principle must be adhered to by obtaining customers' consent prior to the collection and use of their personal information. Within the context of Islamic finance, which prioritises integrity and openness, this principle assumes an even greater level of significance. Before using customer information to promote a new product or service, for instance, financial institutions are required to obtain the customer's explicit consent.

Concerning the utilisation and analysis of data, Generative AI must be ethically programmed and operated. This entails guaranteeing that the analysis of data is conducted in adherence to Islamic finance law and in good faith. For instance, in the context of generating investment recommendations using generative AI, it is imperative that the decision adheres to the tenets of Islam. Following the principles of justice and equality as they pertain to Islamic finance, AI programmes should additionally abstain from discrimination and bias against particular groups. Data protection is a high priority. It is imperative that generative AI systems are outfitted with sufficient technological security protocols, such as data encryption (Aggarwal, 2021; Lee, 2020; Pálmai et al., 2021), safeguards against unauthorised access, and routine security updates. An ongoing commitment to security requires the implementation of a resilient monitoring system capable of promptly identifying and addressing security concerns. In general, these ethical principles must guide the implementation of Generative AI in Islamic finance in order to preserve customer confidence and adhere to the strict standards of Islamic finance.

## Bias in Training Data

Generative models, which are frequently implemented in the field of artificial intelligence, have a tendency to mirror biases that are inherent in the training exercise data (Aggarwal, 2021; Lee, 2020; Pálmai et al., 2021). Because the model learns from the information and patterns contained in the dataset utilised during training, this is the case. In the event that exercise data exhibits specific inequalities or biases, generative models may internalise and perpetuate said biases through the output they generate. In the realm of finance, for instance, recommendations or decisions generated by generative models trained on historical datasets that may reflect inequities in financing decision-making could be unjust to particular groups. The significance of meticulous and discerning data processing becomes tangible when training generative models, as endeavours are made to detect and mitigate the biases that may be inherent in the datasets (Banerjee et al., 2023; Faghani et al., 2022; K, 2023; Pagano et al., 2023). Furthermore, it is critical to maintain ongoing monitoring subsequent to model implementation in order to identify and rectify any biases that may emerge while the model is being applied in practical scenarios. Therefore, it is crucial to maintain ethical and equitable standards in the application of these technologies by being cognizant of the potential for bias in generative models.

The implementation of generative models within the financial sector can exert a substantial influence on the susceptibility to bias. In the context of financing decision-making, for instance, generative models may inadvertently reinforce and reflect biases present in historical data, thereby generating financing recommendations that are unjust to particular groups, such as socioeconomic or ethnic minorities. Furthermore, generative models utilised in investment risk assessment and trained on datasets that exhibit sector-specific biases may generate recommendations that fail to adequately account for the entirety of the sector or investment type. Generative models in algorithm-based portfolio management that rely on unrepresentative trends or patterns may result in an imbalanced distribution of assets and compromise the objectivity of investment choices.

This potential impact extends to the insurance industry as well, where generative models may generate policy pricing that fail to accurately reflect the true risk, provided that exercise data reveals biases in risk assessment towards specific groups. A final consideration is that generative models trained with datasets that predominantly consist of crime cases involving specific groups may exhibit reduced efficacy when it comes to detecting financial crimes, such as money laundering or fraud. Recognising the possibility of bias in generative models is crucial for ensuring that the financial decisions they produce adhere to ethical and fair standards. This necessitates proactive measures in data processing and ongoing oversight throughout the model's life cycle.

Envision an investment firm that generates investment recommendations through the utilisation of generative AI models, employing historical data analysis and market trend monitoring. Although these models have the capability to generate informed recommendations by utilising training data, they encounter the most significant vulnerability when confronted with unforeseen market fluctuations. Extreme market fluctuations caused by unforeseen economic or political events, for instance, can render investment decisions that were previously prudent futile. Additional unanticipated consequences may result from bias in exercise data. When generative AI models are trained using datasets that contain biases that are known to exist in investment decision making, those biases may be reinforced in the model's recommendations. This may lead to unjust investment decisions and may also be detrimental to specific groups or industries. Furthermore, it is possible that generative AI models lack the capability to account for social and environmental factors when making investment decisions. Strictly financial considerations may overshadow the social or environmental ramifications of an investment, leading to inadvertent consequences for the environment or society. The potential consequence of investment decisions produced by generative AI models being significant losses or adverse effects on clients or society is a potential risk to the investment management firm's reputation. Additionally, legal and social accountability for the decisions made by AI systems must be given due consideration. Amidst this intricacy, it is imperative for organisations and financial professionals to not solely concentrate on the efficacy of their generative AI models, but also proactively oversee and assess the ethical ramifications that might emerge. Simultaneously, they should advance the development of optimal methodologies for incorporating generative AI technologies into financial decision making.

The potential exists for a generative AI model designed for financial risk analysis within the framework of Islamic finance to produce undesirable recommendations that deviate from Islamic principles (Darwiesh et al., 2023; Guo et al., 2023; Nikookar, 2023). An essential factor to bear in mind in Islamic finance is the avoidance of riba (interest), a practise that is deemed haram. In the event that the model lacks a comprehensive comprehension of the principles of Islamic finance or is not trained with data that incorporates riba, it may generate investment recommendations pertaining to financial instruments that incorporate usury elements. Customers who place their trust in these recommendations may unwittingly partake in financial activities that are contrary to the principles of Islam. As a consequence, their investment choices may be adversely affected in an unanticipated way. Hence, it is critical to ensure that generative AI models intended for Islamic finance possess a comprehensive understanding of and strictly adhere to the ethical tenets of the faith, specifically with regard to the avoidance of usury.

## Ethical Challenges and Uncertainty in Content Authenticity

A comprehension of creativity and authenticity within the framework of generative AI generates a substantial domain of ethical inquiries, particularly when these technologies are implemented in the financial sector. It is possible for generative models to generate content that blurs the distinction between manipulation and authenticity. Ethical concerns arise in relation to information honesty and intellectual property rights as a result of this.

The act of fabricating a sound or image that is nearly indistinguishable from the authentic version presents a substantial ethical dilemma(Ballesteros et al., 2021; X. Li et al., 2021). Trust and transparency may be influenced by this within the financial domain. As an illustration, in the event that a generative model is capable of producing an imperceptible visual or auditory representation, it may be employed to manipulate perceptions or disseminate inaccurate information in business presentations or financial statements.

Particularly in financial contexts, clear guidelines regarding the use of works generated by generative AI are required to resolve these ethical challenges. This necessitates careful consideration of intellectual property rights, a commitment to transparency concerning the implementation of generative technologies, and the creation of mechanisms for verification or validation to ascertain the genuineness of the information produced. Furthermore, the inclusion of stakeholders and legal professionals in the policy development process can contribute to the formulation of a more holistic ethical framework that aligns with current financial standards.

The utilisation of generative artificial intelligence (AI) within the realm of finance presents prospects for generating content that may engender doubts concerning its genuineness or manipulation. This is demonstrated through the development of marketing materials, financial statements, and business presentations. Consider a financial institution that generates graphs and visualisations to accompany its annual financial statements using generative AI. Advanced generative models are capable of generating visually stunning and remarkably lifelike images. Nevertheless, a degree of uncertainty emerges when evaluating whether the financial performance depicted in the statistics is an accurate reflection of reality or if it has been altered to present a more favourable image.

When applied to investment strategies, generative AI models have the capability to generate exceptionally innovative approaches (Ballesteros et al., 2021; X. Li et al., 2021). Nevertheless, ethical concerns emerge when it becomes necessary to ascertain whether the strategy merely demonstrates patterns evident in the exercise data or whether it truly demonstrates an in-depth comprehension of financial markets. Could the ingenuity of the model be likened to a dependable investment strategy? When generative AI is implemented in marketing content, such as promotional materials or advertisements, ethical concerns arise concerning the veracity and integrity of the generated information. When the model produces customer testimonials or narratives that are exceptionally persuasive, it becomes critical to ascertain the information's validity and veracity. What measures can be taken to guarantee that content produced by generative models remains ethically sound in the context of marketing endeavours?

An additional illustration is present within the framework of a business presentation. A financial company executive may employ generative AI to generate a visual presentation that depicts anticipated expansion of the company. These models are capable of generating convincing graphs and diagrams. However, ethical concerns emerge when we speculate on the degree to which these outcomes accurately represent reality and whether they were "savoured" to bolster the allure or endorsement of a specific narrative. Generative AI has the capability to generate visually appealing advertisements and content for financial products and services that are effective in marketing materials. However, an ethical concern arises regarding the veracity of the narratives or testimonies that could potentially be produced by generative models. Are the testimonials fabricated to enhance the company's image or are they based on genuine consumer experiences? The primary obstacle is that, despite the seemingly genuine appearance of the results generated by generative models, the automated nature of the process and the models' capacity to discern trends and preferences can cast doubt on whether those results accurately represent reality or are the result of intentional manipulations.

It is crucial to establish ethical guidelines and best practises regulating the use of generative AI in the production of financial content in order to address these issues. Prioritise transparency by providing consumers and other stakeholders with comprehensive information regarding the processes involved in content production and the level of trustworthiness associated with its authenticity. Additionally, rigorous oversight and stringent regulation of financial marketing practises utilising generative technology can contribute to the preservation of honesty and integrity in financial communications.

## MINIMIZING ETHICAL CHALLENGES

Given the potential ethical ramifications associated with the implementation of generative artificial intelligence (AI) in the banking industry, it is imperative to adopt proactive strategies that mitigate risks and uphold ethical standards in decision-making. Listed below are some additional detailed measures you can take:

### Deep Understanding of Algorithms and Models

A comprehensive comprehension of algorithms and generative models is a fundamental requirement for the ethical implementation of artificial intelligence (AI). It is imperative for organisations to thoroughly examine model parameters, acquiring a comprehensive comprehension of weights and altered variables throughout the model training phase. It is critical to comprehend the model's utilised features, including financial data and transaction history, in order to assess the possible ethical ramifications and partiality that could result from feature selection. Furthermore, it is critical to comprehend the decision logic of models, encompassing algorithm structures and activation functions, in order to detect and rectify any biases or inequities that may exist in the decision-making process. To the intricacy of the model training procedure and the interplay among variables, particular care must be taken to comprehend potential hazards and optimise model performance in an ethical manner. Evaluation of model performance utilising pertinent metrics becomes crucial in determining the model's success or failure, taking into account any ethical and fairness considerations that may be applicable. Armed with this profound comprehension, organisations are capable of identifying and resolving potential biases or ethical concerns that may emerge in the course of employing generative models, thereby guaranteeing that decisions made adhere to established ethical principles.

## Meticulous Data Processing

A critical and meticulous processing of data is an essential prerequisite for the implementation of generative artificial intelligence. Organisations must undertake a comprehensive examination of the data intended for utilisation, comprehending its origin, configuration, and portrayal of the broader populace. Data cleansing and normalisation procedures are required to eliminate sources of bias such as impure data and scale variations. It is critical to identify and address pre-existing biases in data, particularly when historical data indicates discrimination against particular groups. Ensuring that the dataset encompasses a broad spectrum of customer situations and profiles that mirror the diversity of the market or community being served is a primary objective.

It is imperative to rectify any class or group imbalances present in the data in order to mitigate the model's potential to grant unjust preference to the predominant group. The data quality verification process ensures that the information is precise and pertinent, whereas a comprehensive comprehension of the field or sector in which the model is implemented aids in identifying particularised elements that necessitate scrutiny. By implementing these measures, organisations can mitigate the potential for bias in generative models, thereby establishing a robust framework for training models with datasets that accurately represent diversity and equity. Predicated on the outcomes, more precise and equitable judgements are anticipated within the banking domain, thereby guaranteeing that the implementation of artificial intelligence upholds moral principles and impartiality.

## Dataset Diversification

Dataset diversification plays a crucial role in mitigating potential bias and ensuring generative artificial intelligence (AI) in the financial sector is able to provide fair and balanced decisions. The importance of dataset diversification can be elaborated in detail, starting with the need to involve various situations and contexts in the exercise data. Datasets that cover diverse scenarios and conditions allow generative models to understand wide variations in real-world circumstances.

In the context of finance, diversification of datasets is key to avoid overfocusing on specific groups. If exercise data is too centralized on one customer group or one transaction type, the model can internalize those biases and produce outputs that tend to favor the dominant group. Therefore, diversification helps ensure that models are not only trained on one perspective or context, but rather have a more holistic understanding of the various situations that may be encountered in the company's operations.

By involving diverse datasets, companies can reduce the risk of generative models adopting narrow views or distorting decisions based on disproportionately representative exercise data. Diversifying datasets creates a stronger foundation for training models that are more responsive and accurate to customer diversity, market conditions, and other business challenges. As a result, the resulting generative model will be better able to provide fair and balanced decisions, reflecting the ethical and fair values that are the cornerstones of the banking sector.

## Continuous Surveillance

Constant oversight is an essential measure to guarantee the ethical and impartial execution of generative artificial intelligence (AI) within the financial industry. It is imperative to execute a surveillance system with meticulousness, commencing with consistent monitoring of model performance. It is imperative for organisations to guarantee that generative models function in accordance with anticipated outcomes and do not generate decisions that may be misconstrued or biassed. This scrutiny is primarily concerned with identifying potential biases or inequalities in the decisions produced by the model. Ethical testing is a component of this procedure that assesses the soundness of model decisions with respect to ethical principles and impartiality. Additionally, real-time monitoring is a successful tactic that enables organisations to identify and address potential issues immediately.

Furthermore, routine audits serve as an essential preventive measure to comprehensively assess the performance of the model. These audits may encompass a scrutiny of the utilised data, model parameters, and decision outputs in order to ascertain the absence of detrimental or discriminatory biases that may disadvantage a specific group.

Continuous supervision serves as a mechanism for delivering feedback and corrections to its primary function of evaluation. Following the identification of potential bias, monitoring should be followed by appropriate and prompt corrective action. Therefore, ongoing oversight serves as both a control mechanism and a guarantee that the generative model maintains adherence to the ethical and equitable principles upheld by banking institutions. By establishing and maintaining an efficient supervisory system, organisations can mitigate the potential for inadvertent ethical repercussions and guarantee that decisions produced by models adhere to rigorous ethical principles, are transparent, and fair.

## Human Involvement in Important Decisions

Participation of individuals in critical decision-making processes is crucial for financial institutions to maintain operational integrity and ensure long-term viability. While generative AI models have the capability to generate recommendations by analysing identified patterns and data, human decision-making is still indispensable in certain crucial domains. To uphold ethical values and business principles, for instance, it is necessary for humans to assess whether model recommendations align with the moral stances and policies of the organisation. In markets and unique circumstances, uncertainty necessitates human intervention to evaluate circumstances and implement essential modifications to mitigate risk. Furthermore, it is incumbent upon individuals to ensure that the decisions they make are in accordance with relevant regulations and policies when legal and compliance considerations arise. In the realms of data identification and correction, social impact decision-making, and consumer trust consideration, human presence contributes to a more profound comprehension. In order to achieve sustainable and well-balanced decisions, significant financial institution decisions must incorporate non-quantitative aspects, including ethical considerations, legal compliance, and social impact. This is achieved through the participation of humans.

## Commitment to Transparency

Ensuring transparency in the implementation of technology, particularly generative models, is an essential cornerstone for organisations. This transparency encompasses the comprehensive disclosure of the methods employed to implement generative models, the criteria that inform the decision-making process, and the approaches taken to mitigate any ethical concerns that may emerge. A company implementing an artificial intelligence system for financial decision making, for instance, is obligated to provide comprehensive documentation regarding the model's data processing methodology, the criteria employed to generate recommendations, and the application of data security policies. Adherence to transparency principles not only fosters enhanced comprehension among stakeholders but also bolsters public confidence in the technology's operation.

### Quick Response to Ethical Issues

It is crucial to provide a prompt response to ethical concerns, which aligns with the dedication to transparency. Organisations ought to establish and maintain responsive mechanisms to detect, evaluate, and address ethical concerns that may emerge in the application of generative models. For instance, should concerns emerge pertaining to discrimination or bias in the output of models, organisations ought to establish unambiguous protocols to promptly assess and rectify such concerns. In order to ensure that corrective and preventive measures are openly explained to affected parties, such as customers, employees, and the broader community, a responsive response to ethical issues also entails proactive communication with these stakeholders. Through the integration of a steadfast dedication to transparency and an expeditious reaction to ethical concerns, organisations can establish a robust groundwork for the advancement and implementation of technologies that are both more ethical and sustainable.

## CONCLUSION

The advent of artificial intelligence (AI) technology propelled Islamic finance to the forefront of sustainable financial inclusion during the era of digital transformation. While the implementation of generative AI presents significant prospects for the growth and enhanced availability of Islamic financial services, it also gives rise to ethical dilemmas. A primary emphasis in the realm of Islamic finance is the observance of Sharia principles; however, the interaction between generative AI and such principles may give rise to ethical concerns.

Islamic finance is a pivotal component in addressing societal needs; nevertheless, it is critical to bear in mind that ethical dilemmas necessitate significant consideration to ensure that the beneficial prospects of this technology are actualized without compromising foundational values. A rigorous framework that guarantees generative AI adheres to the principles of Islamic finance is required to address ethical dilemmas. The implementation of technology in a transparent manner, a comprehensive comprehension of algorithms, and the diversification of datasets are crucial. Furthermore, rapid response to ethical concerns and human participation in decision-making are fundamental for mitigating risks and preserving the integrity of Sharia principles.

The research underscores a dearth of understanding in the field of Islamic finance concerning the ethical ramifications of generative artificial intelligence. Further investigation is warranted to enhance comprehension regarding the convergence of AI technology and the tenets of Islamic finance. Research areas that hold potential value include the examination of the practical ramifications of integrating generative AI into Islamic financial products and services, and the development of more comprehensive ethical frameworks. This study establishes a fundamental basis for subsequent investigations to ensure that Islamic finance can persistently adapt to the digital age while upholding its ethical values and fundamental principles.

This research emphasises the significance of digital transformation in elevating Islamic finance to the forefront of sustainable financial inclusion. This is accomplished through the implementation of artificial intelligence (AI) technology. Despite the fact that the implementation of generative AI presents enormous opportunities for the expansion and accessibility of Islamic financial services, there are ethical concerns that must be resolved. A primary emphasis in the realm of Islamic finance is the observance of Sharia principles; however, the interaction between generative AI and such principles may give rise to ethical concerns. In order to attain sustainable financial inclusion, Islamic finance must acquire a comprehensive understanding of the ethical ramifications associated with the implementation of generative AI. While the integration of AI technology may facilitate the growth of Islamic financial services, it is imperative to uphold the sustainability and integrity of Shariah principles. Islamic finance is of paramount importance in delivering solutions that address societal needs; however, in order to fully harness the positive potential of this technology while upholding fundamental values, significant concern must be given to ethical challenges. In order to confront ethical quandaries, it is imperative to implement a stringent framework that guarantees the adherence of generative AI to the tenets of Islamic finance. Diversification of datasets, transparency in technology implementation, and an in-depth comprehension of algorithms are crucial. Incorporating human participation in the decision-making process and promptly addressing ethical concerns are fundamental in order to mitigate risks and uphold the integrity of Sharia principles.

This publication significantly enhances comprehension regarding the effects of digital transformation on Islamic finance through the implementation of artificial intelligence (AI) technology, specifically generative AI. This study provides a comprehensive analysis of the ethical dilemmas that emerge in relation to the implementation of these technologies, with a particular focus on the observance of Shariah principles. By identifying potential ethical risks, the research lays the groundwork for the construction of a robust framework that guarantees the integrity and long-term viability of Islamic finance principles. Practical suggestions, including the implementation of transparency measures, a comprehensive comprehension of algorithms, and the diversification of datasets, are unveiled as crucial components in resolving ethical quandaries and preserving the integrity of Shariah principles. Additionally, the research underscores a dearth of understanding in the field of Islamic finance literature concerning the ethical ramifications of generative AI. This calls for additional investigation into the intersection of AI technology and Islamic finance principles. This publication establishes principles for the sustainable advancement of Islamic financial products and services by examining ethical risks and proposing strategies to mitigate them. Therefore, this study not only offers significant perspectives but also practical recommendations to direct forthcoming policies, business operations, and research endeavours towards preserving the ethical values and principles of Islamic finance amidst the digital revolution.

This study has substantial implications across multiple domains. Initially, policymakers and regulators may be aided in the development of regulatory frameworks that facilitate the implementation of artificial intelligence (AI) technologies in the Islamic finance industry by the results of this study. Policy formulation ought to prioritise ethical considerations and adherence to Sharia principles in order to safeguard the foundational values of Islamic finance while fostering technological innovation. Secondly, these findings can serve as a guide for Islamic finance industry participants to develop sustainable business practises. Incorporating transparency, acquiring a comprehensive comprehension of algorithms, and diversifying datasets were recognised as critical approaches to preserve the integrity of Shariah principles while capitalising on the capabilities of AI technology. This can contribute to the establishment of a commercial milieu that adheres to the ethical requirements of Islamic finance. Moreover, these results establish the necessity for improved Islamic finance education and training concerning AI technology. It is imperative to provide industry participants and experts with enhanced knowledge regarding the ethical ramifications of generative AI and prudent management strategies within the framework of Islamic finance.

The results increased the urgency for additional research. Further investigation may be warranted to refine the ethical framework and examine the practical ramifications of integrating generative AI into Islamic financial products and services. Moreover, participants in the Islamic finance sector may draw inspiration from these discoveries in order to create novel products and services that not only exploit the capabilities of AI technology but also uphold the ethical standards of Sharia. Therefore, the aforementioned discoveries have favourable ramifications for the long-term expansion and progress of Islamic finance during the period of digital revolution.

## REFERENCES

Abbas, K., & Hafeez, M. (2021). Wıll Artıfıcıal Intellıgence Rejuvenate Islamıc Fınance? A Versıon of World Academıa. *Hitit Theology Journal*, *20*(3), 311–324. doi:10.14395/hid.931401

Aggarwal, N. (2021). The norms of algorithmic credit scoring. *The Cambridge Law Journal*, *80*(1), 42–73. doi:10.1017/S0008197321000015

Al Shehab, N., & Hamdan, A. (2021). Artificial intelligence and women empowerment in bahrain. In Studies in Computational Intelligence, 954, 101–121. doi:10.1007/978-3-030-72080-3_6

Ala'raj, M., Abbod, M. F., & Majdalawieh, M. (2021). Modelling customers credit card behaviour using bidirectional LSTM neural networks. *Journal of Big Data*, *8*(1), 69. doi:10.1186/s40537-021-00461-7

Albalawee, N., & Al Fahoum, A. S. (2023). Islamic legal perspectives on digital currencies and how they apply to Jordanian legislation. *F1000 Research*, *12*, 97. doi:10.12688/f1000research.128767.2 PMID:37868298

Ali, M. M., Devi, A., Furqani, H., & Hamzah, H. (2020). Islamic financial inclusion determinants in Indonesia: An ANP approach. *International Journal of Islamic and Middle Eastern Finance and Management*, *13*(4), 727–747. doi:10.1108/IMEFM-01-2019-0007

Allam, Z., & Dhunny, Z. A. (2019). On big data, artificial intelligence and smart cities. *Cities (London, England)*, *89*, 80–91. doi:10.1016/j.cities.2019.01.032

Alonso, R. S., Sittón-Candanedo, I., García, Ó., Prieto, J., & Rodríguez-González, S. (2020). An intelligent Edge-IoT platform for monitoring livestock and crops in a dairy farming scenario. *Ad Hoc Networks*, *98*, 102047. doi:10.1016/j.adhoc.2019.102047

Amjad, M. S., Rafique, M. Z., Khan, M. A., Khan, A., & Bokhari, S. F. (2022). Blue Ocean 4.0 for sustainability–harnessing Blue Ocean Strategy through Industry 4.0. *Technology Analysis and Strategic Management*. doi:10.1080/09537325.2022.2060072

Anand, S., & Mishra, K. (2022). Identifying potential millennial customers for financial institutions using SVM. *Journal of Financial Services Marketing*, *27*(4), 335–345. doi:10.1057/s41264-021-00128-7

Azmat, S., & Ghaffar, H. (2021). Ethical Commitments and Credit Market Regulations. *Journal of Business Ethics, 171*(3), 421–433. doi:10.1007/s10551-019-04391-6

Baber, H. (2020). Financial inclusion and FinTech: A comparative study of countries following Islamic finance and conventional finance. *Qualitative Research in Financial Markets, 12*(1), 24–42. doi:10.1108/QRFM-12-2018-0131

Ballesteros, D. M., Rodriguez-Ortega, Y., Renza, D., & Arce, G. (2021). Deep4SNet: Deep learning for fake speech classification. *Expert Systems with Applications, 184*, 115465. Advance online publication. doi:10.1016/j.eswa.2021.115465

Bandara, R., Fernando, M., & Akter, S. (2020). Privacy concerns in E-commerce: A taxonomy and a future research agenda. *Electronic Markets, 30*(3), 629–647. doi:10.1007/s12525-019-00375-6

Banerjee, I., Bhattacharjee, K., Burns, J. L., Trivedi, H., Purkayastha, S., Seyyed-Kalantari, L., Patel, B. N., Shiradkar, R., & Gichoya, J. (2023). "Shortcuts" Causing Bias in Radiology Artificial Intelligence: Causes, Evaluation, and Mitigation. *Journal of the American College of Radiology, 20*(9), 842–851. doi:10.1016/j.jacr.2023.06.025 PMID:37506964

Borges, A. F. S., Laurindo, F. J. B., Spínola, M. M., Gonçalves, R. F., & Mattos, C. A. (2021). The strategic use of artificial intelligence in the digital era: Systematic literature review and future research directions. *International Journal of Information Management, 57*, 102225. doi:10.1016/j.ijinfomgt.2020.102225

Calder, R. (2019). How religio-economic projects succeed and fail: The field dynamics of Islamic finance in the Arab Gulf states and Pakistan, 1975–2018. *Socio-economic Review, 17*(1), 167–193. doi:10.1093/ser/mwz015

Canhoto, A. I., & Clear, F. (2020). Artificial intelligence and machine learning as business tools: A framework for diagnosing value destruction potential. *Business Horizons, 63*(2), 183–193. doi:10.1016/j.bushor.2019.11.003

Checco, A., Bracciale, L., Loreti, P., Pinfield, S., & Bianchi, G. (2021). AI-assisted peer review. *Humanities & Social Sciences Communications, 8*(1), 25. doi:10.1057/s41599-020-00703-8

Chen, C., Lin, K., Rudin, C., Shaposhnik, Y., Wang, S., & Wang, T. (2022). A holistic approach to interpretability in financial lending: Models, visualizations, and summary-explanations. *Decision Support Systems, 152*, 113647. Advance online publication. doi:10.1016/j.dss.2021.113647

Chowdhury, M. A. F., Sultan, Y., & Mahmudul Haque, M. (2020). Conventional futures: A review of major issues from the Islamic finance perspective. *International Journal of Pluralism and Economics Education, 11*(2), 201–210. doi:10.1504/IJPEE.2020.111258

Dachs, B., Amoroso, S., Castellani, D., Papanastassiou, M., & von Zedtwitz, M. (2023). The internationalisation of R&D: Past, present and future. *International Business Review*. doi:10.1016/j.ibusrev.2023.102191

Darwiesh, A., El-Baz, A. H., Abualkishik, A. Z., & Elhoseny, M. (2023). Artificial Intelligence Model for Risk Management in Healthcare Institutions: Towards Sustainable Development. *Sustainability (Basel), 15*(1), 420. doi:10.3390/su15010420

Du, S., & Xie, C. (2021). Paradoxes of artificial intelligence in consumer markets: Ethical challenges and opportunities. *Journal of Business Research*, *129*, 961–974. doi:10.1016/j.jbusres.2020.08.024

Durán, J. M., & Jongsma, K. R. (2021). Who is afraid of black box algorithms? On the epistemological and ethical basis of trust in medical AI. *Journal of Medical Ethics*, *47*(5), 329–335. doi:10.1136/medethics-2020-106820 PMID:33737318

Faghani, S., Khosravi, B., Zhang, K., Moassefi, M., Jagtap, J. M., Nugen, F., Vahdati, S., Kuanar, S. P., Rassoulinejad-Mousavi, S. M., Singh, Y., Vera Garcia, D. V., Rouzrokh, P., & Erickson, B. J. (2022). Mitigating Bias in Radiology Machine Learning: 3. Performance Metrics. *Radiology. Artificial Intelligence*, *4*(5), e220061. doi:10.1148/ryai.220061 PMID:36204539

Floridi, L., & Strait, A. (2020). Ethical Foresight Analysis: What it is and Why it is Needed? *Minds and Machines*, *30*(1), 77–97. doi:10.1007/s11023-020-09521-y

Gao, B. (2022). The Use of Machine Learning Combined with Data Mining Technology in Financial Risk Prevention. *Computational Economics*, *59*(4), 1385–1405. doi:10.1007/s10614-021-10101-0

González, M. O., Jareño, F., & El Haddouti, C. (2019). Sector portfolio performance comparison between Islamic and conventional stock markets. *Sustainability (Basel)*, *11*(17), 4618. doi:10.3390/su11174618

Gu, G., & Zhu, W. (2021). Time-varying transmission effects of internet finance under economic policy uncertainty and internet consumers' behaviors: Evidence from China. *Journal of Advanced Computational Intelligence and Intelligent Informatics*, *25*(5), 554–562. doi:10.20965/jaciii.2021.p0554

Guo, J., Kang, W., & Wang, Y. (2023). Option pricing under sub-mixed fractional Brownian motion based on time-varying implied volatility using intelligent algorithms. *Soft Computing*, *27*(20), 15225–15246. doi:10.1007/s00500-023-08647-2

Ibrahim, M. H. (2015). Issues in Islamic banking and finance: Islamic banks, Shari'ah-compliant investment and sukuk. *Pacific-Basin Finance Journal*, *34*, 185–191. doi:10.1016/j.pacfin.2015.06.002

Jorzik, P., Yigit, A., Kanbach, D. K., Kraus, S., & Dabic, M. (2023). Artificial Intelligence-Enabled Business Model Innovation: Competencies and Roles of Top Management. *IEEE Transactions on Engineering Management*, 1–13. doi:10.1109/TEM.2023.3275643

K, R. (2023). Algorithm justice – moving towards equitable artificial intelligence. *CASE Journal, 19*(6), 788–799. doi:10.1108/TCJ-11-2021-0211

Kalayci, C. B., Polat, O., & Akbay, M. A. (2020). An efficient hybrid metaheuristic algorithm for cardinality constrained portfolio optimization. *Swarm and Evolutionary Computation*, *54*, 100662. doi:10.1016/j.swevo.2020.100662

Kharbat, F. F., Alshawabkeh, A., & Woolsey, M. L. (2021). Identifying gaps in using artificial intelligence to support students with intellectual disabilities from education and health perspectives. *Aslib Journal of Information Management*, *73*(1), 101–128. doi:10.1108/AJIM-02-2020-0054

Kraus, K., Kraus, N., Hryhorkiv, M., Kuzmuk, I., & Shtepa, O. (2022). Artificial Intelligence in Established of Industry 4.0. *WSEAS Transactions on Business and Economics*, *19*, 1884–1900. doi:10.37394/23207.2022.19.170

Kusiak, A. (2018). Smart manufacturing. *International Journal of Production Research*, *56*(1–2), 508–517. doi:10.1080/00207543.2017.1351644

Lăzăroiu, G., Androniceanu, A., Grecu, I., Grecu, G., & Neguriță, O. (2022). Artificial intelligence-based decision-making algorithms, Internet of Things sensing networks, and sustainable cyber-physical management systems in big data-driven cognitive manufacturing. *Oeconomia Copernicana*, *13*(4), 1047–1080. doi:10.24136/oc.2022.030

Lee, J. (2020). Access to Finance for Artificial Intelligence Regulation in the Financial Services Industry. *European Business Organization Law Review*, *21*(4), 731–757. doi:10.1007/s40804-020-00200-0

Li, D., Lai, J., Wang, R., Li, X., Vijayakumar, P., Gupta, B. B., & Alhalabi, W. (2023). Ubiquitous intelligent federated learning privacy-preserving scheme under edge computing. *Future Generation Computer Systems*, *144*, 205–218. doi:10.1016/j.future.2023.03.010

Li, X., Yu, N., Zhang, X., Zhang, W., Li, B., Lu, W., Wang, W., & Liu, X. (2021). Overview of digital media forensics technology. *Journal of Image and Graphics*, *26*(6), 1216–1226. doi:10.11834/jig.210081

Liu, D. S., Sawyer, J., Luna, A., Aoun, J., Wang, J., Boachie, L., Halabi, S., & Joe, B. (2022). Perceptions of US Medical Students on Artificial Intelligence in Medicine: Mixed Methods Survey Study. *JMIR Medical Education*, *8*(4), e38325. doi:10.2196/38325 PMID:36269641

Lo Piano, S. (2020). Ethical principles in machine learning and artificial intelligence: Cases from the field and possible ways forward. *Humanities & Social Sciences Communications*, *7*(1), 9. doi:10.1057/s41599-020-0501-9

Luna, J. M., Gennatas, E. D., Ungar, L. H., Eaton, E., Diffenderfer, E. S., Jensen, S. T., Simone, C. B. II, Friedman, J. H., Solberg, T. D., & Valdes, G. (2019). Building more accurate decision trees with the additive tree. *Proceedings of the National Academy of Sciences of the United States of America*, *116*(40), 19887–19893. doi:10.1073/pnas.1816748116 PMID:31527280

Makridakis, S. (2017). The forthcoming Artificial Intelligence (AI) revolution: Its impact on society and firms. *Futures*, *90*, 46–60. doi:10.1016/j.futures.2017.03.006

Mallikarjuna, M., & Rao, R. P. (2019). Evaluation of forecasting methods from selected stock market returns. *Financial Innovation*, *5*(1), 40. doi:10.1186/s40854-019-0157-x

Nikookar, H. (2023). A Risk Analysis of Communication, Navigation and Sensing Satellite Systems Threats. *Journal of Mobile Multimedia*, *19*(1), 277–290. doi:10.13052/jmm1550-4646.19114

Padigar, M., Pupovac, L., Sinha, A., & Srivastava, R. (2022). The effect of marketing department power on investor responses to announcements of AI-embedded new product innovations. *Journal of the Academy of Marketing Science*, *50*(6), 1277–1298. doi:10.1007/s11747-022-00873-8

Pagano, T. P., Loureiro, R. B., Lisboa, F. V. N., Peixoto, R. M., Guimarães, G. A. S., Cruz, G. O. R., Araujo, M. M., Santos, L. L., Cruz, M. A. S., Oliveira, E. L. S., Winkler, I., & Nascimento, E. G. S. (2023). Bias and Unfairness in Machine Learning Models: A Systematic Review on Datasets, Tools, Fairness Metrics, and Identification and Mitigation Methods. *Big Data and Cognitive Computing*, *7*(1), 15. doi:10.3390/bdcc7010015

Pallathadka, H., Mustafa, M., Sanchez, D. T., Sekhar Sajja, G., Gour, S., & Naved, M. (2023). IMPACT OF MACHINE learning ON Management, healthcare AND AGRICULTURE. *Materials Today: Proceedings*, *80*, 2803–2806. doi:10.1016/j.matpr.2021.07.042

Pallathadka, H., Ramirez-Asis, E. H., Loli-Poma, T. P., Kaliyaperumal, K., Ventayen, R. J. M., & Naved, M. (2023). Applications of artificial intelligence in business management, e-commerce and finance. *Materials Today: Proceedings*, *80*, 2610–2613. doi:10.1016/j.matpr.2021.06.419

Pálmai, G., Csernyák, S., & Erdélyi, Z. (2021). Authentic and reliable data in the service of national public data asset. *Public Finance Quarterly*, *66*(Special edition 2021/1), 52–67. doi:10.35551/PFQ_2021_s_1_3

Popkova, E. G., & Sergi, B. S. (2020). Human capital and AI in industry 4.0. Convergence and divergence in social entrepreneurship in Russia. *Journal of Intellectual Capital*, *21*(4), 565–581. doi:10.1108/JIC-09-2019-0224

Prentice, C., & Nguyen, M. (2020). Engaging and retaining customers with AI and employee service. *Journal of Retailing and Consumer Services*, *56*, 102186. Advance online publication. doi:10.1016/j.jretconser.2020.102186

Qudah, H., Malahim, S., Airout, R., Alomari, M., Hamour, A. A., & Alqudah, M. (2023). Islamic Finance in the Era of Financial Technology: A Bibliometric Review of Future Trends. *International Journal of Financial Studies*, *11*(2), 76. doi:10.3390/ijfs11020076

Riedl, M. O. (2019). Human-centered artificial intelligence and machine learning. *Human Behavior and Emerging Technologies*, *1*(1), 33–36. doi:10.1002/hbe2.117

Rodríguez-Espíndola, O., Chowdhury, S., Beltagui, A., & Albores, P. (2020). The potential of emergent disruptive technologies for humanitarian supply chains: The integration of blockchain, Artificial Intelligence and 3D printing. *International Journal of Production Research*, *58*(15), 4610–4630. doi:10.1080/00207543.2020.1761565

Rodríguez-Espíndola, O., Chowdhury, S., Dey, P. K., Albores, P., & Emrouznejad, A. (2022). Analysis of the adoption of emergent technologies for risk management in the era of digital manufacturing. *Technological Forecasting and Social Change*, *178*, 121562. doi:10.1016/j.techfore.2022.121562

Schrettenbrunnner, M. B. (2020). Artificial-Intelligence-Driven Management. *IEEE Engineering Management Review*, *48*(2), 15–19. doi:10.1109/EMR.2020.2990933

Shieh, C.-S., Nguyen, T.-T., Chen, C.-Y., & Horng, M.-F. (2023). Detection of Unknown DDoS Attack Using Reconstruct Error and One-Class SVM Featuring Stochastic Gradient Descent. *Mathematics*, *11*(1), 108. doi:10.3390/math11010108

Shieh, C.-S., Nguyen, T.-T., & Horng, M.-F. (2023). Detection of Unknown DDoS Attack Using Convolutional Neural Networks Featuring Geometrical Metric. *Mathematics*, *11*(9), 2145. doi:10.3390/math11092145

Shin, D. (2021). The effects of explainability and causability on perception, trust, and acceptance: Implications for explainable AI. *International Journal of Human-Computer Studies*, *146*, 102551. Advance online publication. doi:10.1016/j.ijhcs.2020.102551

Sun, Q., Wu, H., & Zhao, B. (2022). Artificial intelligence technology in internet financial edge computing and analysis of security risk. *International Journal of Ad Hoc and Ubiquitous Computing*, *39*(4), 201–210. doi:10.1504/IJAHUC.2022.121654

Tahiri Jouti, A. (2018). Islamic finance: Financial inclusion or migration? *ISRA International Journal of Islamic Finance*, *10*(2), 277–288. doi:10.1108/IJIF-07-2018-0074

Thamik, H., & Wu, J. (2022). The Impact of Artificial Intelligence on Sustainable Development in Electronic Markets. *Sustainability (Basel)*, *14*(6), 3568. doi:10.3390/su14063568

Tlemsani, I., & Matthews, R. (2023). Digitalization and the prospects of cryptocurrency in Islamic finance. *International Journal of Technology Management & Sustainable Development*, *22*(2), 131–152. doi:10.1386/tmsd_00072_1

Villegas-Ch, W., & García-Ortiz, J. (2023). Toward a Comprehensive Framework for Ensuring Security and Privacy in Artificial Intelligence. *Electronics (Basel)*, *12*(18), 3786. doi:10.3390/electronics12183786

Vorzhakova, Y., & Boiarynova, K. (2020). The application of digitalization in enterprises on the basis of multiple criteria selection design. *Central European Management Journal*, *28*(3), 127–148. doi:10.7206/cemj.2658-0845.29

Waliszewski, K., & Warchlewska, A. (2020). Attitudes towards artificial intelligence in the area of personal financial planning: A case study of selected countries. *Entrepreneurship and Sustainability Issues*, *8*(2), 399–420. doi:10.9770/jesi.2020.8.2(24)

Wong, L.-W., Tan, G. W.-H., Ooi, K.-B., Lin, B., & Dwivedi, Y. K. (2022). Artificial intelligence-driven risk management for enhancing supply chain agility: A deep-learning-based dual-stage PLS-SEM-ANN analysis. *International Journal of Production Research*, 1–21. doi:10.1080/00207543.2022.2063089

Xiao, X., Tang, Z., Xiao, B., & Li, K.-L. (2023). A Survey on Privacy and Security Issues in Federated Learning. *Jisuanji Xuebao/Chinese. Journal of Computers*, *46*(5), 1019–1044. doi:10.11897/SP.J.1016.2023.01019

Yigitcanlar, T., & Cugurullo, F. (2020). The sustainability of artificial intelligence: An urbanistic viewpoint from the lens of smart and sustainable cities. *Sustainability (Basel)*, *12*(20), 1–24. doi:10.3390/su12208548

Zhang, B. Z., Ashta, A., & Barton, M. E. (2021). Do FinTech and financial incumbents have different experiences and perspectives on the adoption of artificial intelligence? *Strategic Change*, *30*(3), 223–234. doi:10.1002/jsc.2405

# Compilation of References


Aaron, S. (2023, May 12). *How will generative AI impact healthcare?* WeForum. Https://Www.Weforum.Org/Agenda/2023/05/How-Will-Generative-Ai-Impact-Healthcare/

Abbas, K., & Hafeez, M. (2021). Will Artificial Intelligence Rejuvenate Islamic Finance? A Version of World Academia. *Hitit Theology Journal*, *20*(3), 311–324. doi:10.14395/hid.931401

Abolghasemi, M., Khodakarami, V., & Tehranifard, H. (2015): A new approach for supply chain risk management: Mapping SCOR into Bayesian network. *Journal of Industrial Engineering and Management (JIEM), ISSN 2013-0953, Omnia Science, Barcelona, 8*(1), 280-302. doi:10.3926/jiem.1281

Abramson, A. (2023). How to use ChatGPT as a learning tool. *Monitor on Psychology*, *54*(4). https://www.apa.org/monitor/2023/06/chatgpt-learning-tool

Aggarwal, N. (2021). The norms of algorithmic credit scoring. *The Cambridge Law Journal*, *80*(1), 42–73. doi:10.1017/S0008197321000015

Ahmad, W., & Dethy, E. (2019). Preventing surveillance cities: Developing a set of fundamental privacy provisions. *Journal of Science Policy & Governance*, *15*(1), 1–11.

Ahuja, K., Hada, R., Ochieng, M., Jain, P., Diddee, H., Maina, S., & Sitaram, S. (2023). Mega: Multilingual evaluation of generative ai. *arXiv preprint arXiv:2303.12528*. doi:10.18653/v1/2023.emnlp-main.258

Aita, M., & Richer, M.-C. (2005). Essentials of research ethics for healthcare professionals. *Nursing & Health Sciences*, *7*(2), 119–125. doi:10.1111/j.1442-2018.2005.00216.x PMID:15877688

Akgun, S., & Greenhow, C. (2022). Artificial intelligence in education: Addressing ethical challenges in K-12 settings. *AI and Ethics*, *2*(3), 431–440. doi:10.1007/s43681-021-00096-7 PMID:34790956

Al Shehab, N., & Hamdan, A. (2021). Artificial intelligence and women empowerment in bahrain. In Studies in Computational Intelligence, 954, 101–121. doi:10.1007/978-3-030-72080-3_6

Ala'raj, M., Abbod, M. F., & Majdalawieh, M. (2021). Modelling customers credit card behaviour using bidirectional LSTM neural networks. *Journal of Big Data*, *8*(1), 69. doi:10.1186/s40537-021-00461-7

Albalawee, N., & Al Fahoum, A. S. (2023). Islamic legal perspectives on digital currencies and how they apply to Jordanian legislation. *F1000 Research*, *12*, 97. doi:10.12688/f1000research.128767.2 PMID:37868298

Ali, M. M., Devi, A., Furqani, H., & Hamzah, H. (2020). Islamic financial inclusion determinants in Indonesia: An ANP approach. *International Journal of Islamic and Middle Eastern Finance and Management*, *13*(4), 727–747. doi:10.1108/IMEFM-01-2019-0007

Allam, Z., & Dhunny, Z. A. (2019). On big data, artificial intelligence and smart cities. *Cities (London, England)*, *89*, 80–91. doi:10.1016/j.cities.2019.01.032

Almeida, F. (2023). Prospects of Cybersecurity in Smart Cities. *Future Internet*, *15*(9), 285. doi:10.3390/fi15090285

Alonso, R. S., Sittón-Candanedo, I., García, Ó., Prieto, J., & Rodríguez-González, S. (2020). An intelligent Edge-IoT platform for monitoring livestock and crops in a dairy farming scenario. *Ad Hoc Networks*, *98*, 102047. doi:10.1016/j.adhoc.2019.102047

Alsobeh, A., & Woodward, B. (2023). AI as a Partner in Learning: A Novel Student-in-the-Loop Framework for Enhanced Student Engagement and Outcomes in Higher Education. In *Proceedings of the 24th Annual Conference on Information Technology Education*, (pp. 171–172). Research Gate. 10.1145/3585059.3611405

Alter, S. (2022). Understanding artificial intelligence in the context of usage: Contributions and smartness of algorithmic capabilities in work systems. *International Journal of Information Management*, *67*, 102392. doi:10.1016/j.ijinfomgt.2021.102392

Alvero, R. (2023). ChatGPT: Rumors of human providers' demise have been greatly exaggerated. *Fertility and Sterility*, *119*(6), 930–931. doi:10.1016/j.fertnstert.2023.03.010 PMID:36921837

Alyoubi, W. L., Shalash, W. M., & Abulkhair, M. F. (2020). Diabetic retinopathy detection through deep learning techniques: A review. In Informatics in Medicine Unlocked. doi:10.1016/j.imu.2020.100377

Amjad, M. S., Rafique, M. Z., Khan, M. A., Khan, A., & Bokhari, S. F. (2022). Blue Ocean 4.0 for sustainability–harnessing Blue Ocean Strategy through Industry 4.0. *Technology Analysis and Strategic Management*. doi:10.1080/09537325.2022.2060072

Amoroso, D., & Tamburrini, G. (2020). Autonomous weapons systems and meaningful human control: Ethical and legal issues. *Current Robotics Reports*, *1*(4), 187–194. doi:10.1007/s43154-020-00024-3

Anand, S., & Mishra, K. (2022). Identifying potential millennial customers for financial institutions using SVM. *Journal of Financial Services Marketing*, *27*(4), 335–345. doi:10.1057/s41264-021-00128-7

Anderson, L. B., Kanneganti, D., Houk, M. B., Holm, R. H., & Smith, T. (2023). Generative AI as a tool for environmental health research translation. *GeoHealth*, *7*(7), e2023GH000875.

Angwin, J., Larson, J., Mattu, S., & Kirchner, L. (2016). *Machine bias: There's software used across the country to predict future criminals. And it's biased against blacks*. ProPublica. https://www.propublica.org/article/machine-bias-risk-assessments-in-criminal-sentencing

Anyoha, R. (2017, August 28). The History of Artificial Intelligence. *Science in the News*. https://sitn.hms.harvard.edu/flash/2017/history-artificial-intelligence/

Arango, L., Singaraju, S. P., & Niininen, O. (2023). Consumer Responses to AI-Generated Charitable Giving Ads. *Journal of Advertising*, *52*(4), 486–503. doi:10.1080/00913367.2023.2183285

Araujo, T., Helberger, N., Kruikemeier, S., & De Vreese, C. H. (2020). In AI we trust? Perceptions about automated decision-making by artificial intelligence. *AI & Society*, *35*(3), 611–623. doi:10.1007/s00146-019-00931-w

Armstrong, S., Bostrom, N., & Shulman, C. (2016). Racing to the precipice: A model of artificial intelligence development. *AI & Society*, *31*(2), 201–206. doi:10.1007/s00146-015-0590-y

Aruna, P. (2023, May 6). Generative AI Applications: Episode 10: In Agriculture. *Medium*. Https://Medium.Com/Arunapattam/Generative-Ai-Applications-Episode-10-in-Agriculture-4ac24b6da8ea

Ascione, L. (2024). Here is how AI can increase equity in student success. *eCampus News.*

Atlantic Council. (2023). Annex 6: Learning from Cybersecurity, Preparing for Generative AI. In *Scaling Trust The On Web* (pp. 1–12). Atlantic Council. https://www.jstor.org/stable/resrep51651.26

Austin, T., Rawal, B. S., Diehl, A., & Cosme, J. (2023). *AI for Equity: Unpacking Potential Human Bias in Decision Making in Higher Education.* AI, Computer Science and Robotics Technology.

Avnoon, N., Kotliar, D. M., & Rivnai-Bahir, S. (2023). Contextualizing the ethics of algorithms: A socio-professional approach. *New Media & Society*, 14614448221145728. doi:10.1177/14614448221145728

Azad, N., Davoudpour, H., Saharidis, G. K. D., & Shiripour, M. (2014). A new model to mitigating random disruption risks of facility and transportation in supply chain network design. *International Journal of Advanced Manufacturing Technology*, *70*(9-12), 1757–1774. doi:10.1007/s00170-013-5404-0

Azmat, S., & Ghaffar, H. (2021). Ethical Commitments and Credit Market Regulations. *Journal of Business Ethics*, *171*(3), 421–433. doi:10.1007/s10551-019-04391-6

Baber, H. (2020). Financial inclusion and FinTech: A comparative study of countries following Islamic finance and conventional finance. *Qualitative Research in Financial Markets*, *12*(1), 24–42. doi:10.1108/QRFM-12-2018-0131

Bailey, J. (2022). Why teachers are worried about AI. *Plagiarism Today.* https://www.plagiarismtoday.com/2022/12/07/why-teachers-are-worried-about-ai/

Bailey, J. (2023). AI in education: The leap into a new era of machine intelligence carries risks and challenges, but also plenty of promise. *Education Next*, *23*(4), 28–35. https://www.educationnext.org/a-i-in-education-leap-into-new-era-machine-intelligence-carries-risks-challenges-promises/

Baker, C., & Kirk-Wade, E. (2023). *Mental health statistics: Prevalence, services and funding in England.* Commons Library. https://commonslibrary.parliament.uk/research-briefings/sn06988/

Baldwin, G., & James, R. (2010). *Access and equity in higher education.* Elsevier. . doi:10.1016/B978-0-08-044894-7.00825-3

Ballesteros, D. M., Rodriguez-Ortega, Y., Renza, D., & Arce, G. (2021). Deep4SNet: Deep learning for fake speech classification. *Expert Systems with Applications*, *184*, 115465. Advance online publication. doi:10.1016/j.eswa.2021.115465

Bandara, R., Fernando, M., & Akter, S. (2020). Privacy concerns in E-commerce: A taxonomy and a future research agenda. *Electronic Markets*, *30*(3), 629–647. doi:10.1007/s12525-019-00375-6

Banerjee, I., Bhattacharjee, K., Burns, J. L., Trivedi, H., Purkayastha, S., Seyyed-Kalantari, L., Patel, B. N., Shiradkar, R., & Gichoya, J. (2023). "Shortcuts" Causing Bias in Radiology Artificial Intelligence: Causes, Evaluation, and Mitigation. *Journal of the American College of Radiology*, *20*(9), 842–851. doi:10.1016/j.jacr.2023.06.025 PMID:37506964

Barbanel, J. (2019, April 29). *A look at the proposed Algorithmic Accountability Act of 2019.* IAPP. https://iapp.org/news/a/a-look-at-the-proposed-algorithmic-accountability-act-of-2019/

Bareis, J., & Katzenbach, C. (2022). Talking AI into being: The narratives and imaginaries of national AI strategies and their performative politics. *Science, Technology & Human Values*, *47*(5), 855–881. doi:10.1177/01622439211030007

BarocasS.SelbstA. D. (2016). Big data's disparate impact. *Social Science Research Network.* SSRN. doi:10.2139/ssrn.2477899

Barredo Arrieta, A., Díaz-Rodríguez, N., Del Ser, J., Bennetot, A., Tabik, S., Barbado, A., Garcia, S., Gil-Lopez, S., Molina, D., Benjamins, R., Chatila, R., & Herrera, F. (2020). Explainable Artificial Intelligence (XAI): Concepts, taxonomies, opportunities and challenges toward responsible AI. *Information Fusion*, *58*, 82–115. doi:10.1016/j.inffus.2019.12.012

Bartneck, C., Lütge, C., Wagner, A., & Welsh, S. (2021). *An Introduction to Ethics in Robotics and AI*. Springer. doi:10.1007/978-3-030-51110-4

Basu, K., Sinha, R., Ong, A., & Basu, T. (2020). Artificial Intelligence: How is It Changing Medical Sciences and Its Future? *Indian Journal of Dermatology*, *65*(5), 365–370. doi:10.4103/ijd.IJD_421_20 PMID:33165420

Beebe, B. (2017). Bleistein, the problem of aesthetic progress, and the making of American Copyright Law. *Columbia Law Review*, *117*(2). https://columbialawreview.org/content/bleistein-the-problem-of-aesthetic-progress-and-the-making-of-american-copyright-law/

Behmer, E.-J., Chandramouli, K., Garrido, V., Mühlenberg, D., Müller, D., Müller, W., Pallmer, D., Pérez, F. J., Piatrik, T., & Vargas, C. (2019). Ontology Population Framework of MAGNETO for Instantiating Heterogeneous Forensic Data Modalities. In J. MacIntyre, I. Maglogiannis, L. Iliadis, & E. Pimenidis (Eds.), *Artificial Intelligence Applications and Innovations* (pp. 520–531). Springer International Publishing. doi:10.1007/978-3-030-19823-7_44

*Beijing Consensus on artificial intelligence and education*. (2019). Paris: UNESCO. https://unesdoc.unesco.org/ark:/48223/pf0000368303

Benbya, H., Davenport, T. H., & Pachidi, S. (2020). Artificial intelligence in organizations: Current state and future opportunities. *MIS Quarterly Executive*, *19*(4).

Benhamou, Y., & Ferland, J. (2020). *Artificial Intelligence & Damages: Assessing Liability and Calculating the Damages* (SSRN Scholarly Paper 3535387). https://papers.ssrn.com/abstract=3535387

Birhane, A. (2021). The impossibility of automating ambiguity. *Artificial Life*, *27*(1), 44–61. doi:10.1162/artl_a_00336 PMID:34529757

Blake, A. (2023). What is Google's Gemini? Everything you need to know about Google's next-gen AI. *Tech Radar*. https://www.techradar.com/computing/artificial-intelligence/what-is-google-gemini

Blinkoff, E., & Hirsh-pasek, K. (2023). *ChatGPT: Educational friend or foe*? Brookings. https://www.brookings.edu/articles/chatgpt-educational-friend-or-foe/

Boddington, G. (2021). The Internet of Bodies—alive, connected and collective: The virtual physical future of our bodies and our senses. *AI & Society*, 1–17. PMID:33584018

Borges, A. F. S., Laurindo, F. J. B., Spínola, M. M., Gonçalves, R. F., & Mattos, C. A. (2021). The strategic use of artificial intelligence in the digital era: Systematic literature review and future research directions. *International Journal of Information Management*, *57*, 102225. doi:10.1016/j.ijinfomgt.2020.102225

Bostrom, N. (2014). *Superintelligence: Paths, Dangers, Strategies*. Oxford University Press.

Bozkurt, A. (2023). Generative artificial intelligence (AI) powered conversational educational agents: The inevitable paradigm shift. *Asian Journal of Distance Education*, *18*(1).

Braun, M., Hummel, P., Beck, S., & Dabrock, P. (2020). Primer on an ethics of AI-based decision support systems in the clinic. *Journal of Medical Ethics*, *47*(12), e3. doi:10.1136/medethics-2019-105860 PMID:32245804

Brey, P. A. E. (2012). Anticipatory Ethics for Emerging Technologies. *NanoEthics*, *6*(1), 1–13. doi:10.1007/s11569-012-0141-7

Brittain, B. (2023, February 6). *Getty Images lawsuit says Stability AI misused photos to train AI*. Reuters. https://www.reuters.com/legal/getty-images-lawsuit-says-stability-ai-misused-photos-train-ai-2023-02-06/

Bryman, A. (2018). *Social Research Methods* (3rd ed.). OUP Oxford.

Brynjolfsson, E., Li, D., & Raymond, L. R. (2023). Generative AI at work (No. w31161). National Bureau of Economic Research.

Budhwar, P., Malik, A., De Silva, M. T. T., & Thevisuthan, P. (2022). Artificial intelligence – challenges and opportunities for international HRM: A review and research agenda. *International Journal of Human Resource Management*, *33*(6), 1065–1097. doi:10.1080/09585192.2022.2035161

Buiten, M., de Streel, A., & Peitz, M. (2023). The law and economics of AI liability. *Computer Law & Security Report*, *48*, 105794. doi:10.1016/j.clsr.2023.105794

Bulgakova, E., Bulgakov, V., Trushchenkov, I., Vasiliev, D. V., & Kravets, E. (2019a). Big Data in Investigating and Preventing Crimes. In *Studies in Systems* (pp. 61–69). Decision and Control. doi:10.1007/978-3-030-01358-5_6

Buolamwini, J. (2018). *Gender Shades: Intersectional accuracy Disparities in commercial gender classification*. PMLR. https://proceedings.mlr.press/v81/buolamwini18a.html.

Buolamwini, J., & Gebru, T. (2018, January). Gender shades: Intersectional accuracy disparities in commercial gender classification. In *Conference on fairness, accountability, and transparency*, (pp. 77-91).

Burrell, J., & Fourcade, M. (2021). The society of algorithms. *Annual Review of Sociology*, *47*(1), 213–237. doi:10.1146/annurev-soc-090820-020800

Calder, R. (2019). How religio-economic projects succeed and fail: The field dynamics of Islamic finance in the Arab Gulf states and Pakistan, 1975–2018. *Socio-economic Review*, *17*(1), 167–193. doi:10.1093/ser/mwz015

Caliskan, A., Bryson, J. J., & Narayanan, A. (2017). *Semantics derived automatically from language corpora contain human-like biases*. ResearchGate. https://www.researchgate.net/publication/316973825

Canhoto, A. I., & Clear, F. (2020). Artificial intelligence and machine learning as business tools: A framework for diagnosing value destruction potential. *Business Horizons*, *63*(2), 183–193. doi:10.1016/j.bushor.2019.11.003

Cardona, M. A., Rodriguez, R. J., & Ishmael, K. (2023). *Artificial Intelligence and the future of teaching and learning: Insights and recommendations*. Department of Education – Office of Educational Technology. https://www2.ed.gov/documents/ai-report/ai-report.pdf

Carrera-Rivera, A., Ochoa, W., Larrinaga, F., & Lasa, G. (2022). How-to conduct a systematic literature review: A quick guide for computer science research. *MethodsX*, *9*, 101895. doi:10.1016/j.mex.2022.101895 PMID:36405369

Cataleta, M. S. (2020). *Humane Artificial Intelligence: The Fragility of Human Rights Facing AI*. East-West Center. https://www.jstor.org/stable/resrep25514

Cath, C. (2018). Governing artificial intelligence: ethical, legal and technical opportunities and challenges. *Philosophical Transactions: Mathematical, Physical and Engineering Sciences, 376*(2133), 1–8. https://www.jstor.org/stable/26601838

Chakrabarti, R., & Sanyal, K. (2020). Towards a 'Responsible AI': Can India take the lead? *South Asia Economic Journal*, *21*(1), 158–177. doi:10.1177/1391561420908728

Chan, C. K. Y. (2023). A comprehensive AI policy education framework for university teaching and learning. *International Journal of Educational Technology in Higher Education*, *20*(38), 1–25. doi:10.1186/s41239-023-00408-3

ChanC. K. Y.HuW. (2023). *Students' voices on Generative AI: Perceptions, benefits, and challenges in higher education.* Cornell University. https://arxiv.org/abs/2305.00290

Chang, C.-Y., Chien, L.-C., Kuo, E.-C., & Hwan, Y.-S. (2019b). Designing Intelligence system of Image Processing and Mining in Cloud-Example of New Taipei City Police Department. *IEEE International Conference on Knowledge Innovation and Invention*, (pp. 293–295). IEEE. 10.1109/ICKII46306.2019.9042641

Chase, J., Du, J., Fu, N., Le, T. V., & Lau, H. C. (2017). Law enforcement resource optimization with response time guarantees. *2017 IEEE Symposium Series on Computational Intelligence (SSCI)*, (pp. 1–7). IEEE. 10.1109/SSCI.2017.8285326

Chaudhry, M. A., & Kazim, E. (2022). Artificial Intelligence in Education (AIEd): A high-level academic and industry note 2021. *AI and Ethics*, *2*(1), 157–165. doi:10.1007/s43681-021-00074-z PMID:34790953

Checco, A., Bracciale, L., Loreti, P., Pinfield, S., & Bianchi, G. (2021). AI-assisted peer review. *Humanities & Social Sciences Communications*, *8*(1), 25. doi:10.1057/s41599-020-00703-8

Cheddadi, S., & Bouache, M. (2021, August). Improving equity and access to higher education using artificial intelligence. In *2021 16th International Conference on Computer Science & Education (ICCSE)* (pp. 241-246). IEEE. 10.1109/ICCSE51940.2021.9569548

Chen, L., & Wong, G. (2018). *Transcriptome Informatics*. Science Direct. doi:10.1016/B978-0-12-809633-8.20204-5

Chen, C., Lin, K., Rudin, C., Shaposhnik, Y., Wang, S., & Wang, T. (2022). A holistic approach to interpretability in financial lending: Models, visualizations, and summary-explanations. *Decision Support Systems*, *152*, 113647. Advance online publication. doi:10.1016/j.dss.2021.113647

Cheng, L., Varshney, K. R., & Liu, H. (2021). Socially responsible AI algorithms: Issues, purposes, and challenges. *Journal of Artificial Intelligence Research*, *71*, 1137–1181. doi:10.1613/jair.1.12814

Chen, J., Sun, J., & Wang, G. (2022). From unmanned systems to autonomous intelligent systems. *Engineering (Beijing)*, *12*, 16–19. doi:10.1016/j.eng.2021.10.007

Chin, C. (2023). *Navigating the Risks of Artificial Intelligence on the Digital News Landscape*. Center for Strategic and International Studies (CSIS). https://www.jstor.org/stable/resrep53077

Choi, R. Y., Coyner, A. S., Kalpathy-Cramer, J., Chiang, M. F., & Campbell, J. P. (2020). Introduction to Machine Learning, Neural Networks, and Deep Learning. *Translational Vision Science & Technology*, *9*(2), 14. doi:10.1167/tvst.9.2.14 PMID:32704420

Chopra, A. (2017, March 10). *How far has India come in its digitization journey?* Mint. https://www.livemint.com/Industry/nAIcrfPTv5G1yGLGQ54LzN/EmTech-India-2017-How-far-has-India-come-in-its-digitizatio.html

Choudhary, V. & Ali, Aamir S. M. (2023). ChatGPT and Copyright Concerns. *Economic and Political Weekly*, *58*(16), 4–5.

Chowdhury, J. (2012). Hacking Health: Bottom-up Innovation for Healthcare. *Technology Innovation Management Review*, 31–35.

Chowdhury, M. A. F., Sultan, Y., & Mahmudul Haque, M. (2020). Conventional futures: A review of major issues from the Islamic finance perspective. *International Journal of Pluralism and Economics Education*, *11*(2), 201–210. doi:10.1504/IJPEE.2020.111258

Chui, M., Hazan, E., Roberts, R., Singla, A., & Smaje, K. (2023). *The economic potential of generative AI.*

Clelia Casciola, C. (2022). Artificial Intelligence and Health Care: Reviewing the Algorithmic Accountability Act in Light of the European Artificial Intelligence Act. *Vermont Law Review*, *47*(1), 127–155. https://lawreview.vermontlaw.edu/wp-content/uploads/2023/03/06_Casciola_Book1_Final-copy.pdf

Coffey, L. (2023). *Students outrunning faculty in AI use*. Inside Higher Ed. https://www.insidehighered.com/news/tech-innovation/artificial-intelligence/2023/10/31/most-students-outrunning-faculty-ai-use

Colaner, N. (2022). Is explainable artificial intelligence intrinsically valuable? *AI & Society*, 1–8.

Collins, N. (2011). Trading Faures: Virtual Musicians and Machine Ethics. *Leonardo Music Journal*, *21*, 35–39. https://www.jstor.org/stable/41416821. doi:10.1162/LMJ_a_00059

Complete College America. (2023). *Attainment with AI: Making a real difference in college completion with Artificial Intelligence*. Complete College America. https://completecollege.org/wp-content/uploads/2023/11/CCA_AttainmentwithAI.pdf

Complete College America. (2023). *The AI Divide: Equitable Application of AI in Higher Education to advance the completion Agenda*. Complete College. https://completecollege.org/wp-content/uploads/2023/11/CCA_The_AI_Divide.pdf

Contardo, P., Sernani, P., Falcionelli, N., & Dragoni, A. F. (2021b, July 16). Deep learning for law enforcement: A survey about three application domains. *RTA-CSIT 2021*. IEEE.

Cools, H., Van Gorp, B., & Opgenhaffen, M. (2022). Where exactly between utopia and dystopia? A framing analysis of AI and automation in US newspapers. *Journalism*, •••, 14648849221122647.

Cooper, G. (2023). Examining science education in chatgpt: An exploratory study of generative artificial intelligence. *Journal of Science Education and Technology*, *32*(3), 444–452. doi:10.1007/s10956-023-10039-y

Cornell University Center for Teaching Innovation. (2023). *Ethical AI for teaching and learning*. Cornell University. https://teaching.cornell.edu/generative-artificial-intelligence/ethical-ai-teaching-and-learning

Corrigan, J. (2019, April 11). *Lawmakers Introduce Bill to Curb Algorithmic Bias*. Nextgov.Com. https://www.nextgov.com/artificial-intelligence/2019/04/lawmakers-introduce-bill-curb-algorithmic-bias/156237/

Cossette-Lefebvre, H., & Maclure, J. (2023). AI's fairness problem: Understanding wrongful discrimination in the context of automated decision-making. *AI and Ethics*, *3*(4), 1255–1269. doi:10.1007/s43681-022-00233-w

Crawford, K., & Calo, R. (2016). There is a blind spot in AI research. *Nature*, *538*(7625), 311–313. doi:10.1038/538311a PMID:27762391

D'agostino, S. (2023). *How AI tools both help and hinder equity in higher ed*. Inside higher Ed. https://www.insidehighered.com/news/tech-innovation/artificial-intelligence/2023/06/05/how-ai-tools-both-help-and-hinder-equity

D'Agostino, S. (2023). *How AI tools both help and hinder equity*. Inside Higher Ed. https://www.insidehighered.com/news/tech-innovation/artificial-intelligence/2023/06/05/how-ai-tools-both-help-and-hinder-equity

Dachs, B., Amoroso, S., Castellani, D., Papanastassiou, M., & von Zedtwitz, M. (2023). The internationalisation of R&D: Past, present and future. *International Business Review*. doi:10.1016/j.ibusrev.2023.102191

Daly, A., Devitt, S. K., & Mann, M. (2021). AI Ethics Needs Good Data. In P. Verdegem (Ed.), *AI for Everyone?: Critical Perspectives* (pp. 103–122). University of Westminster Press. https://www.jstor.org/stable/j.ctv26qjjhj.9 doi:10.16997/book55.g

Darwiesh, A., El-Baz, A. H., Abualkishik, A. Z., & Elhoseny, M. (2023). Artificial Intelligence Model for Risk Management in Healthcare Institutions: Towards Sustainable Development. *Sustainability (Basel)*, *15*(1), 420. doi:10.3390/su15010420

Das, P., & Das, A. (2019). *Application of Classification Techniques for Prediction and Analysis of Crime in India.*, doi:10.1007/978-981-10-8055-5_18

Dastin, J. (2018). *Amazon scraps secret AI recruiting tool that showed bias against women.* Reuters. https://www.reuters.com/article/us-amazon-com-jobs-automation-insight/amazon-scraps-secret-ai-recruiting-tool-that-showed-bias-against-women-idUSKCN1MK08G

Dastin, J. (2018, October 9). Amazon scraps secret AI recruiting tool that showed bias against women. *Reuters.* https://www.reuters.com/article/idUSKCN1MK0AG/

Dastin, J. (2022). Amazon Scraps Secret AI Recruiting Tool that Showed Bias against Women. In *Ethics of Data and Analytics.* Auerbach Publications. doi:10.1201/9781003278290-44

Dauvergne, P. (2021). The globalization of artificial intelligence: Consequences for the politics of environmentalism. *Globalizations, 18*(2), 285–299. doi:10.1080/14747731.2020.1785670

Davenport, T. H., & Nitin, M. (2022, November 14). *How Generative AI Is Changing Creative Work.* HBR. Https://Hbr.Org/2022/11/How-Generative-Ai-Is-Changing-Creative-Work

Davenport, T., & Kalakota, R. (2019). The potential for artificial intelligence in healthcare. *Future Healthcare Journal, 6*(2), 94–98. doi:10.7861/futurehosp.6-2-94 PMID:31363513

Davies, L. J. P., & Howard, R. M. (2016). Plagiarism and the Internet: Fears, Facts, and Pedagogies. In T. Bretag (Ed.), *Handbook of Academic Integrity* (pp. 591–606). Springer., doi:10.1007/978-981-287-098-8_16

Dawson, D., Schleiger, E., Horton, J., McLaughlin, J., Robinson, C., Quezada, G., Scowcroft, J., & Hajkowicz, S. (2019). *Artificial Intelligence: Australia's ethics framework - a discussion paper.* Commonwealth Scientific and Industrial Research Organisation, Department of Industry, Innovation and Science (Australia). https://apo.org.au/node/229596

De Angelis, L., Baglivo, F., Arzilli, G., Privitera, G. P., Ferragina, P., Tozzi, A. E., & Rizzo, C. (2023). ChatGPT and the rise of large language models: The new AI-driven infodemic threat in public health. *Frontiers in Public Health, 11,* 1567. doi:10.3389/fpubh.2023.1166120 PMID:37181697

De Bruijn, H., Warnier, M., & Janssen, M. (2022). The perils and pitfalls of explainable AI: Strategies for explaining algorithmic decision-making. *Government Information Quarterly, 39*(2), 101666. doi:10.1016/j.giq.2021.101666

Deac, A. (2018). Regulation (Eu) 2016/679 Of The European Parliament And Of The Council On The Protection Of Individuals With Regard To The Processing Of Personal Data And The Free Movement Of These Data. *Perspectives of Law and Public Administration, 7*(2), 151–156. https://ideas.repec.org//a/sja/journl/v7y2018i2p151-156.html

Deep, D. (2023, July 10). *The Transformative Power of Generative AI in Education Technology.* Express Computer. Https://Www.Expresscomputer.in/Artificial-Intelligence-Ai/the-Transformative-Power-of-Generative-Ai-in-Education-Technology/100809/

Department for Digital. Cultural Media and Sport (DCMS), (2019, March 20). Investigation launched into potential for #AI bias in algorithmic decision-making in society. *FE News.* https://www.fenews.co.uk/skills/investigation-launched-into-potential-for-ai-bias-in-algorithmic-decision-making-in-society/

Dewey, J. (2016). *Ethics.* Read Books Ltd.

Dixit, P. (2023, March 29). This chatbot will use the n-word and teach you how to build a bomb. *BuzzFeed News.* https://www.buzzfeednews.com/article/pranavdixit/freedomgpt-ai-chatbot-test

Dolan, D., & Yasin, E. (2023). *A guide to generative AI policy making*. Inside Higher Ed. https://www.insidehighered.com/views/2023/03/22/ai-policy-advice-administrators-and-faculty-opinion

Döring, H. (2019). Public perceptions of the proper role of the state. In *The State in Western Europe Retreat or Redefinition?* (pp. 12–31). Routledge. doi:10.4324/9781315037479-2

Duffourc, M. N., & Gerke, S. (2023). The proposed EU Directives for AI liability leave worrying gaps likely to impact medical AI. *NPJ Digital Medicine*, *6*(1), 1. doi:10.1038/s41746-023-00823-w PMID:37100860

Du, H., Xu, Z., Yan, Z., & Gao, S. (2018). *Intelligent Video Analysis Technology of Public Security Standard Sets of Data and Measurements.*, doi:10.1007/978-981-10-7398-4_47

Duivenvoorde, B. (2022). The Liability of Online Marketplaces under the Unfair Commercial Practices Directive, the E-commerce Directive and the Digital Services Act. *Journal of European Consumer and Market Law, 11*(2). https://kluwerlawonline.com/api/Product/CitationPDFURL?file=Journals\EuCML\EuCML2022009.pdf

Dul, C. (2022). Facial Recognition Technology vs Privacy: The Case of Clearview AI. *QMLJ, 1.*

Durán, J. M., & Jongsma, K. R. (2021). Who is afraid of black box algorithms? On the epistemological and ethical basis of trust in medical AI. *Journal of Medical Ethics*. doi:10.1136/medethics-2020-106820

Du, S., & Xie, C. (2021). Paradoxes of artificial intelligence in consumer markets: Ethical challenges and opportunities. *Journal of Business Research, 129*, 961–974. doi:10.1016/j.jbusres.2020.08.024

Dwivedi, Y. K., Hughes, L., Ismagilova, E., Aarts, G., Coombs, C., Crick, T., Duan, Y., Dwivedi, R., Edwards, J., Eirug, A., Galanos, V., Ilavarasan, P. V., Janssen, M., Jones, P., Kar, A. K., Kizgin, H., Kronemann, B., Lal, B., Lucini, B., & Williams, M. D. (2021). Artificial Intelligence (AI): Multidisciplinary perspectives on emerging challenges, opportunities, and agenda for research, practice and policy. *International Journal of Information Management, 57*, 101994. doi:10.1016/j.ijinfomgt.2019.08.002

Ebers, M., Hoch, V. R. S., Rosenkranz, F., Ruschemeier, H., & Steinrötter, B. (2021). The European Commission's Proposal for an Artificial Intelligence Act—A Critical Assessment by Members of the Robotics and AI Law Society (RAILS). *J, 4*(4), 4. doi:10.3390/j4040043

Edelson, L. (2022, April 29). *Platform Transparency Legislation: The Whos, Whats and Hows*. Lawfare. https://www.lawfaremedia.org/article/platform-transparency-legislation-whos-whats-and-hows

Ellen, G., & Matthew, U. (2023, June 26). *15 Popular AI Video Generators*. BuiltIn. Https://Builtin.Com/Artificial-Intelligence/Ai-Video-Generator

Elliott, A. (2019). *The culture of AI: Everyday life and the digital revolution*. Routledge. doi:10.4324/9781315387185

Engelke, P. (2020). *AI, Society, and Governance: An Introduction*. Atlantic Council. https://www.jstor.org/stable/resrep29327

Enriquez, F., Soria Morillo, L., Alvarez-Garcia, J., Caparrini, F., Velasco-Morente, F., Deniz, O., & Vállez, N. (2019). *Vision and Crowdsensing Technology for an Optimal Response in Physical-Security.*, doi:10.1007/978-3-030-22750-0_2

Epstein, Z., Hertzmann, A., Akten, M., Farid, H., Fjeld, J., Frank, M. R., Groh, M., Herman, L., Leach, N., Mahari, R., Pentland, A. S., Russakovsky, O., Schroeder, H., & Smith, A. (2023). Art and the science of generative AI. *Science, 380*(6650), 1110–1111. doi:10.1126/science.adh4451 PMID:37319193

EU Agency for Fundamental Rights. (2018). *Fundamental Rights and Artificial Intelligence: Key Fundamental Rights considerations for the future regulatory framework*. FRA. https://fra.europa.eu/sites/default/files/fra_uploads/fra-2018-focus-ai-rights_en.pdf

European Commission. (2019, April 8). *Ethics guidelines for trustworthy AI.* EC. https://digital-strategy.ec.europa.eu/en/library/ethics-guidelines-trustworthy-ai

European Commission. (2021). *Proposal for a regulation of the European Parliament and of the Council laying down harmonized rules on artificial intelligence (artificial intelligence act) and amending certain union legislative acts.* European Commission. https://eur-lex.europa.eu/resource.html?uri=cellar:e0649735-a372-11eb-9585-01aa75ed71a1.0001.

Extance, A. (2023). ChatGPT has entered the classroom: How LLMs could transform education. *Nature*. https://www.nature.com/articles/d41586-023-03507-3

Faghani, S., Khosravi, B., Zhang, K., Moassefi, M., Jagtap, J. M., Nugen, F., Vahdati, S., Kuanar, S. P., Rassoulinejad-Mousavi, S. M., Singh, Y., Vera Garcia, D. V., Rouzrokh, P., & Erickson, B. J. (2022). Mitigating Bias in Radiology Machine Learning: 3. Performance Metrics. *Radiology. Artificial Intelligence*, *4*(5), e220061. doi:10.1148/ryai.220061 PMID:36204539

Farrow, E. (2021). Mindset matters: How mindset affects the ability of staff to anticipate and adapt to Artificial Intelligence (AI) future scenarios in organisational settings. *AI & Society*, *36*(3), 895–909. doi:10.1007/s00146-020-01101-z PMID:33223620

Federspiel, F., Mitchell, R., Asokan, A., Umana, C., & McCoy, D. (2023). Threats by artificial intelligence to human health and human existence. *BMJ Global Health*, *8*(5), e010435. doi:10.1136/bmjgh-2022-010435 PMID:37160371

Fernandes, M., Vieira, S. M., Leite, F., Palos, C., Finkelstein, S., & Sousa, J. M. C. (2020). Clinical Decision Support Systems for Triage in the Emergency Department using Intelligent Systems: A Review. *Artificial Intelligence in Medicine*, *102*, 101762. doi:10.1016/j.artmed.2019.101762 PMID:31980099

Feuerriegel, S., Hartmann, J., & Janiesch, C. (2023). Generative AI. *Business & Information Systems Engineering*. doi:10.1007/s12599-023-00834-7

Finkel, M., & Krämer, N. C. (2022). Humanoid robots–artificial. Human-like. Credible? Empirical comparisons of source credibility attributions between humans, humanoid robots, and non-human-like devices. *International Journal of Social Robotics*, *14*(6), 1397–1411. doi:10.1007/s12369-022-00879-w

Finlayson-Brown, J., & Bossotto, L. (2021, August 9). *Italian data protection supervisory authority fines two food delivery companies for non-compliant algorithmic processing.* Allen Overy. https://www.allenovery.com/en-gb/global/blogs/data-hub/italian-data-protection-supervisory-authority-fines-two-food-delivery-companies-for-non-compliant-algorithmic-processing

Fischer, I., Mirbahai, L., Beer, L., Buxton, D., Grierson, S., Griffin, L., & Gupta, N. (2023). *Transforming Higher Education: How we can harness AI in teaching and assessments and uphold academic rigour and integrity.* Warwick International Higher Education Academy (WIHEA), University of Warwick. https://warwick.ac.uk/fac/cross_fac/academy/activities/learningcircles/future-of-learning/ai__education_12-7-23.pdf

Floridi, L., & Cowls, J. (2022). A unified framework of five principles for AI in society. *Machine learning and the city: Applications in architecture and urban design*, 535-545

Floridi, L., & Strait, A. (2020). Ethical Foresight Analysis: What it is and Why it is Needed? *Minds and Machines*, *30*(1), 77–97. doi:10.1007/s11023-020-09521-y

Flowerdew, J., & Li, Y. (2007). Plagiarism and second language writing in an electronic age. *Annual Review of Applied Linguistics*, *27*, 161–183. doi:10.1017/S0267190508070086

Fontes, C., Hohma, E., Corrigan, C. C., & Lütge, C. (2022). AI-powered public surveillance systems: Why we (might) need them and how we want them. *Technology in Society*, *71*, 102137. doi:10.1016/j.techsoc.2022.102137

Fortney, C., & Steward, D. (2017). A Qualitative study of nurse observation of symptoms in infants at end-of-life in the neonatal intensive care unit. *Intensive & Critical Care Nursing*, *40*, 57–68. doi:10.1016/j.iccn.2016.10.004 PMID:28189383

Foster, D. (2022). *Generative deep learning*. O'Reilly Media, Inc.

Franke, U. (2019). *Harnessing Artificial Intelligence*. European Council on Foreign Relations. https://www.jstor.org/stable/resrep21491

Frankenfield, J. (2021). *Artificial Intelligence: What It Is and How It Is Used*. Investopedia. https://www.investopedia.com/terms/a/artificial-intelligence-ai.asp

Fuchs, K. (2023). *Exploring the opportunities and challenges of NLP models in higher education: Is Chat GPT a blessing or a curse?* Frontiers. https://www.frontiersin.org/articles/10.3389/feduc.2023.1166682/full

Füller, J., Hutter, K., Wahl, J., Bilgram, V., & Tekic, Z. (2022). How AI revolutionizes innovation management – Perceptions and implementation preferences of AI-based innovators. *Technological Forecasting and Social Change*, *178*, 121598. doi:10.1016/j.techfore.2022.121598

Gabison, G. (2016). Policy considerations for the blockchain technology public and private applications. *SMU Sci. & Tech. L. Rev.*, *19*, 327.

Ganapini, M. B., Campbell, M., Fabiano, F., Horesh, L., Lenchner, J., Loreggia, A., & Venable, K. B. (2023). Thinking fast and slow in AI: The role of metacognition. In *Machine Learning, Optimization, and Data Science: 8th International Conference, LOD 2022,* (pp. 502-509). Cham: Springer Nature Switzerland.

Gao, B. (2022). The Use of Machine Learning Combined with Data Mining Technology in Financial Risk Prevention. *Computational Economics*, *59*(4), 1385–1405. doi:10.1007/s10614-021-10101-0

GaoC. A.HowardF. M.MarkovN. S.DyerE. C.RameshS.LuoY.PearsonA. T. (2022). Comparing scientific abstracts generated by ChatGPT to original abstracts using an artificial intelligence output detector, plagiarism detector, and blinded human reviewers (p. 2022.12.23.521610). bioRxiv. doi:10.1101/2022.12.23.521610

García-Peñalvo, F. J. (2023). The perception of Artificial Intelligence in educational contexts after the launch of ChatGPT: Disruption or panic? *Education in the Knowledge Society*, *24*, 1–9. doi:10.14201/eks.31279

Gartner. (2023). *Generative AI isn't just a technology or a business case*. Gartner. Https://Www.Gartner.Com/En/Topics/Generative-Ai

*Generative AI global weekly search trends on Google 2023*. (2023). Statista. Https://Www.Statista.Com/Statistics/1367868/Generative-Ai-Google-Searches-Worldwide/. https://www.statista.com/statistics/1367868/generative-ai-google-searches-worldwide/

George, L. (2023). *Generative AI Ethics: 8 Biggest Concerns and Risks*. Tech Target. https://www.techtarget.com/searchenterpriseai/tip/Generative-AI-ethics-8-biggest-concerns

Gerlick, J. A., & Liozu, S. M. (2020). Ethical and legal considerations of artificial intelligence and algorithmic decision-making in personalized pricing. *Journal of Revenue and Pricing Management*, *19*(2), 85–98. doi:10.1057/s41272-019-00225-2

Ghose, A. & Ali, Aamir S. M. (2023). Amplifying Music with Artificial Intelligence. *Economic and Political Weekly*, *58*(17), 4–6.

Ghotbi, N. (2023). The ethics of emotional Artificial Intelligence: A mixed method analysis. *Asian Bioethics Review, 15*(4), 417–430. doi:10.1007/s41649-022-00237-y PMID:37808444

Gil, R., Virgili-Gomà, J., López-Gil, J.-M., & García, R. (2023). Deepfakes: Evolution and trends. *Soft Computing, 27*(16), 11295–11318. doi:10.1007/s00500-023-08605-y

Gipson Rankin, S. M. (2021). Technological tethereds: Potential impact of untrustworthy artificial intelligence in criminal justice risk assessment instruments. *Washington and Lee Law Review, 78*, 647.

Godinho, I., Flores, C., & Marques, N. (2021). CONSULTATION ON THE WHITE PAPER ON ARTIFICIAL INTELLIGENCE - A EUROPEAN APPROACH. *ULP Law Review, 14*, 157–167. doi:10.46294/ulplr-rdulp.v14i1.7475

Goh, Y., & Leon, N. R. (2020). The innovation of Singapore's AI ethics model framework. In L. Hui & B. Tse (Eds.), *AI Governance in 2019: A Year in Review (Observations of 50 Experts in the World* (pp. 77–78). Shanghai Institute for Science of Science.

González, M. O., Jareño, F., & El Haddouti, C. (2019). Sector portfolio performance comparison between Islamic and conventional stock markets. *Sustainability (Basel), 11*(17), 4618. doi:10.3390/su11174618

Google DeepMind. (2023). *Welcome to the Gemini era*. Google. https://deepmind.google/technologies/gemini/#introduction

Google. (2023). *Hands-on with Gemini: Interacting with multimodal AI* [Video]. YouTube. https://www.youtube.com/watch?v=UIZAiXYceBI

Google. (2023). *NotebookLM onboarding and FAQ*. Internal training document.

Goralski, M. A., & Tan, T. K. (2020). Artificial intelligence and sustainable development. *International Journal of Management Education, 18*(1), 100330. doi:10.1016/j.ijme.2019.100330

Goutzamanis, Y. (2023). Closing the Floodgates on Privacy Class Actions: Lloyd v Google LLC. *The Modern Law Review, 86*(1), 249–262. doi:10.1111/1468-2230.12744

Gozalo-BrizuelaR.Garrido-MerchánE. C. (2023). *A survey of Generative AI Applications*. http://arxiv.org/abs/2306.02781

Gu, G., & Zhu, W. (2021). Time-varying transmission effects of internet finance under economic policy uncertainty and internet consumers' behaviors: Evidence from China. *Journal of Advanced Computational Intelligence and Intelligent Informatics, 25*(5), 554–562. doi:10.20965/jaciii.2021.p0554

Gulbahar, K. (2023, April 10). *Speech Recognition: Everything You Need to Know in 2023*. AI Multiple. Https://Research.Aimultiple.Com/Speech-Recognition/

Guo, J., Kang, W., & Wang, Y. (2023). Option pricing under sub-mixed fractional Brownian motion based on time-varying implied volatility using intelligent algorithms. *Soft Computing, 27*(20), 15225–15246. doi:10.1007/s00500-023-08647-2

Gupta, M., Parra, C. M., & Dennehy, D. (2022). Questioning racial and gender bias in AI-based recommendations: Do espoused national cultural values matter? *Information Systems Frontiers, 24*(5), 1465–1481. doi:10.1007/s10796-021-10156-2 PMID:34177358

Hacker, P. (2022). *The European AI Liability Directives – Critique of a Half-Hearted Approach and Lessons for the Future* (*SSRN* Scholarly Paper 4279796). doi:10.2139/ssrn.4279796

HackerP. (2023). What's Missing from the EU AI Act: Addressing the Four Key Challenges of Large Language Models. *Verfassungsblog*. doi:10.17176/20231214-111133-0

Hajian, S., Bonchi, F., & Castillo, C. (2016). Algorithmic Bias. *Algorithmic Bias: From Discrimination Discovery to Fairness-aware Data Mining*. ACM. . doi:10.1145/2939672.2945386

Haluza, D., & Jungwirth, D. (2023). Artificial Intelligence and Ten Societal Megatrends: An Exploratory Study Using GPT-3. *Systems*, *11*(3), 120. doi:10.3390/systems11030120

Hamet, P., & Tremblay, J. (2017). Artificial intelligence in medicine. *Metabolism: Clinical and Experimental*, *69S*, S36–S40. doi:10.1016/j.metabol.2017.01.011 PMID:28126242

Hao, K. (2019, February 4). This is how AI bias really happens – and why it's so hard to fix. *MIT Technology Review*. https://www.technologyreview.com/2019/02/04/137602/this-is-how-ai-bias-really-happensand-why-its-so-hard-to-fix/

Harish, K. (2023, July 20). *Key Use Cases for Generative AI in Media and Entertainment industry*. LinkedIn. Https://Www.Linkedin.Com/Pulse/Key-Use-Cases-Generative-Ai-Media-Entertainment-Harish-Kotadia

Harper, R. H. (2021). AI: The social future of intelligence. In *Routledge Handbook of Social Futures* (pp. 52–58). Routledge. doi:10.4324/9780429440717-4

Hatherley, J. J. (2020). Limits of trust in medical AI. *Journal of Medical Ethics*, *46*(7), 478–481. doi:10.1136/medethics-2019-105935 PMID:32220870

Haug, C. J., & Drazen, J. M. (2023). Artificial intelligence and machine learning in clinical medicine, 2023. *The New England Journal of Medicine*, *388*(13), 1201–1208. doi:10.1056/NEJMra2302038 PMID:36988595

Haugh, B. A., Kaminski, N. J., Madhavan, P., McDaniel, E. A., Pavlak, C. R., Sparrow, D. A., Tate, D. M., & Williams, B. L. (2018). Strategy 3 Proposed Changes. In *RFI Response: National Artificial Intelligence Research and Development Strategic Plan* (pp. 3–6). Institute for Defense Analyses. https://www.jstor.org/stable/resrep22865.6

Hawking, S. (2016). *The best or worst thing to happen to humanity. The launch of the Leverhulme Centre for the Future of Intelligence*. CAM. https://www.cam.ac.uk/research/news/the-best-or-worst-thing-to-happen-to-humanity-stephen-hawking-launches-centre-for-the-future-of

Helm, J. M., Swiergosz, A. M., Haeberle, H. S., Karnuta, J. M., Schaffer, J. L., Krebs, V. E., Spitzer, A. I., & Ramkumar, P. N. (2020). Machine learning and artificial intelligence: Definitions, applications, and future directions. *Current Reviews in Musculoskeletal Medicine*, *13*(1), 69–76. doi:10.1007/s12178-020-09600-8 PMID:31983042

Henin, C., & Le Métayer, D. (2021). Beyond explainability: Justifiability and contestability of algorithmic decision systems. *AI & Society*, 1–14.

Henley, T. B. (1990). Natural Problems and Artificial Intelligence. *Behavior and Philosophy*, *18*(2), 43–56. https://www.jstor.org/stable/27759223

He, Q., Zheng, H., Ma, X., Wang, L., Kong, H., & Zhu, Z. (2022). Artificial intelligence application in a renewable energy-driven desalination system: A critical review. *Energy and AI*, *7*, 100123. doi:10.1016/j.egyai.2021.100123

Herrmann, H. (2023). What's next for responsible artificial intelligence: A way forward through responsible innovation. *Heliyon*, *9*(3), e14379. doi:10.1016/j.heliyon.2023.e14379 PMID:36967876

Hickman, E., & Petrin, M. (2021). Trustworthy AI and corporate governance: The EU's ethics guidelines for trustworthy artificial intelligence from a company law perspective. *European Business Organization Law Review*, *22*(4), 593–625. doi:10.1007/s40804-021-00224-0

Hie, A., & Thouary, C. (2023). *How AI is reshaping higher education*. AACSB. https://www.aacsb.edu/insights/articles/2023/10/how-ai-is-reshaping-higher-education

Hlávka, J. P. (2020). Security, privacy, and information-sharing aspects of healthcare artificial intelligence. In *Artificial Intelligence in Healthcare* (pp. 235–270). Academic Press. doi:10.1016/B978-0-12-818438-7.00010-1

Hoffman, R. (2023). *AI in the classroom: The potential drawbacks and benefits of AI in education.* Greylock. https://greylock.com/greymatter/ai-in-the-classroom/

Holmes, W., Porayska-Pomsta, K., Holstein, K., Sutherland, E., Baker, T., Shum, S. B., ... Koedinger, K. R. (2021). Ethics of AI in education: Towards a community-wide framework. *International Journal of Artificial Intelligence in Education*, 1–23.

Hong, J.-W., & Williams, D. (2019). Racism, Responsibility and Autonomy in HCI: Perceptions of an AI Agent. *Computers in Human Behavior, 100*, 79–84. doi:10.1016/j.chb.2019.06.012

Honigsfeld, A., & Dove, M. G. (2023). *5 collaborative teaching practices for teacher learning.* TESOL International Association. https://www.tesol.org/blog/posts/5-collaborative-teaching-practices-for-teacher-learning

Hoofnagle, C. J., Kesari, A., & Perzanowski, A. (2019). The Tethered Economy. *Geo. Wash. L. Rev., 87*, 783.

How, M. L., Cheah, S. M., Chan, Y. J., Khor, A. C., & Say, E. M. P. (2020). Artificial intelligence-enhanced decision support for informing global sustainable development: A human-centric AI-thinking approach. *Information (Basel), 11*(1), 39. doi:10.3390/info11010039

Hrastinski, S., Olofsson, A. D., Arkenback, C., Ekström, S., Ericsson, E., Fransson, G., Jaldemark, J., Ryberg, T., Öberg, L., Fuentes, A., Gustafsson, U., Humble, N., Mozelius, P., Sundgren, M., & Utterberg, M. (2019). Critical imaginaries and reflections on artifcial intelligence and robots in post-digital K-12 education. *Post Digit. Science Education., 1*(2), 427–445. doi:10.1007/s42438-019-00046-x

Huang, R., Liu, D., Chen, Y., Adarkwah, M. A., Zhang, X. L., Xiao, G. D., Li, X., Zhang, J. J., & Da, T. (2023). *Learning for All with AI? 100 Influential Academic Articles of Educational Robots.* Beijing: Smart Learning Institute of Beijing Normal University. https://sli.bnu.edu.cn/uploads/soft/230413/1_1744328351.pdf

Huang, S., & Siddarth, D. (2023). Generative AI and the digital commons. *arXiv preprint arXiv:2303.11074.*

Huang, J., Li, J., Huang, X., & Cai, W. (2020). Ethical considerations in artificial intelligence and healthcare. *Journal of Healthcare Engineering, 2020*, 1–10. oi:10.1155/2020/8845224

Humble, K. (2023). Artificial Intelligence, International Law and the Race for Killer Robots in Modern Warfare. In *Artificial Intelligence, Social Harms and Human Rights* (pp. 57–76). Springer International Publishing. doi:10.1007/978-3-031-19149-7_3

Humerick, M. (2018). Taking AI Personally: How the E.U. Must Learn to Balance the Interests of Personal Data Privacy & Artificial Intelligence. *Santa Clara High-Technology Law Journal, 34*(4), 393. https://digitalcommons.law.scu.edu/chtlj/vol34/iss4/3

Hurel, L. M. (2018). *Architectures of security and power: IoT platforms as technologies of government.* [MSc diss., London School of Economics and Political Science].

Hwang, H., & Park, M. H. (2020). The Threat of AI and Our Response: The AI Charter of Ethics in South Korea. *Asian Journal of Innovation & Policy, 9*(1).

Hyden, H. (2022). Regulation of AI: Problems and Options. *The Swedish Law and Informatics Research Institute*, (pp. 295–314). Law Pub. doi:10.53292/208f5901.9118259e

IBM. (2024). *What are AI hallucinations?* IBM. https://www.ibm.com/topics/ai-hallucinations

Ibrahim, M. H. (2015). Issues in Islamic banking and finance: Islamic banks, Shari'ah-compliant investment and sukuk. *Pacific-Basin Finance Journal, 34*, 185–191. doi:10.1016/j.pacfin.2015.06.002

Igna, I., & Venturini, F. (2023). The determinants of AI innovation across European firms. *Research Policy, 52*(2), 104661. doi:10.1016/j.respol.2022.104661

Ionescu, B., Ghenescu, M., Răstoceanu, F., Roman, R., & Buric, M. (2020). Artificial Intelligence Fights Crime and Terrorism at a New Level. *IEEE MultiMedia, 27*(2), 55–61. doi:10.1109/MMUL.2020.2994403

Islam, S. R., Russell, I., Eberle, W., & Dicheva, D. (2022). Instilling conscience about bias and fairness in automated decisions. *Journal of Computing Sciences in Colleges, 37*(8), 22–31.

Jaar, D., & Zeller, P. E. (2009). Canadian Privacy Law: The Personal Information Protection and Electronic Documents Act (PIPEDA). *International In-house Counsel Journal, 2*(7), 1135-1146. https://www.iicj.net/subscribersonly/09june/iicj4jun-dataprotection-patrickzeller-guidancesoftware-USA.pdf

Jack, C. (2023, August 15). *Generative-AI.* Scribbr. Https://Www.Scribbr.Com/Ai-Tools/Generative-Ai/

Jacobs, M., & Simon, J. (2023). Reexamining computer ethics in light of AI systems and AI regulation. *AI and Ethics, 3*(4), 1203–1213. doi:10.1007/s43681-022-00229-6

Jagreet, K. (2023, July 6). *Generative AI in Education Industry | Benefits and Future Trends.* Xenon Stack. Https://Www.Xenonstack.Com/Blog/Generative-Ai-Education#:~:Text=Educators%20can%20use%20generative%20AI,%2C%20motivation%2C%20and%20academic%20achievement

Jie, Y., Liu, C. Z., Li, M., Choo, K.-K. R., Chen, L., & Guo, C. (2020). Game theoretic resource allocation model for designing effective traffic safety solution against drunk driving. *Applied Mathematics and Computation, 376*, 125142. doi:10.1016/j.amc.2020.125142

Jindal, S., & Sharma, K. (2018a). Intend to analyze Social Media feeds to detect behavioral trends of individuals to proactively act against Social Threats. *Procedia Computer Science, 132*, 218–225. doi:10.1016/j.procs.2018.05.191

Jin, K. W., Li, Q., Xie, Y., & Xiao, G. (2023). Artificial intelligence in mental healthcare: A scoping review. *The British Journal of Radiology, 96*(1150), 20230213. doi:10.1259/bjr.20230213 PMID:37698582

Jirout, J., Hirsh-pasek, K., & Evans, N. (2023). What ChatGPT can't do: Educating for curiosity and creativity. *Fortune Magazine.* https://www.brookings.edu/articles/what-chatgpt-cant-do-educating-for-curiosity-and-creativity

Jo, A. (2023). The promise and peril of generative AI. *Nature, 614*(1), 214–216.

Jorzik, P., Yigit, A., Kanbach, D. K., Kraus, S., & Dabic, M. (2023). Artificial Intelligence-Enabled Business Model Innovation: Competencies and Roles of Top Management. *IEEE Transactions on Engineering Management*, 1–13. doi:10.1109/TEM.2023.3275643

Josefina, R. (2023, June 20). *Enhancing Patient Engagement with Generative AI in Healthcare.* Light IT. Https://Lightit.Io/Blog/Enhancing-Patient-Engagement-with-Generative-Ai-in-Healthcare/.

Jovanovic, M., & Campbell, M. (2022). Generative artificial intelligence: Trends and prospects. *Computer, 55*(10), 107–112. doi:10.1109/MC.2022.3192720

Jurewicz, I. (2015). Mental health in young adults and adolescents – supporting general physicians to provide holistic care. *Clinical Medicine, 15*(2), 151–154. doi:10.7861/clinmedicine.15-2-151 PMID:25824067

K, R. (2023). Algorithm justice – moving towards equitable artificial intelligence. *CASE Journal, 19*(6), 788–799. doi:10.1108/TCJ-11-2021-0211

Kakatkar, C., Bilgram, V., & Füller, J. (2018). *Innovation Analytics: Leveraging Artificial Intelligence in the Innovation Process* (SSRN Scholarly Paper 3293533). doi:10.2139/ssrn.3293533

Kakuma, R., Minas, H., van Ginneken, N., Dal Poz, M. R., Desiraju, K., Morris, J. E., Saxena, S., & Scheffler, R. M. (2011). Human resources for mental health care: Current situation and strategies for action. *Lancet*, *378*(9803), 1654–1663. doi:10.1016/S0140-6736(11)61093-3 PMID:22008420

Kalayci, C. B., Polat, O., & Akbay, M. A. (2020). An efficient hybrid metaheuristic algorithm for cardinality constrained portfolio optimization. *Swarm and Evolutionary Computation*, *54*, 100662. doi:10.1016/j.swevo.2020.100662

Karnow, C. E. A. (1996). Liability for Distributed Artificial Intelligences. *Berkeley Technology Law Journal*, *11*(1), 147–204. https://www.jstor.org/stable/24115584

Kather, J. N., Ghaffari Laleh, N., Foersch, S., & Truhn, D. (2022). Medical domain knowledge in domain-agnostic generative AI. *NPJ Digital Medicine*, *5*(1), 90. doi:10.1038/s41746-022-00634-5 PMID:35817798

Katyal, S. K. (2022). Democracy & Distrust in an Era of Artificial Intelligence. *Daedalus*, *151*(2), 322–334. https://www.jstor.org/stable/48662045. doi:10.1162/daed_a_01919

Kaur, B., Ahuja, L., & Kumar, V. (2019). Decision tree Model. *Predicting Sexual Offenders on the Basis of Minor and Major Victims*, *193–197*, 193–197. doi:10.1109/AICAI.2019.8701276

Kaynak, O. (2021). The golden age of Artificial Intelligence. *Discover Artificial Intelligence*, *1*(1), 1. doi:10.1007/s44163-021-00009-x

Kazim, E., & Koshiyama, A. S. (2021). A high-level overview of AI ethics. *Patterns (New York, N.Y.)*, *2*(9), 100314. doi:10.1016/j.patter.2021.100314 PMID:34553166

Kazim, E., Koshiyama, A. S., Hilliard, A., & Polle, R. (2021). Systematizing audit in algorithmic recruitment. *Journal of Intelligence*, *9*(3), 46. doi:10.3390/jintelligence9030046 PMID:34564294

KenneallyE.DittrichD. (2012). The menlo report: Ethical principles guiding information and communication technology research. SSRN 2445102. doi:10.2139/ssrn.2445102

Kesavan, N. H. (2023, April 14). *Generative AI – Enterprise Use Cases for a Manufacturing Organization*. LinkedIn. Https://Www.Linkedin.Com/Pulse/Generative-Ai-Enterprise-Use-Cases-Manufacturing-Harikrishnan/

Khairuddin, A., Alwee, R., & Haron, H. (2020). A Comparative Analysis of Artificial Intelligence Techniques in Forecasting Violent Crime Rate. *IOP Conference Series. Materials Science and Engineering*, *864*(1), 012056. doi:10.1088/1757-899X/864/1/012056

Khan, S. M. (2008). Copyright, Data Protection, and Privacy with Digital Rights Management and Trusted Systems: Negotiating a Compromise between Proprietors and Users. *ISJLP*, *5*, 603.

Kharbat, F. F., Alshawabkeh, A., & Woolsey, M. L. (2021). Identifying gaps in using artificial intelligence to support students with intellectual disabilities from education and health perspectives. *Aslib Journal of Information Management*, *73*(1), 101–128. doi:10.1108/AJIM-02-2020-0054

Khogali, H. O., & Mekid, S. (2023). The blended future of automation and AI: Examining some long-term societal and ethical impact features. *Technology in Society*, *73*, 102232. doi:10.1016/j.techsoc.2023.102232

Kile, F. (2013). Artificial intelligence and society: A furtive transformation. *AI & Society*, *28*(1), 107–115. doi:10.1007/s00146-012-0396-0

King, T. C., Aggarwal, N., Taddeo, M., & Floridi, L. (2020). Artificial Intelligence Crime: An Interdisciplinary Analysis of Foreseeable Threats and Solutions. *Science and Engineering Ethics*, *26*(1), 89–120. doi:10.1007/s11948-018-00081-0 PMID:30767109

Kiritchenko, S., & Mohammad, S. M. (2018). *Examining gender and race bias in two hundred sentiment analysis systems*. arXiv preprint arXiv:1805.04508. doi:10.18653/v1/S18-2005

Kitsos, P. (2020, September 2). The Limits of Government Surveillance: Law Enforcement in the Age of Artificial Intelligence. *11th EETN Conference on Artificial Intelligence (SETN 2020)*. IEEE.

Kleinberg, J., Mullainathan, S., & Raghavan, M. (2016). Inherent Trade-Offs in the Fair Determination of Risk Scores. https://doi.org//arXiv.1609.05807. doi:10.48550

Knox, D., & Pardos, Z. (2022). *Toward Ethical and Equitable AI in Higher Education*. Inside Higher Ed. https://www.insidehighered.com/blogs/beyond-transfer/toward-ethical-and-equitable-ai-higher-education

Koch, B. A., Borghetti, J.-S., Machnikowski, P., Pichonnaz, P., Ballell, T. R. de las H., Twigg-Flesner, C., & Wendehorst, C. (2022). Response of the European Law Institute to the Public Consultation on Civil Liability – Adapting Liability Rules to the Digital Age and Artificial Intelligence. *Journal of European Tort Law*, *13*(1), 25–63. doi:10.1515/jetl-2022-0002

Kokciyan, N., Sassoon, I., Sklar, E., Modgil, S., & Parsons, S. (2021). Applying Metalevel Argumentation Frameworks to Support Medical Decision Making. *IEEE Intelligent Systems*, *36*(2), 64–71. doi:10.1109/MIS.2021.3051420

Kostka, I., & Toncelli, R. (2023). Exploring applications of ChatGPT to English language teaching: Opportunities, challenges, and recommendations. *The Electronic Journal for English as a Second Language*, *27*(3), 1–19. doi:10.55593/ej.27107int

Kraus, K., Kraus, N., Hryhorkiv, M., Kuzmuk, I., & Shtepa, O. (2022). Artificial Intelligence in Established of Industry 4.0. *WSEAS Transactions on Business and Economics*, *19*, 1884–1900. doi:10.37394/23207.2022.19.170

Kusiak, A. (2018). Smart manufacturing. *International Journal of Production Research*, *56*(1–2), 508–517. doi:10.1080/00207543.2017.1351644

Lambert, J., & Stevens, M. (2023). *AI ChatGPT: Academia Disruption and Transformation*. In T. Bastiaens (Ed.), *Proceedings of EdMedia + Innovate Learning* (pp. 1498-1504). Association for the Advancement of Computing in Education (AACE). https://www.learntechlib.org/primary/p/222671/

Lanagan, S., & Choo, K.-K. R. (2021). On the need for AI to triage encrypted data containers in U.S. law enforcement applications. *Forensic Science International Digital Investigation*, *38*, 301217. doi:10.1016/j.fsidi.2021.301217

Larson, D. B., Harvey, H., Rubin, D. L., Irani, N., Justin, R. T., & Langlotz, C. P. (2021). Regulatory frameworks for development and evaluation of artificial intelligence–based diagnostic imaging algorithms: Summary and recommendations. *Journal of the American College of Radiology*, *18*(3), 413–424. doi:10.1016/j.jacr.2020.09.060 PMID:33096088

Lathrop, B. (2019). The Inadequacies of the Cybersecurity Information Sharing Act of 2015 in the Age of Artificial Intelligence. *The Hastings Law Journal*, *71*, 501.

Laux, J., Wachter, S., & Mittelstadt, B. (2022). *Trustworthy Artificial Intelligence and the European Union AI Act: On the Conflation of Trustworthiness and the Acceptability of Risk* (SSRN Scholarly Paper 4230294). doi:10.2139/ssrn.4230294

Lăzăroiu, G., Androniceanu, A., Grecu, I., Grecu, G., & Neguriță, O. (2022). Artificial intelligence-based decision-making algorithms, Internet of Things sensing networks, and sustainable cyber-physical management systems in big data-driven cognitive manufacturing. *Oeconomia Copernicana*, *13*(4), 1047–1080. doi:10.24136/oc.2022.030

Lee, D., & Yoon, S. N. (2021). Application of Artificial Intelligence-Based Technologies in the Healthcare Industry: Opportunities and Challenges. *International Journal of Environmental Research and Public Health*, *18*(1), 271. doi:10.3390/ijerph18010271 PMID:33401373

Lee, E. E., Torous, J., De Choudhury, M., Depp, C. A., Graham, S. A., Kim, H.-C., Paulus, M. P., Krystal, J. H., & Jeste, D. V. (2021). Artificial Intelligence for Mental Healthcare: Clinical Applications, Barriers, Facilitators, and Artificial Wisdom. *Biological Psychiatry: Cognitive Neuroscience and Neuroimaging*, *6*(9), 856–864. doi:10.1016/j.bpsc.2021.02.001 PMID:33571718

Lee, J. (2020). Access to Finance for Artificial Intelligence Regulation in the Financial Services Industry. *European Business Organization Law Review*, *21*(4), 731–757. doi:10.1007/s40804-020-00200-0

Lee, J. D., & See, K. A. (2004). Trust in Automation: Designing for Appropriate Reliance. *Human Factors*, *46*(1), 50–80. doi:10.1518/hfes.46.1.50.30392 PMID:15151155

Lei, L., Li, J., & Li, W. (2023). Assessing the role of artificial intelligence in the mental healthcare of teachers and students. *Soft Computing*, •••, 1–11. doi:10.1007/s00500-023-08072-5 PMID:37362257

Leiser, M. R., & Dechesne, F. (2020). Governing machine-learning models: challenging the personal data presumption. *International data privacy law*, *10*(3), 187-200.

Lepri, B., Oliver, N., Letouzé, E., Pentland, A., & Vinck, P. (2017). Fair, transparent, and accountable algorithmic decision-making processes. *Philosophy & Technology*, *31*(4), 611–627. doi:10.1007/s13347-017-0279-x

Li, X. (2023) Artificial intelligence applications in finance: a survey. *Journal of Management Analytics*. ACM. . doi:10.1080/23270012.2023.2244503

Li, D., Lai, J., Wang, R., Li, X., Vijayakumar, P., Gupta, B. B., & Alhalabi, W. (2023). Ubiquitous intelligent federated learning privacy-preserving scheme under edge computing. *Future Generation Computer Systems*, *144*, 205–218. doi:10.1016/j.future.2023.03.010

Lim, W. M., Gunasekara, A., Pallant, J. L., Pallant, J. I., & Pechenkina, E. (2023). Generative AI and the future of education: Ragnarök or reformation? A paradoxical perspective from management educators. *International Journal of Management Education*, *21*(2), 100790. doi:10.1016/j.ijme.2023.100790

Liu, D. S., Sawyer, J., Luna, A., Aoun, J., Wang, J., Boachie, L., Halabi, S., & Joe, B. (2022). Perceptions of US Medical Students on Artificial Intelligence in Medicine: Mixed Methods Survey Study. *JMIR Medical Education*, *8*(4), e38325. doi:10.2196/38325 PMID:36269641

Liu, D., Nanayakkara, P., Sakha, S. A., Abuhamad, G., Blodgett, S. L., Diakopoulos, N., & Eliassi-Rad, T. (2022, July). Examining Responsibility and Deliberation in AI Impact Statements and Ethics Reviews. In *Proceedings of the 2022 AAAI/ACM Conference on AI, Ethics, and Society* (pp. 424-435). AAAI. 10.1145/3514094.3534155

Liu, H., Wang, Y., Fan, W., Liu, X., Li, Y., Jain, S., Liu, Y., Jain, A., & Tang, J. (2022). Trustworthy AI: A computational perspective. *ACM Transactions on Intelligent Systems and Technology*, *14*(1), 1–59. doi:10.1145/3546872

Li, X., Yu, N., Zhang, X., Zhang, W., Li, B., Lu, W., Wang, W., & Liu, X. (2021). Overview of digital media forensics technology. *Journal of Image and Graphics*, *26*(6), 1216–1226. doi:10.11834/jig.210081

Li, Y., & Casanave, C. P. (2012). Two first-year students' strategies for writing from sources: Patchwriting or plagiarism? *Journal of Second Language Writing*, *21*(2), 165–180. doi:10.1016/j.jslw.2012.03.002

Lo Piano, S. (2020). Ethical principles in machine learning and artificial intelligence: Cases from the field and possible ways forward. *Humanities & Social Sciences Communications*, *7*(1), 9. doi:10.1057/s41599-020-0501-9

Lodge, J. M. (2023). *Cheating with generative AI: Shifting focus from means and opportunity to motive [LinkedIn page]*. LinkedIn. https://www.linkedin.com/pulse/cheatinggenerative-ai-shifting-focus-from-means-motive-lodge/

Lombardo, G. (2022). The AI industry and regulation: Time for implementation? In R. Iphofen & D. O'Mathúna (Eds.), *Ethical Evidence and Policymaking* (1st ed., pp. 185–200). Bristol University Press. doi:10.2307/j.ctv2tbwqd5.15

Longoni, C., Fradkin, A., Cian, L., & Pennycook, G. (2022, June). News from generative artificial intelligence is believed less. In *Proceedings of the 2022 ACM Conference on Fairness, Accountability, and Transparency* (pp. 97-106). ACM. 10.1145/3531146.3533077

Loving, T. (2023, March 30). Current AI copyright cases – part 1. *Copyright Alliance*. https://copyrightalliance.org/current-ai-copyright-cases-part-1/

Luna, J. M., Gennatas, E. D., Ungar, L. H., Eaton, E., Diffenderfer, E. S., Jensen, S. T., Simone, C. B. II, Friedman, J. H., Solberg, T. D., & Valdes, G. (2019). Building more accurate decision trees with the additive tree. *Proceedings of the National Academy of Sciences of the United States of America*, *116*(40), 19887–19893. doi:10.1073/pnas.1816748116 PMID:31527280

Machlev, R., Heistrene, L., Perl, M., Levy, K. Y., Belikov, J., Mannor, S., & Levron, Y. (2022). Explainable Artificial Intelligence (XAI) techniques for energy and power systems: Review, challenges and opportunities. *Energy and AI*, *9*, 100169. doi:10.1016/j.egyai.2022.100169

Mac, T. T., Copot, C., Lin, C.-Y., Hai, H. H., & Ionescu, C. M. (2020). Towards The Development of a Smart Drone Police: Illustration in Traffic Speed Monitoring. *Journal of Physics: Conference Series*, *1487*(1), 012029. doi:10.1088/1742-6596/1487/1/012029

Madhav, A. S., & Tyagi, A. K. (2022). The world with future technologies (Post-COVID-19): open issues, challenges, and the road ahead. *Intelligent Interactive Multimedia Systems for e-Healthcare Applications*, 411-452.

Makridakis, S. (2017). The forthcoming Artificial Intelligence (AI) revolution: Its impact on society and firms. *Futures*, *90*, 46–60. doi:10.1016/j.futures.2017.03.006

Maliha, G., Gerke, S., Cohen, G., & Parikh, R. B.MALIHA. (2021). Artificial Intelligence and Liability in Medicine: Balancing Safety and Innovation. *The Milbank Quarterly*, *99*(3), 629–647. doi:10.1111/1468-0009.12504 PMID:33822422

Mallikarjuna, M., & Rao, R. P. (2019). Evaluation of forecasting methods from selected stock market returns. *Financial Innovation*, *5*(1), 40. doi:10.1186/s40854-019-0157-x

Mantello, P., Ho, M. T., Nguyen, M. H., & Vuong, Q. H. (2023). Bosses without a heart: Socio-demographic and cross-cultural determinants of attitude toward Emotional AI in the workplace. *AI & Society*, *38*(1), 97–119. doi:10.1007/s00146-021-01290-1 PMID:34776651

Margetts, H. (2022). Rethinking AI for Good Governance. *Daedalus*, *151*(2), 360–371. https://www.jstor.org/stable/48662048. doi:10.1162/daed_a_01922

Mark, R. (2023, February 4). *Using Generative AI to Enhance Classroom Creativity and Engagement*. LinkedIn. Https://Www.Linkedin.Com/Pulse/Using-Generative-Ai-Enhance-Classroom-Creativity-Mark

Market Research. (2023). *Global Generative AI In Media And Entertainment Market*. Market Research. https://market-research.biz/report/generative-ai-in-media-and-entertainment-market/request-sample/

Marr, B. (2023). A short history of ChatGPT: How we got to where we are today. *Forbes*. https://www.forbes.com/sites/bernardmarr/2023/05/19/a-short-history-of-chatgpt-how-we-got-to-where-we-are-today/?sh=4c973d2c674f

Martin, R., & Johnson, S. (2023). *Introducing NotebookLM.* Google. https://blog.google/technology/ai/notebooklm-google-ai/

Masakowski, Y. R. (2020). Artificial intelligence and the future global security environment. In *Artificial Intelligence and Global Security* (pp. 1–34). Emerald Publishing Limited. doi:10.1108/978-1-78973-811-720201001

Matt, C. (2023, June 9). *AI Music Generators Make You Feel Like a Maestro. Here's How They Work.* Popular mechanics. Https://Www.Popularmechanics.Com/Technology/Audio/A44109081/Ai-Music-Generators-Explained/

Matthew, U., Kazaure, J., Onyebuchi, A., Okey, O., Muhammed, I., & Okafor, N. (2021). Artificial Intelligence Autonomous Unmanned Aerial Vehicle (UAV) System for Remote Sensing in Security Surveillance. *CYBER NIGERIA, 2020*, 1–10. doi:10.1109/CYBERNIGERIA51635.2021.9428862

Mayring, P. (2000). Qualitative Content Analysis. *Forum Qualitative Sozialforschung / Forum: Qualitative. Social Research, 1*(2), 2. doi:10.17169/fqs-1.2.1089

McCarthy, J., Minsky, M. L., Rochester, N., & Shannon, C. E. (1955). A Proposal for the Dartmouth Summer Research Project on Artificial Intelligence: August 31, 1955 - ProQuest. *AI Magazine*.

McKendrick, K. (2019). Artificial intelligence prediction and counterterrorism. London: The Royal Institute of International Affairs-Chatham House.

Mele, C., & Russo-Spena, T. (2023). Artificial Intelligence in Services. Elgar Encyclopedia of Services, 356. Elgar.

MeuwissenM.BollenL. (2021). Transparancy versus Explainability in AI. *ResearchGate*. https://doi.org/ doi:10.13140/RG.2.2.27466.90561

Mikalef, P., Conboy, K., Lundström, J. E., & Popovič, A. (2022). Thinking responsibly about responsible AI and 'the dark side'of AI. *European Journal of Information Systems, 31*(3), 257–268. doi:10.1080/0960085X.2022.2026621

Milivojevic, S. (2021). *Crime and Punishment in the Future Internet: Digital Frontier Technologies and Criminology in the Twenty-First Century*. Routledge. doi:10.4324/9781003031215

Miller, G. J. (2022, March). Artificial Intelligence Project Success Factors—Beyond the Ethical Principles. In *Information Technology for Management: Business and Social Issues: 16th Conference, ISM 2021, and FedCSIS-AIST 2021 Track, Held as Part of FedCSIS 2021,* (pp. 65-96). Cham: Springer International Publishing.

Mirbabaie, M., Brünker, F., Möllmann, N. R., & Stieglitz, S. (2022). The rise of artificial intelligence–understanding the AI identity threat at the workplace. *Electronic Markets, 32*(1), 1–27. doi:10.1007/s12525-021-00496-x

Mittal, A. (2023). ChatGPT and the Legal and Ethical Problems of Copyright in India. *Economic and Political Weekly, 58*(32), 62–63.

Mittelstadt, B., Allo, P., Taddeo, M., Wachter, S., & Floridi, L. (2016). The ethics of algorithms: Mapping the debate. *Big Data & Society, 3*(2), 205395171667967. doi:10.1177/2053951716679679

Mökander, J. (2023). Auditing of AI: Legal, Ethical and Technical Approaches. *Digital Society : Ethics, Socio-Legal and Governance of Digital Technology, 2*(3), 49. doi:10.1007/s44206-023-00074-y

Molina-Molina, J. C., Salhaoui, M., Guerrero-González, A., & Arioua, M. (2021). Autonomous Marine Robot Based on AI Recognition for Permanent Surveillance in Marine Protected Areas. *Sensors (Basel), 21*(8), 8. doi:10.3390/s21082664 PMID:33920075

Morley, J., Machado, C. C. V., Burr, C., Cowls, J., Joshi, I., Taddeo, M., & Floridi, L. (2020). The ethics of AI in health care: A mapping review. *Social Science & Medicine (1982), 260*, 113172. doi:10.1016/j.socscimed.2020.113172

Mount, M., Round, H., & Pitsis, T. S. (2020). Design thinking inspired crowdsourcing: Toward a generative model of complex problem solving. *California Management Review*, *62*(3), 103–120. doi:10.1177/0008125620918626

Mukherjee, S., Coulter, M., Chee, F. Y., Mukherjee, S., & Chee, F. Y. (2023, December 14). Explainer: What's next for the EU AI Act? *Reuters*. https://www.reuters.com/technology/whats-next-eu-ai-act-2023-12-14/

Muller, M., Chilton, L. B., Kantosalo, A., Martin, C. P., & Walsh, G. (2022, April). GenAICHI: generative AI and HCI. In CHI conference on human factors in computing systems extended abstracts (pp. 1-7). ACM. doi:10.1145/3491101.3503719

Müller, V. C. (2023). Ethics of Artificial Intelligence and Robotics. In E. N. Zalta & U. Nodelman (Eds.), *The Stanford Encyclopedia of Philosophy* (Fall 2023). Metaphysics Research Lab, Stanford University. https://plato.stanford.edu/archives/fall2023/entries/ethics-ai/

Myrzashova, R., Alsamhi, S. H., Shvetsov, A. V., Hawbani, A., & Wei, X. (2023). Blockchain meets federated learning in healthcare: A systematic review with challenges and opportunities. *IEEE Internet of Things Journal*, *10*(16), 14418–14437. doi:10.1109/JIOT.2023.3263598

Nader, K., Toprac, P., Scott, S., & Baker, S. (2022). Public understanding of artificial intelligence through entertainment media. *AI & Society*, 1–14. doi:10.1007/s00146-022-01427-w PMID:35400854

Nanos, A. (2023). *Criminal Liability of Artificial Intelligence* (SSRN Scholarly Paper 4623126). doi:10.2139/ssrn.4623126

Newswire, P. R. (2023). Education and tech leaders come together to offer guidance on integrating AI safety into classrooms worldwide. *PR News Wire*. https://www.prnewswire.com/news-releases/education-and-tech-leaders-come-together-to-offer-guidance-on-integrating-ai-safely-into-classrooms-worldwide-301812478.html

Ngai, E. W. T., Peng, S., Alexander, P., & Moon, K. (2014). Decision support and intelligent systems in the textile and apparel supply chain: An academic review of research articles. *Expert Systems with Applications*, *41*(1), 81–91. doi:10.1016/j.eswa.2013.07.013

Nguyen, A., Ngo, H. N., Hong, Y., Dang, B., & Thi Nguyen, B.-P. (2023). Ethical principles for Artificial Intelligence in education. *Education and Information Technologies*, *28*(4), 4221–4241. doi:10.1007/s10639-022-11316-w PMID:36254344

Nichols, R. (2022). The cultural evolution of Chinese morality, and the essential value of multi-disciplinary research in understanding it. In The Routledge International Handbook of Morality, Cognition, and Emotion in China (pp. 19-37). Routledge. doi:10.4324/9781003281566-3

Nikookar, H. (2023). A Risk Analysis of Communication, Navigation and Sensing Satellite Systems Threats. *Journal of Mobile Multimedia*, *19*(1), 277–290. doi:10.13052/jmm1550-4646.19114

Nissenbaum, H. (1996). Accountability in a computerized society. *Science and Engineering Ethics*, *2*(1), 25–42. doi:10.1007/BF02639315

Noble, S. U. (2018). Algorithms of oppression. In *Algorithms of oppression*. New York university press. doi:10.2307/j.ctt1pwt9w5.11

Novelli, C., Taddeo, M., & Floridi, L. (2023). Accountability in artificial intelligence: What it is and how it works. *AI & Society*, 1–12. doi:10.1007/s00146-023-01635-y

NoyS.ZhangW. (2023). Experimental evidence on the productivity effects of generative artificial intelligence. *Available at* SSRN 4375283.

Nyholm, S. (2023). A new control problem? Humanoid robots, artificial intelligence, and the value of control. *AI and Ethics*, *3*(4), 1229–1239. doi:10.1007/s43681-022-00231-y

Obermeyer, Z., Powers, B., Vogeli, C., & Mullainathan, S. (2019). Dissecting racial bias in an algorithm used to manage the health of populations. *Science*, *366*(6464), 447–453. doi:10.1126/science.aax2342 PMID:31649194

Oppy, G., & Dowe, D. (2021). The Turing Test. In E. N. Zalta (Ed.), *The Stanford Encyclopedia of Philosophy* (Winter 2021). Metaphysics Research Lab, Stanford University. https://plato.stanford.edu/archives/win2021/entriesuring-test/

Owe, A., & Baum, S. D. (2021). Moral consideration of nonhumans in the ethics of artificial intelligence. *AI and Ethics*, *1*(4), 517–528. doi:10.1007/s43681-021-00065-0

Padigar, M., Pupovac, L., Sinha, A., & Srivastava, R. (2022). The effect of marketing department power on investor responses to announcements of AI-embedded new product innovations. *Journal of the Academy of Marketing Science*, *50*(6), 1277–1298. doi:10.1007/s11747-022-00873-8

Pagano, T. P., Loureiro, R. B., Lisboa, F. V. N., Peixoto, R. M., Guimarães, G. A. S., Cruz, G. O. R., Araujo, M. M., Santos, L. L., Cruz, M. A. S., Oliveira, E. L. S., Winkler, I., & Nascimento, E. G. S. (2023). Bias and Unfairness in Machine Learning Models: A Systematic Review on Datasets, Tools, Fairness Metrics, and Identification and Mitigation Methods. *Big Data and Cognitive Computing*, *7*(1), 15. doi:10.3390/bdcc7010015

Pallathadka, H., Mustafa, M., Sanchez, D. T., Sekhar Sajja, G., Gour, S., & Naved, M. (2023). IMPACT OF MACHINE learning ON Management, healthcare AND AGRICULTURE. *Materials Today: Proceedings*, *80*, 2803–2806. doi:10.1016/j.matpr.2021.07.042

Pallathadka, H., Ramirez-Asis, E. H., Loli-Poma, T. P., Kaliyaperumal, K., Ventayen, R. J. M., & Naved, M. (2023). Applications of artificial intelligence in business management, e-commerce and finance. *Materials Today: Proceedings*, *80*, 2610–2613. doi:10.1016/j.matpr.2021.06.419

Pálmai, G., Csernyák, S., & Erdélyi, Z. (2021). Authentic and reliable data in the service of national public data asset. *Public Finance Quarterly*, *66*(Special edition 2021/1), 52–67. doi:10.35551/PFQ_2021_s_1_3

Palomares, I., Martínez-Cámara, E., Montes, R., García-Moral, P., Chiachio, M., Chiachio, J., Alonso, S., Melero, F. J., Molina, D., Fernández, B., Moral, C., Marchena, R., de Vargas, J. P., & Herrera, F. (2021). A panoramic view and swot analysis of artificial intelligence for achieving the sustainable development goals by 2030: Progress and prospects. *Applied Intelligence*, *51*(9), 6497–6527. doi:10.1007/s10489-021-02264-y PMID:34764606

Pannu, A. (2015). *Artificial Intelligence and its Application in Different Areas*. Semantic Scholar.https://www.semanticscholar.org/paper/Artificial-Intelligence-and-its-Application-in-Pannu-Student/9a4d9a755134e612854db1897c03adb3983413df

Pasquale, F. (2018). *A Rule of Persons, Not Machines: The Limits of Legal Automation* (SSRN Scholarly Paper 3135549). https://papers.ssrn.com/abstract=3135549

Pasquale, F. (2020). *New laws of robotics*. Harvard University Press.

Pawlicka, A., Choraś, M., Przybyszewski, M., Belmon, L., Kozik, R., & Demestichas, K. (2021). Why Do Law Enforcement Agencies Need AI for Analyzing Big Data? *Computer Information Systems and Industrial Management: 20th International Conference, CISIM 2021*. Springer. 10.1007/978-3-030-84340-3_27

Paykamian, B. (2023). *Who should be regulating AI classroom tools*. Government Technology. https://www.govtech.com/education/higher-ed/who-should-be-regulating-ai-classroom-tools

Pecorari, D. (2001). Plagiarism and international students: How the English-speaking university responds. In D. D. Belcher & A. R. Hirvela (Eds.), *Linking Literacies: Perspectives on L 2 Reading- Writing Connections* (pp. 229–245). University of Michigan Press.

Pedrelli, P., Nyer, M., Yeung, A., Zulauf, C., & Wilens, T. (2015). College Students: Mental Health Problems and Treatment Considerations. *Academic Psychiatry*, *39*(5), 503–511. doi:10.1007/s40596-014-0205-9 PMID:25142250

Pedro, F., Subosa, M., Rivas, A., & Valverde, P. (2019). *Artificial intelligence in education: Challenges and opportunities for sustainable development*. Research Gate.

Peeters, M. M., van Diggelen, J., Van Den Bosch, K., Bronkhorst, A., Neerincx, M. A., Schraagen, J. M., & Raaijmakers, S. (2021). Hybrid collective intelligence in a human–AI society. *AI & Society*, *36*(1), 217–238. doi:10.1007/s00146-020-01005-y

Peltz, J., & Street, A. C. (2020). Artificial intelligence and ethical dilemmas involving privacy. In Artificial Intelligence and Global Security: Future Trends, Threats and Considerations (pp. 95-120). Emerald Publishing Limited. doi:10.1108/978-1-78973-811-720201006

Pennington, M. (2018). Five tools for detecting Algorithmic Bias in AI. *Technomancers – Legal Tech Blog*. https://www.technomancers.co.uk/2018/10/13/fivetools-for-detecting-algorithmic-bias-in-ai/

Pereira, L. M. (2019). Should I kill or rather not? *AI & Society*, *34*(4), 939–943. doi:10.1007/s00146-018-0850-8

Phillips, M., Dove, E. S., & Knoppers, B. M. (2017). Criminal prohibition of wrongful re-identification: Legal solution or minefield for big data? *Journal of Bioethical Inquiry*, *14*(4), 527–539. doi:10.1007/s11673-017-9806-9 PMID:28913771

Pierce, D. (2023). *Google's AI-powered note-taking app is the messy beginning of something great*. ZDNet. https://www.zdnet.com/article/google-launches-its-ai-notebook-notebooklm-heres-what-you-need-to-know/

Pizzi, M., Romanoff, M., & Engelhardt, T. (2020). AI for humanitarian action: Human rights and ethics. *International Review of the Red Cross*, *102*(913), 145–180. doi:10.1017/S1816383121000011

Pobiner, S., & Murphy, T. (2018). *Participatory design in the age of artificial intelligence*. Deloitte Insights. https://www2.deloitte.com/us/en/insights/focus/ cognitive-technologies/participatory-design-artificial-intelligence.html

Popkova, E. G., & Sergi, B. S. (2020). Human capital and AI in industry 4.0. Convergence and divergence in social entrepreneurship in Russia. *Journal of Intellectual Capital*, *21*(4), 565–581. doi:10.1108/JIC-09-2019-0224

Porche, I. (2016). *Emerging cyber threats and implications*. RAND. doi:10.7249/CT453

Porsdam Mann, S., Earp, B. D., Nyholm, S., Danaher, J., Møller, N., Bowman-Smart, H., & Savulescu, J. (2023). Generative AI entails a credit–blame asymmetry. *Nature Machine Intelligence*, 1–4.

Precedence research. (2023). *Generative AI in Travel Market*. Https://Www.Precedenceresearch.Com/Generative-Ai-in-Travel-Market

Prem, E. (2023). From ethical AI frameworks to tools: A review of approaches. *AI and Ethics*, *3*(3), 1–18. doi:10.1007/s43681-023-00258-9

Prentice, C., & Nguyen, M. (2020). Engaging and retaining customers with AI and employee service. *Journal of Retailing and Consumer Services*, *56*, 102186. Advance online publication. doi:10.1016/j.jretconser.2020.102186

Qadir, J. (2023, May). Engineering education in the era of ChatGPT: Promise and pitfalls of generative AI for education. In *2023 IEEE Global Engineering Education Conference (EDUCON)* (pp. 1-9). IEEE. 10.1109/EDUCON54358.2023.10125121

Qiao-Franco, G., & Bode, I. (2023). Weaponised artificial intelligence and Chinese practices of human–machine interaction. *The Chinese Journal of International Politics*, *16*(1), 106–128. doi:10.1093/cjip/poac024

Qudah, H., Malahim, S., Airout, R., Alomari, M., Hamour, A. A., & Alqudah, M. (2023). Islamic Finance in the Era of Financial Technology: A Bibliometric Review of Future Trends. *International Journal of Financial Studies*, *11*(2), 76. doi:10.3390/ijfs11020076

Quinn, P. (2021). Research under the GDPR – a level playing field for public and private sector research? *Life Sciences, Society and Policy*, *17*(1), 1–33. doi:10.1186/s40504-021-00111-z PMID:33397487

Rahman, M., Terano, H. J. R., Rahman, N., Salamzadeh, A., & Rahaman, S. (2023). ChatGPT and academic research: A review and recommendations based on practical examples. *Journal of Education. Management and Development Studies*, *3*(1), 1–12. doi:10.52631/jemds.v3i1.175

Rajamäki, J., Sarlio-Siintola, S., & Simola, J. (2018). *Ethics of Open Source Intelligence Applied by Maritime Law Enforcement Authorities*. Semantic Scholar. https://www.semanticscholar.org/paper/Ethics-of-Open-Source-Intelligence-Applied-by-Law-Rajam%C3%A4ki-Sarlio-Siintola/e76a98d86fadd131646850bb6011ce8473469d58

Rajapakshe, C., Balasooriya, S., Dayarathna, H., Ranaweera, N., Walgampaya, N., & Pemadasa, N. (2019). Using CNNs RNNs and Machine Learning Algorithms for Real-time Crime Prediction. *2019 International Conference on Advancements in Computing (ICAC)*, (pp. 310–316). IEEE. 10.1109/ICAC49085.2019.9103425

Raji, I. D., & Buolamwini, J. (2019). *Actionable auditing: Investigating the impact of publicly naming biased performance results of commercial ai products*. In Proceedings of the 2019 AAAI/ACM Conference on AI, Ethics, and Society, 429-435. 10.1145/3306618.3314244

Raji, I. D., Smart, A., White, R. N., Mitchell, M., Gebru, T., Hutchinson, B., & Barnes, P. (2020). Closing the AI accountability gap: Defining an end-to-end framework for internal algorithmic auditing. In *Proceedings of the 2020 conference on fairness, accountability, and transparency*, (pp. 33-44). ACM. 10.1145/3351095.3372873

Reich, N. (1986). Product safety and product liability—An analysis of the EEC Council Directive of 25 July 1985 on the approximation of the laws, regulations, and administrative provisions of the Member States concerning liability for defective products. *Journal of Consumer Policy*, *9*(2), 133–154. doi:10.1007/BF00380508

Reis, J., Santo, P., & Melão, N. (2020). Artificial Intelligence Research and Its Contributions to the European Union's Political Governance: Comparative Study between Member States. *Social Sciences (Basel, Switzerland)*, *9*(11), 11. doi:10.3390/socsci9110207

Reuters. (2023a). *Artificial Intelligence Liability Directive: Legislation tracker*. Practical Law. https://uk.practicallaw.thomsonreuters.com/w-037-5533?transitionType=Default&contextData=(sc.Default)&firstPage=true

Reuters. (2023b, December 9). *EU clinches deal on landmark AI Act—Reaction*. Reuters. https://www.reuters.com/technology/eu-clinches-deal-landmark-ai-act-2023-12-09/

Richelle, D., Joseph, G. S., & Steve, R. (2023, May 11). *AI-powered marketing and sales reach new heights with generative AI*. McKinsey. Https://Www.Mckinsey.Com/Capabilities/Growth-Marketing-and-Sales/Our-Insights/Ai-Powered-Marketing-and-Sales-Reach-New-Heights-with-Generative-Ai

Richter, D., Wall, A., Bruen, A., & Whittington, R. (2019). Is the global prevalence rate of adult mental illness increasing? Systematic review and meta-analysis. *Acta Psychiatrica Scandinavica*, *140*(5), 393–407. doi:10.1111/acps.13083 PMID:31393996

Riedl, M. O. (2019). Human-centered artificial intelligence and machine learning. *Human Behavior and Emerging Technologies*, *1*(1), 33–36. doi:10.1002/hbe2.117

Rivera, J., & Hare, F. (2014, June). The deployment of attribution agnostic cyberdefense constructs and internally based cyberthreat countermeasures. In *2014 6th international conference on cyber conflict (CyCon 2014)* (pp. 99-116). IEEE. 10.1109/CYCON.2014.6916398

Rodríguez de las Heras Ballell, T. (2023). The revision of the product liability directive: A key piece in the artificial intelligence liability puzzle. *ERA Forum, 24*(2), 247–259. 10.1007/s12027-023-00751-y

Rodríguez-Espíndola, O., Chowdhury, S., Beltagui, A., & Albores, P. (2020). The potential of emergent disruptive technologies for humanitarian supply chains: The integration of blockchain, Artificial Intelligence and 3D printing. *International Journal of Production Research, 58*(15), 4610–4630. doi:10.1080/00207543.2020.1761565

Rodríguez-Espíndola, O., Chowdhury, S., Dey, P. K., Albores, P., & Emrouznejad, A. (2022). Analysis of the adoption of emergent technologies for risk management in the era of digital manufacturing. *Technological Forecasting and Social Change, 178*, 121562. doi:10.1016/j.techfore.2022.121562

Rodriguez, R. (2020). Legal and human rights issues of AI: Gaps, challenges, and vulnerabilities. *Journal of Responsible Technology, 4*, 100005. doi:10.1016/j.jrt.2020.100005

Rose, K., Eldridge, S. D., & Chapin, L. (2015). *THE INTERNET OF THINGS: AN OVERVIEW Understanding the Issues and Challenges of a More Connected World.* Semantic Scholar. https://www.semanticscholar.org/paper/THE-INTERNET-OF-THINGS-%3A-AN-OVERVIEW-Understanding-Rose-Eldridge/6d12bda69e8fcbbf1e9a10471b54e57b15cb07f6

Sachan, S., Yang, J.-B., Xu, D.-L., Benavides, D., & Li, Y. (2019). An Explainable AI Decision-Support-System to Automate Loan Underwriting. *Expert Systems with Applications, 144*, 113100. doi:10.1016/j.eswa.2019.113100

Salahuddin, Z., Woodruff, H. C., Chatterjee, A., & Lambin, P. (2022). Transparency of deep neural networks for medical image analysis: A review of interpretability methods. *Computers in Biology and Medicine, 140*, 105111. doi:10.1016/j.compbiomed.2021.105111 PMID:34891095

Sánchez, J. R., Campo-Archbold, A., Rozo, A. Z., Díaz-López, D., Pastor-Galindo, J., Mármol, F. G., & Díaz, J. A. (2022). On the Power of Social Networks to Analyze Threatening Trends. *IEEE Internet Computing, 26*(02), 19–26. doi:10.1109/MIC.2022.3154712

Sauer, F. (2021). Lethal autonomous weapons systems. In *The Routledge Social Science Handbook of AI* (pp. 237–250). Routledge. doi:10.4324/9780429198533-17

Schiff, D. (2021). Out of the laboratory and into the classroom: The future of artificial intelligence in education. *AI & Society, 36*(1), 331–348. doi:10.1007/s00146-020-01033-8 PMID:32836908

Schiff, D. (2022). Education for AI, *not* AI for Education: The role of education and ethics in national AI policy strategies. *International Journal of Artificial Intelligence in Education, 32*(3), 527–563. doi:10.1007/s40593-021-00270-2

Schrettenbrunnner, M. B. (2020). Artificial-Intelligence-Driven Management. *IEEE Engineering Management Review, 48*(2), 15–19. doi:10.1109/EMR.2020.2990933

Schuett, J. (2019). A Legal Definition of AI. SSRN *Electronic Journal*. doi:10.2139/ssrn.3453632

Schwalbe, N., & Wahl, B. (2020). Artificial intelligence and the future of global health. *Lancet, 395*(10236), 1579–1586. doi:10.1016/S0140-6736(20)30226-9 PMID:32416782

Schwemer, S. F. (2022). *Digital Services Act: A Reform of the e-Commerce Directive and Much More* (*SSRN* Scholarly Paper 4213014). doi:10.2139/ssrn.4213014

Sharma, A., & Singh, B. (2022). Measuring Impact of E-commerce on Small Scale Business: A Systematic Review. *Journal of Corporate Governance and International Business Law*, *5*(1).

Shaw, J., Rudzicz, F., Jamieson, T., & Goldfarb, A. (2019). Artificial intelligence and the implementation challenge. *Journal of Medical Internet Research*, *21*(7), e13659. doi:10.2196/13659 PMID:31293245

Sherer, J., Sterling, N., Burger, L., Banaschik, M., & Taal, A. (2018). *An Investigator's Christmas Carol: Past*. Present, and Future Law Enforcement Agency Data Mining Practices., doi:10.1007/978-3-319-97181-0_12

Shieh, C.-S., Nguyen, T.-T., Chen, C.-Y., & Horng, M.-F. (2023). Detection of Unknown DDoS Attack Using Reconstruct Error and One-Class SVM Featuring Stochastic Gradient Descent. *Mathematics*, *11*(1), 108. doi:10.3390/math11010108

Shieh, C.-S., Nguyen, T.-T., & Horng, M.-F. (2023). Detection of Unknown DDoS Attack Using Convolutional Neural Networks Featuring Geometrical Metric. *Mathematics*, *11*(9), 2145. doi:10.3390/math11092145

Shin, D. (2021). The effects of explainability and causability on perception, trust, and acceptance: Implications for explainable AI. *International Journal of Human-Computer Studies*, *146*, 102551. Advance online publication. doi:10.1016/j.ijhcs.2020.102551

Shneiderman, B. (2020). Bridging the gap between ethics and practice: Guidelines for reliable, safe, and trustworthy human-centered AI systems. [TiiS]. *ACM Transactions on Interactive Intelligent Systems*, *10*(4), 1–31. doi:10.1145/3419764

Siddharth, G. (2023). *How Generative AI Shaping the Media and Entertainment Industry*. QUY Tech. Https://Www.Quytech.Com/Blog/Generative-Ai-in-Media-and-Entertainment-Industry/. https://www.quytech.com/blog/generative-ai-in-media-and-entertainment-industry/

Simonite, T. (2018, January 11). When it comes to gorillas, google photos remains blind. Wired. https://www.wired.com/story/when-it-comes-to-gorillas-google-photos-remains-blind/

Simpson, T. (2021). Real-Time Drone Surveillance System for Violent Crowd Behavior Unmanned Aircraft System (UAS) – Human Autonomy Teaming (HAT). *2021 IEEE/AIAA 40th Digital Avionics Systems Conference (DASC)*. IEEE. https://www.semanticscholar.org/paper/Real-Time-Drone-Surveillance-System-for-Violent-%E2%80%93-Simpson/3d3dad177c3913d46800e286fdc0553f8b2d81d7

Singh, B. (2023). Blockchain Technology in Renovating Healthcare: Legal and Future Perspectives. In Revolutionizing Healthcare Through Artificial Intelligence and Internet of Things Applications (pp. 177-186). IGI Global.

Singh, A., Anand, T., Sharma, S., & Singh, P. (2021). IoT Based Weapons Detection System for Surveillance and Security Using YOLOV4. *ICCES*, *2021*, 488–493. doi:10.1109/ICCES51350.2021.9489224

Singh, B. (2019). Profiling Public Healthcare: A Comparative Analysis Based on the Multidimensional Healthcare Management and Legal Approach. *Indian Journal of Health and Medical Law*, *2*(2), 1–5.

Singh, B. (2020). GLOBAL SCIENCE AND JURISPRUDENTIAL APPROACH CONCERNING HEALTHCARE AND ILLNESS. *Indian Journal of Health and Medical Law*, *3*(1), 7–13.

Singh, B. (2022). COVID-19 Pandemic and Public Healthcare: Endless Downward Spiral or Solution via Rapid Legal and Health Services Implementation with Patient Monitoring Program. *Justice and Law Bulletin*, *1*(1), 1–7.

Singh, B. (2022). Relevance of Agriculture-Nutrition Linkage for Human Healthcare: A Conceptual Legal Framework of Implication and Pathways. *Justice and Law Bulletin*, *1*(1), 44–49.

Singh, B. (2022). Understanding Legal Frameworks Concerning Transgender Healthcare in the Age of Dynamism. *ELECTRONIC JOURNAL OF SOCIAL AND STRATEGIC STUDIES*, *3*(1), 56–65. doi:10.47362/EJSSS.2022.3104

Smith, Z. S. (2022). Self-Driving Car Users Shouldn't Be Held Responsible For Crashes, U.K. Report Says. *Forbes*. https://www.forbes.com/sites/zacharysmith/2022/01/25/self-driving-car-users-shouldnt-be-held-responsible-for-crashes-uk-report-says/

Smith, H. (2021). Clinical AI: Opacity, accountability, responsibility and liability. *AI & Society*, *36*(2), 535–545. doi:10.1007/s00146-020-01019-6

Smith, M., & Miller, S. (2022). The ethical application of biometric facial recognition technology. *AI & Society*, *37*(1), 167–175. doi:10.1007/s00146-021-01199-9 PMID:33867693

Somaya, D., & Varshney, L. R. (2020). Ownership Dilemmas in an Age of Creative Machines. *Issues in Science and Technology*, *36*(2), 79–85. https://www.jstor.org/stable/26949112

Sousa Antunes, H. (2020). *Civil Liability Applicable to Artificial Intelligence: A Preliminary Critique of the European Parliament Resolution of 2020* (*SSRN* Scholarly Paper 3743242). doi:10.2139/ssrn.3743242

Stahl, B. C., & Wright, D. (2018). Ethics and privacy in AI and big data: Implementing responsible research and innovation. *IEEE Security and Privacy*, *16*(3), 26–33. doi:10.1109/MSP.2018.2701164

Stanley, M. (2023, April 18). *Tapping the $6 Trillion Opportunity in AI*. Morgan Stanley. Https://Www.Morganstanley.Com/Ideas/Generative-Ai-Growth-Opportunity

Summers, J. (2009). *Principles of Healthcare Ethics*. Eweb:321396. https://repository.library.georgetown.edu/handle/10822/953367

Sun, J., Liao, Q. V., Muller, M., Agarwal, M., Houde, S., Talamadupula, K., & Weisz, J. D. (2022, March). Investigating explainability of generative AI for code through scenario-based design. In *27th International Conference on Intelligent User Interfaces* (pp. 212-228). ACM. 10.1145/3490099.3511119

Sun, Q., Wu, H., & Zhao, B. (2022). Artificial intelligence technology in internet financial edge computing and analysis of security risk. *International Journal of Ad Hoc and Ubiquitous Computing*, *39*(4), 201–210. doi:10.1504/IJAHUC.2022.121654

Sun, S. (2020). Application of fuzzy image restoration in criminal investigation. *Journal of Visual Communication and Image Representation*, *71*, 102704. doi:10.1016/j.jvcir.2019.102704

Suralkar, S., Gangurde, S., Chintakindi, S., & Chawla, H. (2020). An Autonomous Intelligent Ornithopter. In A. P. Pandian, R. Palanisamy, & K. Ntalianis (Eds.), *Proceeding of the International Conference on Computer Networks, Big Data and IoT (ICCBI - 2019)* (pp. 856–865). Springer International Publishing. 10.1007/978-3-030-43192-1_93

Sweeney, P. (2022). Trusting social robots. *AI and Ethics*, 1–8. PMID:35634257

Tahiri Jouti, A. (2018). Islamic finance: Financial inclusion or migration? *ISRA International Journal of Islamic Finance*, *10*(2), 277–288. doi:10.1108/IJIF-07-2018-0074

Tallman, I., Leik, R. K., Gray, L. N., & Stafford, M. C. (1993). A Theory of Problem-Solving Behavior. *Social Psychology Quarterly*, *56*(3), 157–177. doi:10.2307/2786776

Tariq, A., & Rafi, K. (2012). Intelligent Decision Support Systems- A Framework. *Information and Knowledge Management*. Semantic Scholar. https://www.semanticscholar.org/paper/Intelligent-Decision-Support-Systems-A-Framework-Tariq-Rafi/98250a732c5e5e11f6c7c9d0bf69efdd1ee71f4b

Tasneem, A. (2023). *7 missteps university leaders must avoid in their AI approach*. EAB. https://eab.com/insights/blogs/strategy/missteps-university-leaders-ai-approach/

Taylor, S., Boniface, M., Pickering, B., Anderson, M., Danks, D., Følstad, A., Leese, M., Müller, V., Sorell, T., Winfield, A., & Woollard, F. (2018). *Responsible AI – Key themes, concerns & recommendations for European research and innovation.*

Thamik, H., & Wu, J. (2022). The Impact of Artificial Intelligence on Sustainable Development in Electronic Markets. *Sustainability (Basel), 14*(6), 3568. doi:10.3390/su14063568

Thomas, D. C., Liao, Y., Aycan, Z., Cerdin, J.-L., Pekerti, A. A., Ravlin, E. C., Stahl, G. K., Lazarova, M. B., Fock, H., Arli, D., Moeller, M., Okimoto, T. G., & van de Vijver, F. (2015). Cultural intelligence: A theory-based, short-form measure. *Journal of International Business Studies, 46*(9), 1099–1118. https://www.jstor.org/stable/43653785. doi:10.1057/jibs.2014.67

Thong, J. L. K. (2021, June 29). Mapping Singapore's journey and approach to AI governance. *Digital Asia.* https://medium.com/digital-asia-ii/mapping-singapores-journey-and-approach-to-ai-governance-d01f76bbf5c6

Tlemsani, I., & Matthews, R. (2023). Digitalization and the prospects of cryptocurrency in Islamic finance. *International Journal of Technology Management & Sustainable Development, 22*(2), 131–152. doi:10.1386/tmsd_00072_1

Tlili, A., Shehata, B., Adarkwah, M. A., Bozkurt, A., Hickey, D. T., Huang, R., & Agyemang, B. (2023). What is the devil is my guardian angel: ChatGPT as a case study of using chatbots in education. *Smart Learning Environments, 10*(1), 1–24. doi:10.1186/s40561-023-00237-x

Tobin, S. (2023, June 1). *Getty asks London court to stop UK sales of Stability AI system.* Reuters. https://www.reuters.com/technology/getty-asks-london-court-stop-uk-sales-stability-ai-system-2023-06-01/

*Top Strategic Technology Trends for 2022: Generative AI.* (2021). Gartner. https://www.gartner.com/en/documents/4006921.

ToppiReddy, H. K. R., Saini, B., & Mahajan, G.ToppiReddy. (2018). Crime Prediction & Monitoring Framework Based on Spatial Analysis. *Procedia Computer Science, 132*, 696–705. doi:10.1016/j.procs.2018.05.075

Torres, J. T., & Mayo, C. P. T. (2023). *AI eroding AI? A new era for Artificial Intelligence and academic integrity.* Faculty Focus. https://www.facultyfocus.com/articles/teaching-with-technology-articles/ai-eroding-ai-a-new-era-for-artificial-intelligence-and-academic-integrity/

Traniello, J. F. A., & Bakker, T. C. M. (2016). Intellectual theft: Pitfalls and consequences of plagiarism. *Behavioral Ecology and Sociobiology, 70*(11), 1789–1791. doi:10.1007/s00265-016-2207-y

Uche, O. F. (2023, August 8). *Exploring Generative AI in Media & Entertainment: Virtual News Presenters.* LinkedIn. Https://Www.Linkedin.Com/Pulse/Exploring-Generative-Ai-Media-Entertainment-Virtual-News-Okereke

Ugwudike, P. (2022). AI audits for assessing design logics and building ethical systems: The case of predictive policing algorithms. *AI and Ethics, 2*(1), 199–208. doi:10.1007/s43681-021-00117-5 PMID:35909984

UNESCO (2019). *Artificial intelligence in education: challenges and opportunities for sustainable development.* UNESCO.

UNESCO. (2018). *ICT Competency Framework for Teachers.* UNESCO. https://unesdoc.unesco.org/ark:/48223/pf0000265721

UNESCO. (2021). *Artificial intelligence in education: Guidance for policy makers.* UNESCO. https://en.unesco.org/artificial-intelligence/education.

UNESCO. (2022). *Recommendation on the Ethics of Artificial Intelligence.* UNESCO. https://unesdoc.unesco.org/ark:/48223/pf0000381137.locale=en

UNESCO. (2023). *ChatGPT and artificial intelligence in higher education: Quick start guide*. UNESCO. https://unesdoc.unesco.org/ark:/48223/pf0000385146

Uunona, G. N., & Goosen, L. (2023). Leveraging Ethical Standards in Artificial Intelligence Technologies: A Guideline for Responsible Teaching and Learning Applications. In Handbook of Research on Instructional Technologies in Health Education and Allied Disciplines (pp. 310-330). IGI Global. doi:10.4018/978-1-6684-7164-7.ch014

van der Zant, T., Kouw, M., & Schomaker, L. (2013). *Generative artificial intelligence*. Springer Berlin Heidelberg.

Varkey, B. (2020). Principles of Clinical Ethics and Their Application to Practice. *Medical Principles and Practice*, *30*(1), 17–28. doi:10.1159/000509119 PMID:32498071

VB v Natsionalna agentsia za prihodite, Case C-340/21 (ECJ 2021). https://eur-lex.europa.eu/legal-content/EN/TXT/?uri=CELEX%3A62021CN0340

Veale, M., & Binns, R. (2017). Fairer machine learning in the real world: Mitigating discrimination without collecting sensitive data. *Big Data & Society*, *4*(2), 205395171774353. doi:10.1177/2053951717743530

Villegas-Ch, W., & García-Ortiz, J. (2023). Toward a Comprehensive Framework for Ensuring Security and Privacy in Artificial Intelligence. *Electronics (Basel)*, *12*(18), 3786. doi:10.3390/electronics12183786

Visvizi, A. (2021). Artificial intelligence (AI): Explaining, querying, demystifying. *Artificial Intelligence and Its Contexts: Security, Business and Governance*, 13-26.

Vorzhakova, Y., & Boiarynova, K. (2020). The application of digitalization in enterprises on the basis of multiple criteria selection design. *Central European Management Journal*, *28*(3), 127–148. doi:10.7206/cemj.2658-0845.29

Wagner, G. (2021). *Liability for Artificial Intelligence: A Proposal of the European Parliament* (SSRN Scholarly Paper 3886294). doi:10.2139/ssrn.3886294

Waliszewski, K., & Warchlewska, A. (2020). Attitudes towards artificial intelligence in the area of personal financial planning: A case study of selected countries. *Entrepreneurship and Sustainability Issues*, *8*(2), 399–420. doi:10.9770/jesi.2020.8.2(24)

Waltl, B., & Vogl, R. (2018). Increasing Transparency in Algorithmic- Decision-Making with Explainable AI. *Datenschutz Und Datensicherheit - DuD*, *42*(10), 613–617. doi:10.1007/s11623-018-1011-4

Wang, H., & Ma, S. (2021). Preventing Crimes Against Public Health with Artificial Intelligence and Machine Learning Capabilities. *Socio-Economic Planning Sciences*, *80*, 101043. doi:10.1016/j.seps.2021.101043

Wang, X., Lin, X., & Shao, B. (2023). Artificial intelligence changes the way we work: A close look at innovating with chatbots. *Journal of the Association for Information Science and Technology*, *74*(3), 339–353. doi:10.1002/asi.24621

Weinhardt, M. (2020). Ethical Issues in the Use of Big Data for Social Research. *Historical Social Research. Historische Sozialforschung*, *45*(3), 342–368. https://www.jstor.org/stable/26918416

Weisz, J. D., Muller, M., He, J., & Houde, S. (2023). Toward general design principles for generative AI applications. *arXiv preprint arXiv:2301.05578.*

Weisz, J. D., Muller, M., He, J., & Houde, S. (2023). Toward General Design Principles for Generative AI Applications. *CEUR Workshop Proceedings*, *3359*(1), 130–144.

Wendehorst, C. (2020). Strict Liability for AI and other Emerging Technologies. *Journal of European Tort Law*, *11*(2), 150–180. doi:10.1515/jetl-2020-0140

Wenzlaff, K., & Spaeth, S. (2022). Smarter than Humans? Validating how OpenAI's ChatGPT Model explains crowdfunding, Alternative finance and community finance. SSRN *Scholarly Paper.* doi:10.2139/ssrn.4302443

Werthner, H., Stanger, A., Schiaffonati, V., Knees, P., Hardman, L., & Ghezzi, C. (2023). Digital Humanism: The Time Is Now. *Computer, 56*(1), 138–142. doi:10.1109/MC.2022.3219528

West, D. M. (2018, September 13). *The role of corporations in addressing AI's ethical dilemmas.* Brookings. https://www.brookings.edu/articles/how-to-address-ai-ethical-dilemmas/

Wong, L.-W., Tan, G. W.-H., Ooi, K.-B., Lin, B., & Dwivedi, Y. K. (2022). Artificial intelligence-driven risk management for enhancing supply chain agility: A deep-learning-based dual-stage PLS-SEM-ANN analysis. *International Journal of Production Research,* 1–21. doi:10.1080/00207543.2022.2063089

World Health Organisation. (2016). *Administrative Errors: Technical Series on Safer Primary Care.* WHO *Press.* https://apps.who.int/bookorders.%0Awww.who.int/patientsafety

Wyatt, A. (2021). A Southeast Asian perspective on the impact of increasingly Autonomous systems on subnational relations of power. *Defence Studies, 21*(3), 271–291. doi:10.1080/14702436.2021.1908136

Xiao, X., Tang, Z., Xiao, B., & Li, K.-L. (2023). A Survey on Privacy and Security Issues in Federated Learning. *Jisuanji Xuebao/Chinese. Journal of Computers, 46*(5), 1019–1044. doi:10.11897/SP.J.1016.2023.01019

Xu, Y., Zhou, Y., Sekula, P., & Ding, L. (2021). Machine learning in construction: From shallow to deep learning. *Developments in the Built Environment, 6,* 100045. doi:10.1016/j.dibe.2021.100045

Yamin, M. M., Ullah, M., Ullah, H., & Katt, B. (2021). Weaponized AI for cyber attacks. *Journal of Information Security and Applications, 57,* 102722. doi:10.1016/j.jisa.2020.102722

Yan, D. (2023). Impact of ChatGPT on learners in a L2 writing practicum: An exploratory investigation. *Education and Information Technologies, 28*(11), 13943–13967. doi:10.1007/s10639-023-11742-4

Yeung, K. (2020). Recommendation of the council on artificial intelligence (OECD). *International Legal Materials, 59*(1), 27–34. doi:10.1017/ilm.2020.5

Yigitcanlar, T., & Cugurullo, F. (2020). The sustainability of artificial intelligence: An urbanistic viewpoint from the lens of smart and sustainable cities. *Sustainability (Basel), 12*(20), 1–24. doi:10.3390/su12208548

YounasA.ZengY. (2024) *Proposing Central Asian AI ethics principles: A multilevel approach for responsible AI.* SSRN. https://ssrn.com/abstract=4689770 or doi:10.2139/ssrn.4689770

Young, A. T., Amara, D., Bhattacharya, A., & Wei, M. L. (2021). Patient and general public attitudes towards clinical artificial intelligence: A mixed methods systematic review. *The Lancet. Digital Health, 3*(9), e599–e611. doi:10.1016/S2589-7500(21)00132-1 PMID:34446266

Zajda, J., & Vissing, Y. (Eds.). (2022). *Discourses of globalisation, ideology, and human rights* (Vol. 14). Springer. doi:10.1007/978-3-030-90590-3

Zech, H. (2021). Liability for AI: Public policy considerations. *ERA Forum, 22*(1), 147–158. 10.1007/s12027-020-00648-0

Zeng, J. (2020). Artificial intelligence and China's authoritarian governance. *International Affairs, 96*(6), 1441–1459. doi:10.1093/ia/iiaa172

Zhang, P., & Boulos, M. N. K. (2023). *Generative AI in Medicine and Healthcare : Promises, Opportunities and Challenges.* Research Gate..

Zhang, B. Z., Ashta, A., & Barton, M. E. (2021). Do FinTech and financial incumbents have different experiences and perspectives on the adoption of artificial intelligence? *Strategic Change*, *30*(3), 223–234. doi:10.1002/jsc.2405

Zhang, B., Zhu, J., & Su, H. (2023). Toward the third generation artificial intelligence. *Science China. Information Sciences*, *66*(2), 1–19. doi:10.1007/s11432-021-3449-x

Zhang, R. (2021a). The AI embedding predicts the legal risks of policing and its prevention. *2021 International Conference on Computer Information Science and Artificial Intelligence (CISAI)*, 642–646. 10.1109/CISAI54367.2021.00129

Zhang, Y., Wu, M., Tian, G. Y., Zhang, G., & Lu, J. (2021). Ethics and privacy of artificial intelligence: Understandings from bibliometrics. *Knowledge-Based Systems*, *222*, 106994. doi:10.1016/j.knosys.2021.106994

Zhao, J., Wang, T., Yatskar, M., Ordóñez, V., & Chang, K. (2017). Men Also Like Shopping: Reducing Gender Bias Amplification using Corpus-level Constraints. *Men Also Like Shopping: Reducing Gender Bias Amplification Using Corpus-level Constraints*. ACL. . doi:10.18653/v1/D17-1323

Zimmer, B. (2018). *Towards Privacy by Design: Review of the Personal Information Protection and Electronic Documents Act: Report of the Standing Committee on Access to Information, Privacy and Ethics*. House of Commons, Canada. https://www.ourcommons.ca/Content/Committee/421/ETHI/Reports/RP9690701/ethirp12/ethirp12-e.pdf

Ziosi, M., Mökander, J., Novelli, C., Casolari, F., Taddeo, M., & Floridi, L. (2023). *The EU AI Liability Directive: Shifting the Burden From Proof to Evidence* (SSRN Scholarly Paper 4470725). doi:10.2139/ssrn.4470725

Zittrain, J. L. (2006). The Generative Internet. *Harvard Law Review*, *119*(7), 1974–2040. https://www.jstor.org/stable/4093608

Zohny, H., McMillan, J., & King, M. (2023). Ethics of generative AI. *Journal of Medical Ethics*, *49*(2), 79–80. doi:10.1136/jme-2023-108909 PMID:36693706

Zufferey, R., Tormo-Barbero, J., Feliu-Talegón, D., Nekoo, S. R., Acosta, J. Á., & Ollero, A. (2022). How ornithopters can perch autonomously on a branch. *Nature Communications*, *13*(1), 7713. doi:10.1038/s41467-022-35356-5 PMID:36513661

Zuiderwijk, A., Chen, Y. C., & Salem, F. (2021). Implications of the use of artificial intelligence in public governance: A systematic literature review and a research agenda. *Government Information Quarterly*, *38*(3), 101577. doi:10.1016/j.giq.2021.101577

# About the Contributors

**S. M. Aamir Ali** is presently Assistant Professor at Symbiosis Law School Pune, Symbiosis International (Deemed University), Pune. He is currently pursuing his PhD from West Bengal National University of Juridical Sciences (WBNUJS), Kolkata, and has completed his LL.M from National Law School of India University (NLSIU), Bengaluru, with a specialization in Human Rights Law. Mr Ali has been teaching across specializations in Constitutional Law, Human Rights and Criminal Law at UG and PG levels since 2019.

**A. Anuradha** is an Associate Professor in VIT Business School with 15 years of experience in teaching and research in the emerging field of quantitative marketing and economics (QME) which is the combination of Marketing, Economics and Statistics. She is passionate towards working in the broader areas of marketing which connects consumers, firms and competitors more so the wider range of methodologies covering economic theories and econometrics. Her research focus is to strive to bring in implications to improve (People, Profit and Planet) the standard of living and lifestyle of People, Profits to the business organizations and the protection of Planet through sustainability measures

**B. Sam Paul** is currently pursuing Ph.D in the domain of Marketing. He has completed his MBA (Marketing and HR) and B.E. (Electronics and Communication) in Madras University and Anna University affiliated colleges respectively in Chennai. He is having 3 years of Industry experience in marketing domain and 6 years of teaching experience in Master of Business Administrative Department from a reputed college in Chennai. He has published in International Journals and attended many conferences in the area of marketing. His interest lies in teaching, learning technology marketing and sharing knowledge.

**Astha Chaturvedi** is a Research Scholar at Parul Institute of Law, Parul University Vadodara Gujarat, India. Her research focuses on identifying and understanding key challenges in the regulatory framework for Artificial Intelligence. Identification of these challenges may allow formation of a comprehensive legal framework which would help majorly in core areas like setting up liability, legal personality and ethical and moral grounds for an AI driven technologies.

**Allen Farina** is a full-time faculty member in the Department of Professional Studies, School of Multidisciplinary and Professional Studies, at Purdue University Global. Dr. Farina holds a Master of Education degree from Lakeland University and a Doctor of Education degree with a concentration in Higher Education Leadership from Lynn University. He has worked in higher education settings for twenty years. Instructional experiences include serving on dissertation committees, teaching graduate classes in Leadership, Education, and Organizational Development, and undergraduate courses in diversity, multiculturalism, and general education. Administrative functions include curriculum writing, mentoring, and training online instructors, and course lead. Research interests include best practices in online teaching and learning, DEI in higher education settings, and Autism spectrum disorder (ASD).

**Catherine Hayes** is Professor of Health Professions Pedagogy and Scholarship at the University of Sunderland, UK. She is a UK National Teaching Fellow and Principal Fellow of the UK Higher Education Academy. As a graduate of Podiatric Medicine in 1992, Catherine was a Founding Fellow of the Faculty of Podiatric Medicine at the Royal College of Physicians and Surgeons (Glasgow) in 2012 and was awarded Fellowship of the Royal College of Podiatry (London) in 2010. She is currently Programme Leader of the University of Sunderland's Professional Doctorate pathways for the DBA, EdD, DPM and DProf.

**Zidan Kachhi** is a counselling psychologist and an assistant professor. He has authored two books and published various papers in peer-reviewed journals.

**Early Ridho Kismawadi**, S.E.I, MA is a lecturer at the Department of Islamic Banking, Faculty of Islamic Economics and Business IAIN Langsa, Aceh, Indonesia, he has been a lecturer since 2013, he has completed a doctoral program in 2018 majoring in Sharia economics at the State Islamic University of North Sumatra. He was appointed head of the Islamic economics Law study program (2023) Islamic banking study program (2020) and Islamic financial management study program (2019) at Langsa State Islamic Institute (IAIN Langsa), Aceh, Indonesia. His research interests include financial economics, applied econometrics, Islamic economics, banking, and finance. He has published articles in national and international journals. In addition, he is also a reviewer of several reputable international journals such as Finance Research Letters, Financial Innovation, Cogent Business &; Management, Journal of Islamic Accounting and Business Research. He has also presented his papers at various local and international seminars. Dr. Early Ridho Kismawadi, S.E.I, MA, can be contacted at kismawadi@iainlangsa.ac.id.

**Sadhana Mishra** earned her Ph.D. in management in 2014 from FMS, Banaras Hindu University, Varanasi, India. Following this accomplishment, she embarked on a professional journey that led her to the University of Hail, Saudi Arabia, where she presently holds the position of Assistant Professor in the College of Business Administration. Dr. Mishra's primary research focus revolves around the domains of Retailing and Tourism. With a wealth of experience, she has delved extensively into research areas such as Brand Management, Customer Experience, Brand Engagement, Shopping behavior, Tourism and Hospitality, Medical Tourism, Virtual Tourism and Green Marketing. Beyond her academic endeavors, Dr. Mishra actively contributes to the scholarly community as a reviewer for journals indexed in Scopus and Wiley.

**Carolyn Stevenson** is a full-time faculty member in the Department of Professional Studies, School of Multidisciplinary and Professional Studies. Dr. Stevenson has over 20 years teaching and administrative experience in higher education. She holds a Master of Arts degree in Communication, Master of Business Administration, and Doctor of Education with an emphasis in Higher Education. Recent publications include a chapter entitled: "Leading across Generations: Issues for Higher Education Administrators" published in the Handbook of Research on Transnational Higher Education Management, IGI Global;

Technical Writing: A Comprehensive Resource for Technical Writers at all Levels, (Martinez, Hannigan, Wells, Peterson and Stevenson) Revised and Updated Edition, Kaplan Publishing, Building Online Communities in Higher Education Institutions: Creating Collaborative Experience (with Joanna Bauer), published by IGI Global, and Promoting Climate Change Awareness through Environmental Education (with Lynn Wilson), published by IGI Global.

**Burak Tomak** got his BA from Marmara University in English Language Teaching department in Istanbul, Turkey. He got his MA and Ph.D. degrees at the same department from Middle East Technical University in Turkey. He is currently working at Marmara University as an instructor at the School of Foreign Languages teaching academic English for prep students and as a lecturer in the Faculty of Education training prospective foreign language teachers. His research interests are culture teaching, teacher identity, foreign language learning strategies, teacher education, and teaching writing

**Ayşe Yılmaz Virlan** began at Boğaziçi University, earning her a Master's in English Language Teaching and a Ph.D. in Curriculum and Instruction from Yeditepe University. As a full-time instructor at Marmara University, she led curriculum development and taught language proficiency courses while holding various leadership roles. Simultaneously, she lectured at multiple universities, specializing in ELT, educational philosophy, psychology, and instructional methods. Alongside teaching, she contributed to international publications, authored book chapters, and presented at conferences, dedicated to innovative teaching methods and fostering engaging learning environments.

# Index

## A



## B



## C



## D



## E

## F



## G



## H



## I



## L



## M



## N



## P



## R

## S



## T



## U



## V



## W

www.ingramcontent.com/pod-product-compliance
Lightning Source LLC
LaVergne TN
LVHW080644011026
849160LV00014B/1316